Progress on Meshless Methods

Computational Methods in Applied Sciences

Volume 11

Series Editor

E. Oñate
International Center for Numerical Methods in Engineering (CIMNE)
Technical University of Catalunya (UPC)
Edificio C-1, Campus Norte UPC
Gran Capitán, s/n
08034 Barcelona, Spain
onate@cimne.upc.edu
www.cimne.com

For other titles published in this series, go to
www.springer.com/series/6899

A.J.M. Ferreira • E.J. Kansa
G.E. Fasshauer • V.M.A. Leitão

Editors

Progress on Meshless Methods

 Springer

Editors
A.J.M. Ferreira
Faculty of Engineering
University of Porto
Portugal
ferreira@fe.up.pt

E.J. Kansa
University of California
Davis, California
USA

G.E. Fasshauer
Illinois Institute of Technology
Chicago, Illinois
USA

V.M.A. Leitão
Insituto Superior Técnico
Universidade Tecnica Lisboa
Lisbon
Portugal
vitor@civil.ist.utl.pt

ISBN 978-1-4020-8820-9   e-ISBN 978-1-4020-8821-6

Library of Congress Control Number: 2008933848

Printed on acid-free paper

9 8 7 6 5 4 3 2 1

springer.com

Preface

The term meshless (or meshfree) method refers to a broad class of effective numerical techniques for solving a growing number of science and engineering applications without the dependence of an underlying computational mesh (as required by traditional methods such as finite elements or boundary elements). The variety of problems analysed by these methods is very large, and ranges from fracture mechanics, over fluid mechanics, multiscale problems, and laminated composites, all the way to moving material interfaces, just to name a few.

In recent years, several international conferences have been held to foster the discussion of recent advances, share experiences with meshless methods and promote new ideas in the field. The growing body of literature devoted to this area of research provides clear evidence of the interest in meshless methods.

The objective of this book is to collect state-of-the-art papers contributing to the development of this field of research. The book contains 17 invited papers written by participants of the Second ECCOMAS Thematic Conference on Meshless Methods, held in Porto, Portugal, from July 9th to 11th, 2007. The conference is one of a series of Thematic Conferences sponsored by the European Community on Computational Methods in Applied Sciences.

The list of contributors reveals a mix of highly distinguished authors as well as young researchers who are very active working on promising results.

The editors hope that this book will constitute a valuable reference for researchers in the field of mehless methods as applied to engineering and science problems.

The editors would like to thank all authors for their interesting contribution to this book.

University of Porto, Portugal   *A.J.M. Ferreira*
University of California-Davis, USA   *E.J. Kansa*
Illinois Institute of Technology, USA   *G.E. Fasshauer*
Instituto Superior Tecninco, Portugal   *V.M.A. Leitão*

Contents

Contributors

Carlos J.S. Alves
CEMAT-IST and Departamento de Matemática, Instituto Superior Técnico,
TULisbon, Avenida Rovisco Pais, 1096 Lisboa Codex, Portugal,
e-mail: calves@math.ist.utl.pt

Barbara Bacchelli
Department of Mathematics and Applications, University of Milano-Bicocca,
Via R. Cozzi 53, 20125 Milano, Italy, e-mail: barbara.bacchelli@unimib.it

Jorge Belinha
Researcher, Institute of Mechanical Engineering – IDMEC Rua Dr. Roberto Frias,
4200-465 Porto, Portugal, e-mail: jorge.belinha@fe.up.pt

Francisco Manuel Bernal Martinez
G. Millán Institute for Modeling, Simulation and Industrial Mathematics,
Universidad Carlos III de Madrid, 28911 Leganés, e-mail: fcoberna@math.uc3m.es

Mira Bozzini
Department of Mathematics and Applications, University of Milano-Bicocca,
20133 Milano, Italy

Jiun-Snyam (JS) Chen
Department of Civil and Environmental Engineering, University of Califormia,
Los Angeles, CA 90095-1593, USA, e-mail: jschen@seas.ucla.edu

Andrew Corrigan
George Mason University, Department of Computational and Data Sciences,
MS 6A2, 4400 University Drive, Fairfax, VA 22030-4444, USA,
e-mail: acorriga@gmu.edu

Elías Cueto
Group of Structural Mechanics and Materials Modelling, Aragón Institute of
Engineering Research (I3A), University of Zaragoza, María de Luna, 5, E-50018
Zaragoza, Spain, e-mail: ecueto@unizar.es

Suvranu De
Advanced Computational Research Laboratory, Department of Mechanical,
Aerospace and Nuclear Engineering, Rensselaer Polytechnic Institute, 110,
8th Street, Troy, NY 12180, USA, e-mail: des@rpi.com

Lúcia Maria de Jesus Dinis
Associate Professor, Faculty of Engineering of the University of Porto – FEUP Rua
Dr. Roberto Frias, 4200-465 Porto, Portugal, e-mail: ldinis@fe.up.pt

Armel Djeukou
Airbus Deutschland GmbH, Airbus Allee 1, D-28199 Bremen,
e-mail: armel.djeukou@tu-harburg.de

Manuel Doblaré
Group of Structural Mechanics and Materials Modelling, Aragón Institute
of Engineering Research (I3A), University of Zaragoza, María de Luna, 5,
E-50018 Zaragoza, Spain

Gregory E. Fasshauer
Department of Applied Mathematics, Illinois Institute of Technology, Chicago,
IL 60616, USA, e-mail: fasshauer@iit.edu

António Joaquim Mendes Ferreira
Departamento de Engenharia Mecânica e Gestão Industrial, Faculdade
de Engenharia da Universidade do Porto, Rua Dr. Roberto Frias, 4200-465 Porto,
Portugal, e-mail: ferreira@fe.up.pt

Bruno Figueiredo
DECivil/ICIST, Instituto Superior Técnico, TU Lisbon, Portugal,
e-mail: brunof@civil.ist.utl.pt

Csába Gáspár
Széchenyi István University, P.O. Box 701, H-9007 Győr, Hungary,
e-mail: gasparcs@sze.hu

David González
Group of Structural Mechanics and Materials Modelling, Aragón Institute
of Engineering Research (I3A), University of Zaragoza, María de Luna, 5, E-50018
Zaragoza, Spain

Hsin-Yun Hu
Department of Mathematics, Tunghai University, Taichung 407, Taiwan, R.O.C.

Wei Hu
Department of Civil and Environmental Engineering, University of California,
Los Angeles, CA 90095-1593, USA

Renato Manuel Natal Jorge
Departamento de Engenharia Mecânica e Gestão Industrial, Faculdade
de Engenharia da Universidade do Porto, Rua Dr. Roberto Frias, 4200-465 Porto,
Portugal, e-mail: rnatal@fe.up.pt

Manuel Segura Kindelan
G. Millán Institute for Modeling, Simulation and Industrial Mathematics,
Universidad Carlos III de Madrid, 28911 Leganés, Madrid, Spain,
e-mail: kinde@ing.uc3m.es

Vitor M.A. Leitão
DECivil/ICIST, Instituto Superior Técnico, TU Lisbon, Portugal,
e-mail: vitor@civil.ist.utl.pt

Leevan Ling
Department of Mathematics, Hong Kong Baptist University, Kowloon Tong,
Hong Kong, e-mail: lling@hkbu.edu.hk

Michael Macri
Advanced Computational Research Laboratory, Department of Mechanical,
Aerospace and Nuclear Engineering, Rensselaer Polytechnic Institute, 110,
8th Street, Troy, NY 12180, USA

Nuno F.M. Martins
CEMAT-IST and Departamento de Matemática, Faculdade de Ciências e
Tecnologia, Univ. Nova de Lisboa, Quinta da Torre, 2829-516 Caparica, Portugal,
e-mail: nfm@fct.unl.pt

Herve Morvan
School of Mechanical, Materials and Manufacturing Engineering, Faculty of
Engineering Room B111 Coaks University park, University of Nottingham,
Nottingham, NG7 2RD, UK

Henry Power
School of Mechanical, Materials and Manufacturing Engineering, Faculty of
Engineering Room B111 Coaks University park, University of Nottingham,
Nottingham, NG7 2RD, UK, e-mail: henry.power@nottingham.ac.uk

Zoran Ren
Faculty of Mechanical Engineering, University of Maribor, Smetanova 31, SI-2000
Maribor, Slovenia, e-mail: ren@uni-mb.si

Carla Maria da Cunha Roque
Departamento de Engenharia Mecânica e Gestão Industrial, Faculdade de Engenharia da Universidade do Porto, Rua Dr. Roberto Frias, 4200-465 Porto, Portugal,
e-mail: ferreira@fe.up.pt

David Stevens
School of Mechanical, Materials and Manufacturing Engineering, Faculty of
Engineering Room B111 Coaks University park, University of Nottingham,
Nottingham, NG7 2RD, UK

Matej Vesenjak
University of Maribor, Faculty of Mechanical Engineering, Smetanova 31, SI-2000
Maribor, Slovenia, e-mail: m.vesenjak@uni-mb.si

Otto von Estorff
Institute of Modelling and Computation, Hamburg University of Technology,
Denickestrasse 17, D-21073 Hamburg, Germany

John Wallin
George Mason University, Department of Computational and Data Sciences,
MS 6A2, 4400 University Drive, Fairfax, VA 22030-4444, USA,
e-mail: jwallin@gmu.edu

Jack G. Zhang
Department of Mathematics and Statistics, University of New Mexico,
Albuquerque, NM 87131, USA, e-mail: jackunm@math.unm.edu

Particular Solution of Poisson Problems Using Cardinal Lagrangian Polyharmonic Splines

Barbara Bacchelli(✉) and Mira Bozzini

Abstract In this paper, we propose a method belonging to the class of special meshless kernel techniques for solving Poisson problems. The method is based on a simple new construction of an approximate particular solution of the nonhomogeneous equation in the space of polyharmonic splines. High order Lagrangian polyharmonic splines are used as a basis to approximate the nonhomogeneous term and a closed-form particular solution is given. The coefficients can be computed by convolution products of known vectors. This can be done in all dimensions, without numerical integration nor solution of systems. Numerical experiments are presented in two dimensions and show the quality of the approximations for different test examples.

Keywords: Particular solutions · Poisson equation · fundamental solutions · cardinal Lagrangian polyharmonic splines

1 Introduction

Interpolation by radial basis functions (RBFs) has become a powerful tool in multivariate approximation theory, especially for it allows to work in arbitrary space dimension. During the last decade, there has been an increasing interest and significant progress in applying RBFs for approximating the solutions of partial differential equations (PDEs). It is known that most of the RBFs are globally defined basis functions. This means that the resulting matrix for interpolation is dense and it can be highly ill-conditioned. While, the compactly supported radial basis functions (CS-RBFs) may produce lower approximation results.

In this paper we consider the Dirichlet problem for the Poisson equation in a bounded regular domain $\Omega \subset R^d$. The solution is given as superposition of a

B. Bacchelli and M. Bozzini
Department of Mathematics and Applications, University of Milano-Bicocca, Via R. Cozzi 53, 20125 Milano, Italy, e-mail: barbara.bacchelli@unimib.it

A.J.M. Ferreira et al. (eds.) *Progress on Meshless Methods, Computational Methods in Applied Sciences.*
© Springer Science + Business Media B.V. 2009

particular solution u_P of the Poisson equation $\Delta u = f$, and the solution u_H of the homogeneous problem $\Delta u = 0$ in such a way that the boundary conditions are met.

Several procedures are given in the literature to evaluate u_P. Here we propose a method to obtain closed-form particular solutions (MPSs). Polyharmonic splines are employed as trial space for approximating the solution and the inhomogeneous term by using cardinal Lagrangian polyharmonic splines of order m as basis functions [7]. The goal of the method is a reproducing property of the fundamental solutions v_m of the iterated Laplacian operator Δ^m through the operator Δ, precisely, $\Delta v_m = v_{m-1}$. Then the coefficients of u_P can be evaluated by convolution products of known vectors, forcing interpolatory conditions on a regular mesh of R^d. In this way, no integral equations nor systems are needed to be solved, and the method can be used in all dimensions. Moreover, the computational cost in evaluating u_P in a grid point is O(1).

More precisely, in Sect. 2 we briefly recall polyharmonic splines and some preliminary results. In Sect. 3 we develop the proposed method to derive approximated particular solutions of the Poisson equation and we give rates of convergence to particular solutions for smooth problems in arbitrary dimensions (Theorem 4). In Sect. 4 we present a numerical technique, based on the classical finite difference method (FDM), which is very efficient for solving the homogeneous PDE problem in an hypercube.

Section 5 is devoted to numerical simulation in two dimensions. More precisely, we show the efficiency of the proposed MPS and the quality of the approximations comparing the results with the recent notes [6] and [3]. When Ω is square we couple MPS with FDM using the numerical technique described in Sect. 4. For more general domains we couple MPS with the method of fundamental solutions (MFS). In all the experiments the efficiency of the proposed method is revealed. In particular it turns out that the method is suited to handle near-singular problems too.

For the MFS and for common methods for finding particular solutions, we refer readers to [10] and to references therein for further details.

2 Main Notations and Preliminary Results

For positive integer k, we use the notation Π^k for the class of polynomials in d variables of total degree not exceeding k. $C^k(R^d)$ is the usual space of continuous functions with continuous partial derivatives up to total order k; $C_0^k(R^d)$ is the subspace of $C^k(R^d)$ of functions with compact support. The Fourier transform of a summable function f is defined by $\hat{f}(\omega) := \int_{R^d} e^{-i\omega \cdot x} f(x) dx$, where $\omega \cdot x := \sum_{k=1}^d \omega_k x_k$, and it is extended by duality to all distributions in $S'(R^d)$. $\| \bullet \|_\infty$ and $\| \bullet \|_{\infty, \Omega}$ denote the supremum norm in R^d or in a bounded set Ω of R^d respectively. $|\bullet|$ is the Euclidean norm. In the sequel m will denote a positive integer and h a positive real number.

We use the notation $*$ for the convolution product (when it makes sense), of sequences a, b, $(a * b)_k := \sum_{j \in \mathbb{Z}^d} a_j b_{k-j}$, and for functions f, g, or distributions,

$(f * g)(x) := \int_{\mathbb{R}^d} f(t) g(x - t) dt$. We also denote by $*$ the semi-discrete convolution between a sequence a and a function f: $a * f(\bullet) := \sum_{j \in \mathbb{Z}^d} a_j f(\bullet - j)$. Given a function $f \in C^0(R^d)$ let us denote by $f|_h$ the sequence

$$f|_h := \{f(hj)\}_{j \in \mathbb{Z}^d}.$$

Let $m > d/2$. We denote with $SH^m(R^d)$ the class of m−harmonic cardinal splines. A polyharmonic function is one which is m-harmonic for some $m > d/2$. More precisely, following [7], we can define $SH^m(R^d)$ to be the subspace of $S'(R^d)$ (the usual space of d-dimensional tempered distributions) whose elements s satisfy the following conditions

$$(i) \quad s \in C^{2m-d-1}(R^d)$$
$$(ii) \quad \triangle^m s = 0 \quad \text{on } R^d \backslash Z^d,$$

where $\Delta := \sum_{j=1}^d \partial^2 / \partial x_j^2$ is the Laplace operator and Δ^m is defined iteratively by $\Delta^m s = \Delta(\Delta^{m-1} s)$.

Let v_m be the fundamental solution of Δ^m given by

$$v_m(x) = \begin{cases} c(d,m)|x|^{2m-d} & \text{if } d \text{ is odd,} \\ c(d,m)|x|^{2m-d} \log|x| & \text{if } d \text{ is even,} \end{cases}$$

where the constant $c(d,m)$ depends only on d and m and is chosen so that $\Delta^m v_m = Dirac$ (the unit Dirac distribution at the origin), i.e.,

$$\widehat{v_m}(\omega) = \frac{(-1)^m}{|\omega|^{2m}}.$$

Since $\widehat{\Delta v_m} = -|\omega|^2 \widehat{v_m}$, we derive the property for $m > d/2 + 1$,

$$\Delta v_m = v_{m-1}. \tag{1}$$

The fundamental Lagrangian polyharmonic cardinal spline L_m is the function in $SH^m(R^d)$ which satisfies $L_m(j) = \delta_{0j}, j \in Z^d$, where δ is the usual Kroneker symbol. The Fourier transform of L_m is given by

$$\widehat{L_m}(\omega) = \frac{|\omega|^{-2m}}{\sum_{j \in Z^d} |\omega - 2\pi j|^{-2m}}$$

and L_m is a fast decreasing function, i.e., there are positive constants b and c, depending on d and m but independent of x, such that

$$|L_m(x)| \le c e^{-b|x|}, \text{ for all } x \in R^d.$$

Moreover, L_m has the following representations in terms of v_m:

$$L_m(x) = l^m * v_m(x), \quad x \in R^d, \tag{2}$$

where the sequence l^m is fast decreasing,

$$|l_j^m| \leq c e^{-b|j|}, \text{ for all } j \in Z^d.$$

Note that Lagrangian functions can be represented, and fairly efficiently computed, via the elementary B-splines (see for example [1]). Evaluating (2) in $j \in Z^d$, we derive the relevant property

$$l^m * v_m|_1 = \delta, \tag{3}$$

where $v_m|_1 = \{v_m(j)\}_{j \in Z^d}$.

It is well known that L is a refinable function and the family $\{L_m(\bullet - j)\}_{j \in Z^s}$ is a stable basis of $SH^m(R^d)$ [8].

In [7] it is shown that cardinal polyharmonic splines solve interpolation problems on Z^d. If f is a function of polynomial growth, i.e. there exists $r \in N$ such that $|f(x)| \leq c(1 + |x|^r)$ for all $x \in R^d$, then for every $m > d/2$, there is a (unique) element If in $SH^m(R^d)$ which interpolates f in Z^d, i.e. such that $If(j) = f(j)$ for all $j \in Z^d$:

$$If(x) := f|_1 * L_m(x) = f|_1 * l^m * v_m(x), \quad x \in R^d. \tag{4}$$

We also consider such approximants on a scaled grid where, for positive spacing h,

$$I_h f := f|_h * L_m(h^{-1}\bullet) = f|_h * l^m * v_m(h^{-1}\bullet) \tag{5}$$

We denote by $SH^{m,h}(R^d)$ the class of m-harmonic splines with knots on a scaled grid $\{hZ^d\} := \{hk\}_{k \in Z^d}$. More precisely, a function g is an element in $SH^{m,h}(R^d)$ iff it satisfies $g \in C^{2m-d-1}(R^d)$ and $\triangle^m g = 0$ on $R^d \backslash hZ^d$. Thus $SH^{m,1}(R^d) \equiv SH^m(R^d)$, and s belongs to $SH^m(R^d)$ iff $g := s(h^{-1}\bullet)$ belongs to $SH^{m,h}(R^d)$.

If f is a function of polynomial growth, then the sequence $f|_h = \{f(hj)\}_{j \in Z^d}$ is of polynomial growth and $I_h f$ is the (unique) element in $SH^{m,h}(R^d)$ which interpolates f in hZ^d, i.e. it satisfies $I_h f(hj) = f(hj)$ for all $j \in Z^d$. We can also say that $I_h f$ interpolates the sequence $f|_h$.

Approximation orders of convergence can be founded in [2] that, for our convenience, we resume as follows.

Theorem 1. *The cardinal interpolant $I_1 f$ in $SH^m(R^d)$ is well defined and exact for functions $f \in \Pi^{2m-1}$. For every function $f \in C^{2m}(R^d)$ with bounded $(2m)$th-order partial derivatives, the interpolants (5) satisfy, for $h \to 0$,*

$$\|I_h f - f\|_\infty = O(h^{2m}).$$

We end this section by giving two results on the discrete convolution, which will be relevant for the validity of the construction of the particular solution that we present in the next section.

Theorem 2. *Let u, v be fast decreasing sequences. Then for all $j \in Z^d$ the convolution $(u * v)_j$ is well defined and $u * v$ is fast decreasing.*

Proof. For all $j \in Z^d$,

$$
\begin{aligned}
|(u * v)_j| &\leq \sum_{|k| \leq |j|/2} |v_{j-k}||u_k| + \sum_{|k| > |j|/2} |v_{j-k}||u_k| \\
&\leq \sum_{|k| \leq |j|/2} c_1 e^{-b_1 \frac{|j|}{2}} |u_k| + \sum_{|k| > |j|/2} |v_{j-k}| c_2 e^{-b_2 \frac{|j|}{2}} \\
&\leq C e^{-b \frac{|j|}{2}} (\|u\|_1 + \|v\|_1).
\end{aligned}
$$ □

Theorem 3. *Let u be a fast decreasing sequence and let v be a sequence of polynomial growth. Then for all $j \in Z^d$ the convolution $(u * v)_j$ is well defined and $u * v$ is a sequence of polynomial growth.*

Proof. We have

$$
\begin{aligned}
|v_{j-k}| &\leq c(1 + |j - k|^r) \leq c(1 + (|j| - |k|)^r) \\
&\leq c \, 2^r (1 + |j|^r - |k|^r) \leq c \, 2^r (1 + |j|^r)(1 + |k|^r)
\end{aligned}
$$

Then for all $j \in Z^d$,

$$
\begin{aligned}
|(u * v)_j| &\leq \sum_k |v_{j-k}||u_k| \leq c \, 2^r (1 + |j|^r) \sum_k (1 + |k|^r)|u_k| \\
&\leq C(1 + |j|^r),
\end{aligned}
$$

since the series is convergent.   □

3 The Poisson Equation

Let us consider the Poisson equation in R^d:

$$
\Delta u(x) = f(x), \quad x \in R^d. \tag{6}
$$

In the following lemma we derive a particular solution of this problem when the inhomogeneous term is a polyharmonic spline in $SH^{m-1,h}(R^d)$, for fixed $h > 0$, which is fast decreasing:

$$
f \in SH^{m-1,h}, \quad |f(x)| \leq c e^{-b|x|}, \; x \in R^d.
$$

The solution is a polyharmonic spline in $SH^{m,h}(R^d)$, which is obtained without solving linear systems nor integral equations. The method based on the notable recursive property (1). A simple relation for the coefficients in the solution is derived forcing interpolatory conditions on a regular mesh of R^d. The sequences $f|_h, v_m|_1$ and l^{m-1} are defined as in the previous section.

Lemma 1. *Let* $m > \dfrac{d}{2} + 1$, *and* $h > 0$ *be fixed. Let* f *be a polyharmonic spline in* $SH^{m-1,h}(R^d)$ *and* f *be fast decreasing. We define the sequence* α *as follows*

$$\alpha := h^2 \cdot f|_h * l^{m-1} * v_m|_1, \tag{7}$$

and let s *be the* $m-$*harmonic spline which interpolates the sequence* α *in* $\{hZ^d\}$, *i.e. let* $s|_h := \alpha$ *and*

$$s = s|_h * L_m(h^{-1}\bullet) := h^2 \cdot f|_h * l^{m-1} * v_m(h^{-1}\bullet). \tag{8}$$

Then s *is well defined and it satisfies the Poisson equation in* R^d: $\Delta s(x) = f(x)$, $x \in R^d$.

Proof. Since $f|_h * l^{m-1}$ is fast decreasing and $v_m|_1$ is of polynomial growth, then the sequence $s|_h = \alpha$ defined in (7) is of polynomial growth and s is well defined. By (2), we can express s as

$$s(x) = s|_h * l^m * v_m(h^{-1}x), \quad x \in R^d. \tag{9}$$

The last expression of s in (8) follows substituting $s|_h = \alpha$ (7) into (9), because of the relation $v_m|_1 * l^m = \delta$:

$$s = s|_h * L_m(h^{-1}\bullet) = h^2 \cdot f|_h * l^{m-1} * v_m|_1 * l^m * v_m(h^{-1}\bullet)$$
$$= h^2 \cdot f|_h * l^{m-1} * v_m(h^{-1}\bullet).$$

In order to prove that s is a solution of the Poisson equation in R^d, let us compute the Laplacian of s expressed by (9) by using (1):

$$\Delta s(x) = h^{-2} s|_h * l^m * v_{m-1}(h^{-1}x), \quad x \in R^d. \tag{10}$$

Substituting $s|_h = \alpha$ (7) into (10) we get,

$$\Delta s(x) = f|_h * l^{m-1} * v_m|_1 * l^m * v_{m-1}(h^{-1}x), \quad x \in R^d,$$

and since $v_m|_1 * l^m = \delta$, then, for all x in R^d

$$\Delta s(x) = f|_h * l^{m-1} * v_{m-1}(h^{-1}x) = I_h f(x) = f(x).$$

where the last equality holds since f is a polyharmonic spline in $SH^{m-1,h}(R^d)$. $\square$

The Dirichlet problem for the Poisson equation in a domain $\Omega \subset R^d$ with boundary $\partial\Omega$ has the form

$$\Delta u(x) = f(x), \quad x \in \Omega \tag{11}$$

$$B(u) = g \tag{12}$$

where (12) is the boundary condition $u(x) = g(x), x \in \partial\Omega$.

We assume that this problem is well posed in the sense that the exact solution $u \in C^0(\overline{\Omega}) \cap C^2(\Omega)$ exists and an a-priori inequality

$$\|u\|_{\infty,\overline{\Omega}} \leq C(\|\Delta u\|_{\infty,\Omega} + \|B(u)\|_{\infty,\partial\Omega}) \tag{13}$$

holds in $\overline{\Omega}$, where the constant C depends on Ω but not on u. Such bounds are derived in the literature (see e.g. [5]).

A particular solution u_P satisfies Eq. (11), but does not necessarily satisfies (12). If u_H satisfies the Laplace equation, $\Delta u_H(x) = 0, x \in \Omega$, with boundary conditions $B(u_H) = g - B(u_P)$, then the general solution u of the problem (11), (12) is obtained by superposition, $u = u_H + u_P$.

We suppose Ω to be an open and bounded subset of $[0,1]^d$, having the cone property and a Lipschitz boundary. If $f \in C^k(\Omega)$ has bounded derivatives of total order k, for some $k \geq 0$, then it is possible to extend f to a function $\widetilde{f}$ that belongs to $C_0^k(R^d)$ with compact support given by $I_\delta := [-\delta, 1 + \delta]^d$, $\delta > 0$, i.e. $\widetilde{f}(x) = f(x), x \in \Omega$, and $\widetilde{f}(x) = 0$, $x \in R^d \setminus I_\delta$. Suitable extensions $\widetilde{f}$ can be provided in many cases by the Whitney Extension Theorem [4], although this cannot be used for all domains Ω. In the sequel we still denote by f this extension, simplifying the notation.

In the following Theorem 4 we derive an approximated solution s of the problem (11) in the bounded domain Ω. For fixed $h > 0$, we approximate the extension f by $I_h f$, the $m-1$-harmonic spline that interpolates f in hZ^d. Then s is the element in $SH^{m,h}(R^d)$ that satisfies the Poisson equation $\Delta s_h = I_h f$ in R^d. The existence of this solution follows from the previous lemma. The procedure holds in all dimensions d. The coefficients in the approximation s are expressed by convolution products between known vectors. If f has bounded $2(m-1)$-th order partial derivatives, the error between s and a solution u_P of the problem (11) has order of $h^{2(m-1)}$. The sequences $f|_h, v_m|_1$ and l^{m-1} are defined as in the previous section.

Theorem 4. *Let* $\Omega \subset [0,1]^d$ *be an open subset of* R^d *having the cone property and a Lipschitz boundary and let m be an integer, $m > \dfrac{d}{2}$. Let us consider the Poisson problem (11) in the set Ω, assuming that f belongs to the class $C^{2(m-1)}(\Omega)$ with bounded derivatives of total order $2(m-1)$. Then we can extend f to a function defined on the whole R^d, that we still denote by f, and which belongs to the class $C_0^{2m-2}(R^d)$ with bounded $(2m-2)$th-order partial derivatives. Let $h > 0$ be fixed and we consider the interpolant $I_h f$ of f in $\{hZ^d\}$. Let us define the sequence α as follows:*

$$\alpha := h^2 \cdot f|_h * l^{m-1} * v_m|_1, \tag{14}$$

and let s be the m-harmonic spline which interpolates the sequence α in $\{hZ^d\}$, i.e. $s|_h := \alpha$ and

$$s = s|_h * L_m(h^{-1}\bullet) := h^2 \cdot f|_h * l^{m-1} * v_m(h^{-1}\bullet). \tag{15}$$

Then

(i) s is well defined and it is a solution of the following Poisson equation in R^d

$$\Delta s(x) = I_h f(x), \ x \in R^d; \tag{16}$$

(ii) s is a discrete solution of (11), i.e.,

$$\Delta s(hj) = f(hj), \ hj \in \Omega;$$

(iii) as $h \to 0$,

$$\|\Delta s - f\|_{\infty,\Omega} = O(h^{2m-2});$$

(iv) for every fixed $h > 0$ there exists a particular solution u_p of problem (11) such that, as $h \to 0$,

$$\|s - u_P\|_{\infty,\Omega} = O(h^{2m-2}).$$

Proof. (i) follows from the Lemma. (ii) is an obvious consequence of (i), while (iii) is a consequence of Theorem 1: since Δs is the $(m-1)$-harmonic spline interpolating the function f in the set $\{hZ^d\}$, it follows

$$\|\Delta s - f\|_{\infty,\Omega} = \|\Delta s - f\|_{\infty,\Omega} \le \|\Delta s - f\|_{\infty}$$
$$= \|I_h f - f\|_{\infty} = O(h^{2m-2}), \ h \to 0.$$

In order to prove (iv), we define $g_P := B(s)$ and we consider the Dirichlet problem

$$\begin{cases} \Delta u(x) = f(x), \ x \in \Omega, \\ B(u) = g_P. \end{cases}$$

According to well posedness, there exists a (unique) solution u_P of this problem and by using (13), we can give the following error bound

$$\|s - u_P\|_{\infty,\overline{\Omega}} \le C(\|\Delta(s - u_P)\|_{\infty,\Omega} + \|B(s - u_P)\|_{\infty,\partial\Omega})$$
$$= C(\|I_h f - f\|_{\infty,\Omega} + \|B(s) - g_P\|_{\infty,\partial\Omega})$$
$$= C\|I_h f - f\|_{\infty,\Omega} \le C\|I_h f - f\|_{\infty} = O(h^{2m-2}), \ h \to 0. \qquad \square$$

4 Solving the Dirichlet Problem for the Laplace Equation

Looking at the error and convergence analysis of a numerical solution $\widetilde{u}$ of the problem (11), (12) obtained by superposition, we consider an approximated solution $\widetilde{u}_H$ of the Laplace problem associated with an approximated particular solution s of the problem (11) given by (15). More precisely, let u_H be a solution of the problem: $\Delta u_H(x) = 0, x \in \Omega$, with boundary conditions $B(u_H) = g - B(s) =: g_H$, then we look for an approximation of u_H, say $\widetilde{u}_H$, with small residuals $\|\Delta(\widetilde{u}_H)\|_{\infty}$, and

$\|B(\tilde{u}_H) - g_H\|_{\infty, \partial\Omega}$. Then we use $\tilde{u} := s + \tilde{u}_H$ for our full solution and residuals,

$$\|\Delta\tilde{u} - f\|_{\infty,\Omega} = \|\Delta(s + \tilde{u}_H) - f\|_{\infty,\Omega} \leq \|I_h f - f\|_{\infty,\Omega} + \|\Delta(\tilde{u}_H)\|_{\infty,\Omega},$$
$$\|B(\tilde{u}) - g\|_{\infty,\partial\Omega} = \|B(s + \tilde{u}_H) - g\|_{\infty,\partial\Omega} = \|B(\tilde{u}_H) - g_H\|_{\infty,\partial\Omega}.$$

Then, using (13), there is an a-posteriori error bound

$$\|\tilde{u} - u\|_{\infty,\overline{\Omega}} \leq C(\|\Delta(\tilde{u} - u)\|_{\infty,\Omega} + \|B(\tilde{u} - u)\|_{\infty,\partial\Omega})$$
$$= C(\|I_h f - f\|_{\infty,\Omega} + \|\Delta(\tilde{u}_H)\|_{\infty,\Omega} + \|B(\tilde{u}_H) - g_H\|_{\infty,\partial\Omega}).$$

In particular, if we consider the classical 2nd order FDM with step h_1, $h_1 \leq h$, we choose as homogeneous solution $\tilde{u}_H$ the polyharmonic spline in SH^{m,h_1} which interpolates the approximated nodal values. Note that the value of the residual in the knots belonging to the boundary is zero. In this way the errors are $\|I_h f - f\|_{\infty,\Omega} = O(h^{2m-2})$ due to the particular solution, $\|\Delta(\tilde{u}_H)\|_{\infty,\Omega} = O(h_1^2)$ due to the FDM error, and $\|B(\tilde{u}_H) - g_H\|_{\infty,\partial\Omega} = O(h_1^{2m})$ due to the polyharmonic spline interpolation error.

In this section, we propose a modified version of the 2nd order FDM. The procedure is here presented for a set Ω such that its closure is the hypercube $[0,1]^d$. More precisely, we consider the Laplace equation with given boundary conditions

$$\begin{cases} \Delta u(x) = 0, & x \in (0,1)^d, \\ B(u) = g. \end{cases}$$

Given a uniform mesh of cardinality $N = (n+2)^d$ on $[0,1]^d$, we discretize the operator Δ by the $\Delta_{1,h}$ operator, $\Delta_{1,h} f := \frac{1}{h^2} \sum_{j=1}^d (f(\bullet - he_j) - 2f + f(\bullet + he_j))$, with $h = \frac{1}{n+1}$ (here $e_1, \ldots, e_d \in R^d$ are the coordinate vectors $(e_j)_k := \delta_{jk}, 1 \leq j, k \leq d$). This leads to a linear system of order N

$$Au = b, \tag{17}$$

in which we consider at first the (given) boundary values as unknown, and we assign the new boundary conditions with periodicity, as if Ω where a torus. The so obtained system contains all the equations that one has to write when applying the classical second order method, and the remaining equations are simply identity. What we obtain is equivalent to the classical finite difference system but, due to the special form of the matrix A, it is possible to rewrite it in such a way that we solve a system whose unknowns are only the values corresponding to the knots nearest to the boundary, while the remaining components are obtained by multiplying known values. For example in two dimensions, we have to solve a system of order $4\sqrt{N} - 3$, instead of order N.

Indeed, the matrix A is square, positive semidefinite of order N, and it is a block circulant matrix with circulant blocks. The matrix A has only one null eigenvalue and it is diagonalized by the multivariate Fourier matrix of the eigenvectors of A, $W = F_n \otimes F_n \ldots \otimes F_n$

$$W^T A W = \Lambda$$

where Λ is the eigenvalues matrix of A. Then we can deduce the equivalent system

$$\begin{cases} u = Wz \\ z_i = \dfrac{1}{\lambda_i} \Sigma_k w_{ki} b_k, \quad i = 2, N \\ \Sigma_k b_k = 0 \end{cases}$$

and the last equation derives from the null eigenvalue λ_1. Substituting the boundary values into u and b when they occur, we note that it is possible to extract a system of order $M + 1$, where $M = n^d - (n-2)^d$,

$$Kx = \widetilde{g},$$

which is solved for example by Gaussian elimination.

The structures of the matrix K and of its inverse are particular, as the graphs for the case $d = 2$ show, in Fig. 1.

These structures allow us to formulate a stability conjecture which is confirmed by the numerical evidence. Indeed, we note that the plot of the condition number $k_2(K)$ versus n tends to a line with a slope about 2, as it is shown in Fig. 2.

We note that in the implementation of the method, the matrices A and K can be computed before the experiments, and then reloaded, when they are needed.

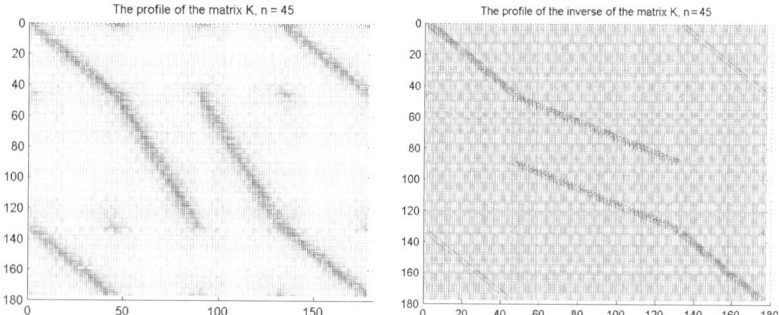

Fig. 1 The profile of the matrix K and of its inverse, $d = 2, n = 45$

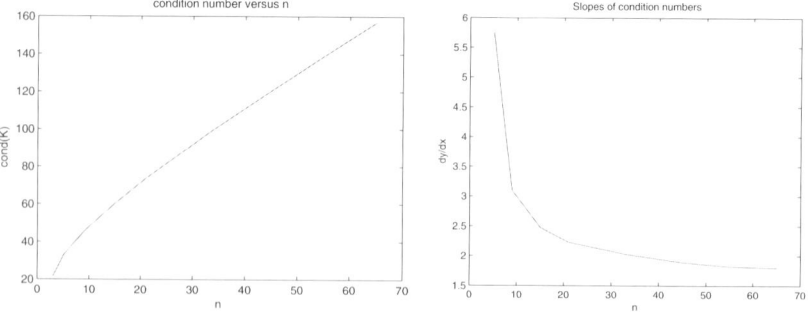

Fig. 2 Condition numbers of the matrix K

5 Numerical Results

We have considered different experiments in two dimensions in order to test the performance of the proposed method. Here we present only four significant cases, where we choose the parameter m equal to tree.

We remark that the computation of the approximated particular solution with MPS needs some care. Precisely, in order to overcome the Gibbs phenomena and since the method works with convolution products, it is necessary to extend the right hand-hand side $f(x,y)$ of problem (11) in a square domain which strictly contains the assigned domain Ω, as we said in Sect. 3.

In the following, we will precise the extended domain and the different procedures of construction of the extension of f that we apply in the experiments.

The computations are performed with an Intel Core 2 Duo T7300 processor (2 GHz, cache L2 4 MB, FSB 800 MHz).

The first three experiments are concerned with a square domain Ω, and we couple MPS with the modified 2nd order FDM of step $h_1 = h$, where h is the grid space, as described in Sect. 4.

The matrices A and K are computed before the experiments and then reloaded when necessary.

In all these cases, $\Omega \cup \partial\Omega = [a,b]^2$ is extended to the square $I_h := [a-h,b+h]^2$, and we compute in I_h the values of the extension of the inhomogeneous term $f(x,y)$ of problem (11), according to the values of the lines in the horizontal/vertical directions, evaluated by using the two points nearest to the boundary $\partial\Omega$. The evaluations in the "corners", which are intersection of the horizontal/vertical directions, is made by taking the mean of the resulting adjacent values.

Example 1. Kansa's test function.

The well known function is used for testing numerical methods by several authors. For comparison we focus in particular our attention on the recent work [6].

The Poisson problem (11), (12) is given in the domain $\Omega \cup \partial\Omega = [1,2]^2$ and the solution is

$$u(x,y) = \sin\frac{\pi x}{6}\sin\frac{7\pi x}{4}\sin\frac{3\pi y}{4}\sin\frac{5\pi y}{4}, \quad (x,y) \in [1,2]^2.$$

The profiles of the solution $u(x,y)$ and of the inhomogeneous term $f(x,y)$ are shown in Fig. 3.

We compute the approximated solution for the series of data-size N as shown in Table 1. The errors are obtained on 94×94 regular gridded points.

The results show that the rate of convergence is analogous of the one presented in [6], but a satisfactory accuracy is reached with a much smaller size of the data. Indeed, in [6], the accuracy of 5.37×10^{-4} relevant to the maximum error is reached when the cardinality N of the sample is equal to 5,329.

Comparing the increment of the accuracy between $N = 1,089$ and $N = 2,025$, with the increment of the CPU, this show a decrease of efficiency, so then we stopped increasing the size of the data.

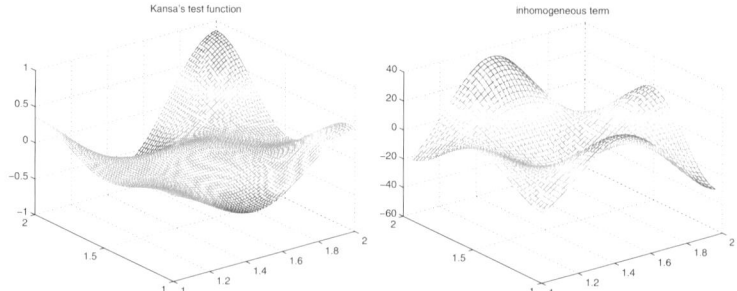

Fig. 3 Example 1: solution and right-hand side

Table 1 Kansa's test function, numerical results

N	MAX	MSE	CPU
49	1.0×10^{-2}	3.1×10^{-3}	0.03 sec
225	1.2×10^{-3}	2.7×10^{-4}	0.14 sec
441	3.8×10^{-4}	9.9×10^{-5}	0.31 sec
1,089	1.7×10^{-4}	6.1×10^{-5}	1.06 sec
2,025	1.2×10^{-4}	4.2×10^{-5}	3.68 sec

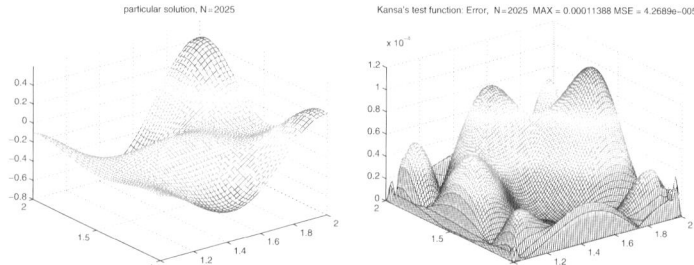

Fig. 4 Example 1: particular solution and errors in the approximated solution, $N = 2,025$

The profiles of the particular solution computed by MPS with $N = 2,025$, and the errors in the approximated solution are shown in Fig. 4.

The following two examples concern with near singular problems, which are considered in [3]. In general, most of the near singular or singular problems cannot be solved directly by standard numerical methods. As a result, special treatments are required before applying these standard methods. With MPS good accuracy is reached with no special treatments.

Example 2. Sharp spikes inhomogeneous source term.

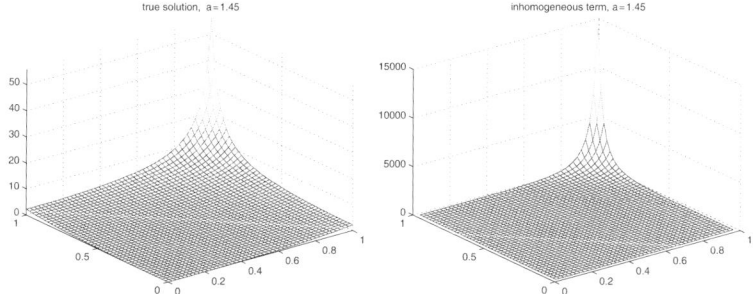

Fig. 5 Example 2: solution and right-hand side, $a = 1.45$

Table 2 Sharp spikes test function, $N = 2{,}025$

a	Range of f	MAX	Relative error	MSE
1.6	[2.5, 850]	0.0080	7.4×10^{-4}	0.0012
1.5	$[2.67, 7.34 \times 10^3]$	0.039	1.7×10^{-3}	0.0044
1.45	$[2.75, 9.29 \times 10^4]$	0.12	2.2×10^{-3}	0.01
1.415	$[2.82, 8.23 \times 10^9]$	3.38	1.3×10^{-3}	0.09

We consider here the Poisson problem (11), (12) where $\Omega \cup \partial\Omega = [0,1]^2$ and the solution is given by

$$u(x,y) = \frac{(x^2 + y^2)}{a - \sqrt{x^2 + y^2}}, \qquad (x,y) \in [0,1]^2.$$

The right-hand side is given by

$$f(x,y) = \frac{-4a^2 + 3a\sqrt{x^2 + y^2} - (x^2 + y^2)}{(\sqrt{x^2 + y^2} - a)^3}.$$

If $a = \sqrt{2} \simeq 1.4142$, the solution $u(x,y)$ and the right-hand $f(x,y)$ have a singularity in $(1,1)$. The profiles of these functions are shown in Fig. 5, when $a = 1.45$.

We test the effectiveness of the method to escape the singularity by assuming different values of a and N. The results obtained for $N = 2{,}025$ grid points are shown in Table 2, where the errors are computed on the same points. We see that in any case we get a significant result, also when we are very near to the singular case ($a = 1.415$). In addition, the accuracy is of the same order, as we deduce by looking at the relative error.

We point out that the approach does not require knowing the explicit analytic behavior of the inhomogeneous term in the neighborhood of singularity. The profile of the approximated particular solution is shown in Fig. 6, when $a = 1.5$ and $N = 2{,}025$.

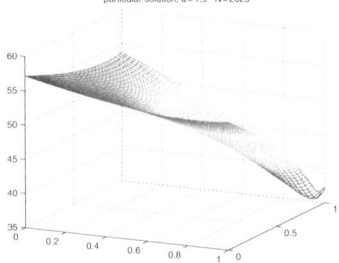

Fig. 6 Example 2: particular solution, $a = 1.5, N = 2{,}025$

Table 3 Sharp spikes test function, $N = 441$

a	MAX	Relative error
1.5	0.05	2.1×10^{-3}
1.415	2.05	8.0×10^{-4}

The results are satisfactory even with a smaller set of data points, as it is seen in Table 3, for $N = 441$.

Example 3. Mild near-singular inhomogeneous source term.

In this case we solve the problem (11), (12) with $\Omega \cup \partial\Omega = [0,1]^2$ and the solution is given by

$$u(x,y) = \frac{r^2\sqrt{a-r}}{(r+b)} - \frac{500(x-y)^4}{24}, \qquad (x,y) \in [0,1]^2,$$

where $r = \sqrt{x^2 + y^2}$. The right-hand side is given by

$$f(x,y) = \frac{9r^4 + Ar^3 + Br^2 + Cr + D}{4(r+b)^3(a-r)^{3/2}} - 500(x-y)^2,$$

$$A = 26b - 14a, \ B = 4a^2 - 40ab - 25b^2, \ C = ab(12a - 42b), \ D = 16a^2b^2.$$

Here, we consider the case when $a = 1.415$ and $b = 0.02$. We observe that there are two sharp spikes near the points at (0,0) and (1,1). Since $f(0,0) = 237$ and $f(1,1) = -15{,}861$, these spikes are relatively milder than the one in example 2, and no singularity is present in the solution.

In this case we choose $N = 2{,}025$ grid points and the resulting maximum error computed on the same points was $MAX = 5.38 \times 10^{-3}$. This result is comparable with those obtained with the scheme presented in [3], where the authors use a smoothing technique to cut off the spike behavior in the inhomogeneous term.

The following experiment is concerned with a non convex domain, in order to check MPS in this more general case.

Example 4. Non convex domain.

As we said at the beginning of this section, we need to extend the rigth-hand side of Eq. (11) in a square domain which strictly contains the assigned domain Ω, in order to compute the particular solution with MPS. If Ω, is non convex and bounded into a square $I = [a,b]^2$, then we extend f in the square $I_h = [a-h,b+h]^2$, and we evaluate the extension in a uniform lattice in I_h. It is clear that when the function f is known and well defined in I_h, we can use its values as extended values.

Otherwise, a procedure of construction of the extension is to extrapolate the polyharmonic spline which interpolate f in a set of points in Ω (extrapolation procedure).

As a significant experiment, we couple MPS with MFS (method of fundamental solution) in the following problem:

$$\Delta u(x,y) = -\frac{5\pi^2}{4}\sin \pi x \cos \frac{\pi y}{2}, \text{ in } \Omega,$$

$$u(x,y) = \sin \pi x \cos \frac{\pi y}{2}, \text{ on } \partial\Omega,$$

where $\Omega = (0,1)^2 \backslash [0.5,1]^2$ which is three-quarters of the unit square. The exact solution is given by $u(x,y) = \sin \pi x \cos \frac{\pi y}{2}$.

In the MFS, we use 16 evenly distributed points on the boundary $\partial\Omega$ and the same number of source points on a circle with center at $(0.5,0.5)$ and radius 5.

In the computation of the approximated particular solution with the extrapolation procedure, we interpolate the inhomogeneous term at 65 evenly space grid points in $\Omega \subset [0,1]^2$, and we evaluate the extended function in 1,225 grid points in $I_h = [-h,1+h]^2$. The resulting maximum error evaluated in a grid in Ω with step $1/91$ is $MAX = 4.0 \times 10^{-3}$. While, taking the true function f in the whole I_h, the resulting maximum error is $MAX = 1.4 \times 10^{-4}$.

In Fig. 7 it is shown the plot of the errors in the approximated solution in the case of the extrapolation procedure. We note that the behavior of the approximated solution in a neighborhood of the inner non convex zone is very good (the error near the point $(0.5,0.5)$ is of order 10^{-6}).

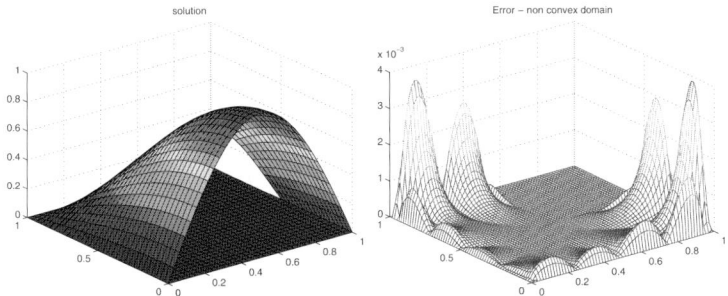

Fig. 7 Example 4: solution and errors in the approximated solution

6 Conclusions

The examples here presented show that the proposed method is simple and very effi-cient, also in presence of spikes, where no special treatments are required. Moreover, a good accuracy is reached also with a small set of data points.

References

1. Bacchelli, B., M. Bozzini, and C. Rabut, Polyharmonic wavelets based on Lagrangean functions, in *Curve and Surface Fitting: Avignon 2006*, A. Cohen, J.L. Merrien, and L.L. Schumaker (eds.), Nashboro Press, Brentwood, TN, 2007, 11–20.
2. Buhmann, M.D., Multivariate cardinal interpolation with radial basis functions, Constr. Approx. **6** (1990) 225–255.
3. Chen, C.S., G. Kuhn, J. Ling, and G. Mishuris, Radial basis functions for solving near singular Poisson problems, Commun. Numer. Meth. Eng. **19** (2003), 333–347.
4. Hörmander, L., The Analysis of Linear Partial Differential Operators, Springer, Berlin, Vol. 1, 1983.
5. Ladyzhenskaja, O.A., The boundary value problems of Mathematical Physics, Applied Mathematical Sciences, Springer, New York, Vol. 49, 1985.
6. Ling, L., and E.J. Kansa, A least-squares preconditioner for radial basis functions collocation methods, Adv. Comput. Math. **28** (2005), 31–54.
7. Madych, W.R., and S.A. Nelson, Polyharmonic cardinal splines, J. Approx. Theory **60** (1990), 141–156.
8. Micchelli, C., C. Rabut, and F. Utreras, Using the refinement equation for the construction of pre-wavelets, III: Elliptic splines, Numer. Algorithms **1** (1991), 331–352.
9. Rabut, C., Elementary polyharmonic cardinal B-splines, Numer. Algorithms **2** (1992), 39–46.
10. Schaback, R., and H. Wendland, Kernel techniques: from machine learning to meshless methods, Acta Numerica (2006), 1–97.

A Meshless Solution to the p-Laplace Equation

Francisco Manuel Bernal Martinez(✉) and Manuel Segura Kindelan

Abstract The p-Laplace equation is a non-linear elliptic PDE which plays an important role in the modeling of many phenomena in areas such as glaciology, non-Newtonian rheology or edge-preserving image deblurring. We have linearized it and applied a scheme introduced by G. Fasshauer which allows to solve it in the framework of Kansa's method. In order to confirm the validity of the approach, a 2D example (the pressure distribution in Hele-Shaw flow) has been numerically solved. The convergence and accuracy of the method are discussed, and an improvement based on smoothing up the linearized PDE is suggested.

Keywords: p-Laplace · Non-linear methods · Kansa Method

1 Introduction

The motivation for this work is the simulation of injection molding, a process of industrial relevance whereby molten polymer is driven into a cavity (the mold) in order to manufacture small plastic parts. If the polymer viscosity obeys a power law and the mold is thin compared to its planar dimensions, the classical mathematical model of injection molding is the Hele-Shaw approximation ([13]). In the remainder of this paper, we will restrict ourselves to isothermal Hele-Shaw flows, which physically arise whenever the fluid viscosity does not depend on temperature. In this case it suffices to solve the following 2D, non-linear, elliptic equation

$$div(|\nabla u|^{\gamma}\nabla u) = 0 \tag{1}$$

F.M.B. Martinez
G. Millán Institute for Modeling, Simulation and Industrial Mathematics, Universidad Carlos III de Madrid, 28911 Leganés, e-mail: fcoberna@math.uc3m.es

M.S. Kindelan
G. Millán Institute for Modeling, Simulation and Industrial Mathematics, Universidad Carlos III de Madrid, 28911 Leganés, Madrid, Spain, e-mail: kinde@ing.uc3m.es

A.J.M. Ferreira et al. (eds.) *Progress on Meshless Methods, Computational Methods in Applied Sciences.*
© Springer Science + Business Media B.V. 2009

whose solution yields the pressure distribution $u(x,y)$ in the filled region of the mold. Exponent γ completely characterizes the polymer rheology, and is typically $\gamma \approx 1/2$. We will assume dimensionless units: the independent variables in this problem have been scaled in such a way that the maximal fluid velocity is one (the reasons for this criterion will be evident in Section 3). If the pressure profile (p_{IN}) is set along the injection gates by the injection machine, the boundary conditions are

$$u = p_{IN} \ (injection) \qquad \partial u/\partial n = 0 \ (walls) \qquad u = 0 \ (front) \qquad (2)$$

From this pressure field, the average planar velocity $< \vec{v} >$ can be computed and the location of the advancing front can be updated:

$$< \vec{v} >= -|\nabla u|^{\gamma} \nabla u \qquad (3)$$

Although the boundary conditions (BCs) (2) are the most usual in commercial software, the rate Q at which the polymer is fed can be alternatively used. In this case, assuming that there is a single injection segment Γ, and always in dimensionless units, $\oint_{\Gamma} |\nabla u|^{\gamma}(\partial u/\partial n)dl = Q$. Instead, we will make the further assumption that the profile of $< \vec{v} >$ along the injection segment, q_{IN}, is known. The proper BCs for this situation are therefore non-linear,

$$-|\nabla u|^{\gamma}\partial u/\partial n = q_{IN} \ (injection) \qquad \partial u/\partial n = 0 \ (walls) \qquad u = 0 \ (front) \quad (4)$$

We will refer to (2) and (4) as *Dirichlet-injection* and *Neumann-injection* BCs, respectively.

The numerical simulation of the Hele-Shaw flow requires coupling (a) some method for solving Eq. (1) at every time step with (b) some technique to advance the front to its new position, until the mold domain has been completely filled. In the state-of-the-art approach (a) is accomplished through finite elements (FEM), whereas for (b), either the volume-of-flow (VoF) method is used, or the nodes along the front are tracked to their new positions. The latter option entails remeshing around the front at every time step, while the former avoids it at the price of forgoing a sharp frontline. Another disadvantage of FEM is the fact that the numerical interpolant is not differentiable along element borders, so that an averaging, upwind process is required in order to compute gradients on element nodes. In [4], an alternative, meshless framework was proposed for solving this problem combining the method of asymmetric Radial Basis Function (RBF) collocation for pressure with Level Sets for capturing the front motion. We believe that this approach has the potential to overcome some difficulties inherent to the FEM formulation.

Equation (1) is a *p-Laplace* (also called *p-harmonic*) equation of index $p = \gamma+2$. The p-Laplace operator

$$-\triangle_p u := div(|\nabla u|^{p-2} \nabla u) \qquad (5)$$

has been regarded as a counterpart to the Laplace operator for non-linear phenomena. The equation $\triangle_p u = 0$ has been the subject of extensive mathematical research for its own sake (see for instance [1]). In the remainder of this paper, we will consider the 2D case only (i.e. $u \in \mathbb{R}^d$ with $d = 2$), and will report on our recent progress in solving the non-linear elliptic PDE (1) in a meshless numerical environment. We believe that this approach could also be applied to other topical two-dimensional PDEs involving the p-Laplace operator as the core nonlinearity, such as the Perona-Malik equation for non-linear (edge-preserving) image denoising or the problem of finding the minimal surface resting on a given boundary.

2 Asymmetric RBF Collocation

2.1 Kansa's Method

The idea of using RBFs to solve PDEs was first introduced by Kansa [14, 15]. Consider the boundary-value problem (BVP) $L(u) = f(\vec{x})$ in domain Ω with boundary conditions along $\partial\Omega$ given by $G(u) = g(\vec{x})$, where L and G are linear operators. Ω is discretized into a set of $N = N_I + N_B$ scattered nodes (called *centers*) $\chi = \{\vec{x}_i \in \Omega, \ i = 1...N_I\} \cup \{\vec{x}_j \in \partial\Omega, \ j = N_I + 1...N_I + N_B\}$ and an approximate solution to the PDE is sought in the form of a linear combination of RBFs $\{\phi_k(\vec{x}), \ k = 1...N\}$ centered at each of them,

$$u(\vec{x}) = \sum_{k=1}^{N} \alpha_k \, \phi_k(\vec{x}), \qquad \phi_k(\vec{x}) \equiv \phi(\|\vec{x} - \vec{x}_k\|) \qquad (6)$$

Having L and G operate on the RBF, the unknown coefficients α_k are determined by appropriate collocation of either the PDE or the BC on N points, which usually – but not necessarily – are the same set of centers:

$$\sum_{k=1}^{N} \alpha_k \, L\phi_k(\vec{x}_i) = f(\vec{x}_i), \qquad i = 1, \ldots, N_I \qquad (7)$$

$$\sum_{k=1}^{N} \alpha_k \, G\phi_k(\vec{x}_j) = g(\vec{x}_j), \qquad j = N_I + 1, \ldots, N_I + N_B \qquad (8)$$

Inversion of the linear system (7–8) is guaranteed for positive-definite RBFs. In the event of a non-positive-definite RBF, positive-definiteness can be restored by adding a low order polynomial to (7–8) plus suitable constraints for the additional coefficients. However, even if the system (7–8) is formally solvable, it may be extremely ill-conditioned in practice. This drawback is compounded by 'Schaback's uncertanty principle', which establishes a trade-off between accuracy and ill-condition in RBF collocation.

   PDE collocation on boundary (PDEBC). Accuracy can be greatly improved (especially in presence of Neumann BCs) by enforcing the PDE on the boundary nodes also [12]. In this case expansion (6) must be supplemented with N_B extra RBF centers $\{\vec{x}_m, m = N + 1, \ldots, N + N_B\}$ in order to match the N_B new collocation equations. Since these extra centers are not to be collocated on, they may lie outside the PDE domain. With them, the RBF interpolant takes on the form

$$u(\vec{x}) = \sum_{k=1}^{N} \alpha_k \, \phi_k(\vec{x}) + \sum_{m=N+1}^{N+N_B} \alpha_m \, \phi_m(\vec{x}) \tag{9}$$

2.2 Nonlinear Equations

In the event of a non-linear PDE the collocation equations (7–8) give rise to a non-linear system of algebraic equations. An entirely different approach is the *operator-Newton method*, which was first introduced by Fasshauer in the context of meshless methods [8, 9], and which is sketched below:

Algorithm 1

- Let $Hu = 0$ be an elliptic non-linear PDE in Ω and L a linearization of it
- Pick an initial guess $u_{k=0}$ of solution. We seek v such that $H(u+v) = 0$
- *For $k = 1, 2...$ until convergence*

  - Compute residual $R_k = -Hu_{k-1}$
  - Solve $L_k v_k = R_k$ by Kansa's method, where $L_k = L(u_{k-1})$
  - Perform the smoothing of the Newton update, $\tilde{v}_k = S_k v_k$
  - Update the previous iterate, $u_k = u_{k-1} + \tilde{v}_k$

The idea behind the operator-Newton method is therefore to recast the nonlinear elliptic PDE into a succession of linear elliptic problems, whose coefficients are determined by the previous iterate. The smoothing step is optional (if $S = I$, smoothing is skipped) and designed to counter the phenomenon of *lack of derivatives*, which takes place whenever the numerical solution of the linearized PDE prevents the Newton iterations from achieving full quadratic convergence rate [10, 11]. Since Kansa's method has been shown to possess spectral convergence [7] in the solution of linear elliptic PDEs, Algorithm 1 is well suited to the RBF collocation framework, especially when the solution domain Ω has an irregular shape (such as that of an expanding fluid).

3 Linearization of the p-Laplace Equation

3.1 Linearization of the PDE

Let us define *flow fluidity* as $S(u) := |\nabla u|^\gamma$ so that Eq. (1) may be rewritten as $H(u) := div(S(u)\nabla u) = 0$. In order to linearize this equation, we assume that

$$|\nabla u| \gg |\nabla v| \tag{10}$$

and expand $S(u)$ in a Taylor series around the origin, retaining only terms to first order in $|\nabla v|/|\nabla u|$,

$$S(u+v) \approx S(u) + \gamma |\nabla u|^{\gamma-2} (\nabla u \cdot \nabla v) \qquad (11)$$

Therefore the truncation criterion is $\left(\frac{|\nabla v|}{|\nabla u|}\right)^2 \approx 0$. It will be useful to define $\vec{K}_u :=$ $\gamma |\nabla u|^{\gamma-2}\nabla u$, so that $S(u+v) \approx S(u) + \vec{K}_u \cdot \nabla v$. Now

$$H(u+v) \approx H(u) + div[S(u)\nabla v] + div[(\vec{K}_u \cdot \nabla v)\nabla u] + div[(\vec{K}_u \cdot \nabla v)\nabla v] \qquad (12)$$

The only remaining term which is not linear in v is the last one. Notice that

$$|(\vec{K}_u \cdot \nabla v)\nabla v| \le |\vec{K}_u| \cdot |\nabla v|^2 = \gamma|\nabla u|^{\gamma+1}\left(\frac{|\nabla v|}{|\nabla u|}\right)^2 = \gamma|<\vec{v}>|\left(\frac{|\nabla v|}{|\nabla u|}\right)^2 \approx 0 \qquad (13)$$

and therefore

$$div[(\vec{K}_u \cdot \nabla v)\nabla v] \approx 0 \qquad (14)$$

as long as $|<\vec{v}>| = O(1)$. Hence the convenience for normalizing the velocity field when recasting the independent variables into dimensionless form. The third term may be rewritten as

$$div[(\vec{K}_u \cdot \nabla v)\nabla u] = div[(\nabla u \cdot \nabla v)\vec{K}_u] \qquad (15)$$

It is also worthwhile noticing that

$$\vec{K}_u \cdot (\nabla v \cdot \nabla)\nabla u = \nabla v \cdot (\vec{K}_u \cdot \nabla)\nabla u = \nabla v \cdot \nabla S(u) \qquad (16)$$

After some manipulation we finally arrive at

$$H(u+v) \approx H(u) + S(u)\nabla^2 v + \nabla v \cdot \nabla S(u) + \vec{K}_u \cdot (\nabla u \cdot \nabla)\nabla v$$
$$+ \nabla S(u) \cdot \nabla v - \frac{(2-\gamma)\nabla u \cdot \nabla S(u) - \gamma S(u)\nabla^2 u}{|\nabla u|^2}(\nabla u \cdot \nabla v) \qquad (17)$$

Now particularize to $\mathbb{R}^2$ and define

$$A(x,y) = S\left[1 + \frac{\gamma}{|\nabla u|^2}\left(\frac{\partial u}{\partial x}\right)^2\right] \qquad (18)$$

$$B(x,y) = \frac{2\gamma S}{|\nabla u|^2}\left(\frac{\partial u}{\partial x}\right)\left(\frac{\partial u}{\partial y}\right) \qquad (19)$$

$$C(x,y) = S\left[1 + \frac{\gamma}{|\nabla u|^2}\left(\frac{\partial u}{\partial y}\right)^2\right] \qquad (20)$$

$$\vec{D}(x,y) = 2\nabla S + \frac{(\gamma-2)\nabla u \cdot \nabla S + \gamma S\nabla^2 u}{|\nabla u|^2}\nabla u \qquad (21)$$

where we have dropped the u in $S(u)$. Then, in order for $H(u+v) = 0$, the correction v must obey the following *linear* PDE:

$$A\frac{\partial^2 v}{\partial x^2} + B\frac{\partial^2 v}{\partial x \partial y} + C\frac{\partial^2 v}{\partial y^2} + \vec{D} \cdot \nabla v = R \tag{22}$$

where $R(x,y) := -H(v)$ is the residual of the current iterate u to the nonlinear PDE (1). We will refer to R as *nonlinear residual* (to distinguish it from the *linear residual* r, i.e. the residual of the RBF approximation to (22), which will appear in the ensuing discussion). Since $\triangle = B^2 - 4AC = -4(1+\gamma)S^2 < 0$, (22) is elliptic everywhere and can be always reduced to canonical form. As a guess of the solution (iteration 0), it seems natural to solve a Laplace equation (which would correspond to $\gamma = 0$) with the same BCs as in (2) or (4).

3.2 Linearization of the BCs

Dirichlet-injection case. In this case no linearization is needed and the BCs for the correction are homogeneous if the guess complies with (2),

$$v = p_{IN} - u \ (injection) \qquad \frac{\partial v}{\partial n} = -\frac{\partial u}{\partial n} \ (walls) \qquad v = -u \ (front) \tag{23}$$

Neumann-injection case. Here, the BC operator $Gu := |\nabla u|^\gamma \frac{\partial u}{\partial n}$ must be linearized

$$G(u+v) \approx |\nabla u|^\gamma \frac{\partial u}{\partial n} + \gamma \frac{\partial u}{\partial n} |\nabla u|^{\gamma-2}(\nabla u \cdot \nabla v) \tag{24}$$

Therefore, the BCs for v are

$$|\nabla u|^\gamma \frac{\partial v}{\partial n} + \gamma \frac{\partial u}{\partial n} |\nabla u|^{\gamma-2}(\nabla u \cdot \nabla v) = q_{IN} - |\nabla u|^\gamma \frac{\partial u}{\partial n} := R_{BC} \ (injection) \tag{25}$$

$$\frac{\partial v}{\partial n} = -\frac{\partial u}{\partial n} \ (walls) \qquad\qquad v = -u \ (front) \tag{26}$$

where R_{BC} is the residual to the non-linear BC. As the iterations progress and both $R \to 0$ and $R_{BC} \to 0$, v becomes the solution of a Laplace PDE with homogeneous BCs (if u_0 complies with the BCs of the non-linear problem) and vanishes.

There are a number of qualitative differences between this problem and the test problem analyzed in [8]. First, the present nonlinearity is a *differential operator* rather than a *function* of the solution. Secondly, both Dirichlet and Neumann BCs must be enforced, instead of only Dirichlet. Finally, the highest gradients take place along the boundary. In order to meet the latter two features, we have slightly modified Fasshauer's Algorithm 1 to incorporate PDEBC.

4 The Test Problem

4.1 Motivation

In order to validate the proposed method, we have solved (1) in a non-trivial domain. It is a square box $[0,1] \times [-\frac{1}{2}, \frac{1}{2}]$ with an elliptical insert $(\frac{x-1/2}{a})^2 + (\frac{y}{b})^2 = 1$, with $a = 2$, $b = 3$. The meshless discretization of such a domain is shown (for the injection-Dirichlet case) in Fig. 1. Most of the nodes stem from an $n \times n = 30 \times 30$ grid from which those inside the insert have been removed and $2n = 60$ further nodes have been added in order to model the internal, elliptical wall. There is also an additional layer of nodes at half-grid-constant distance of the boundary, which is intended to improve the accuracy of the RBF approximation close to the boundary, especially along the portions of it with Neumann BCs. As long as there are no coincidental nodes, no minimal distance among nodes has been enforced (it happens to be $h_{min} = 0.0018$). The set of RBF centers is the same as that of collocation nodes plus a set of extra centers required for PDEBC, which are placed outside the mold at a distance $h = 1/29$ (the grid constant) along the outward vector. The exponent γ is 0.6 (which models polyethylene). The mold has been taken from [4] with minor modifications, the most important of which is that the side $x = 1$ has been removed to be replaced by a freely moving frontline.

The RBF that we have used is the scaled Matérn function $M_{2,11,c}$ (see the Appendix for the definition). Since it is positive definite, there is no need for including polynomial terms into the RBF interpolant. The same RBF was used in [17] and found to significantly outperform MQs. Here, we have chosen to restrict c below the ill-condition threshold and found a nearly optimal value of $c = 0.1$, which gives rise (in our collocation point set) to a condition number of about 10^{13}. It performs indeed better than a MQ RBF with an equal or lower condition number (in the best case, the MQ delivers half of the accuracy of $M_{2,11,0.1}$) and, more importantly, is more stable with respect to the condition number. However, the Matérn RBF is more expensive to compute and has two tunable parameters, instead of just one as the MQ.

The injection gate is the segment $x = 0, -\xi \le y \le \xi, \xi = 0.15$, highlighted with an arrow in Fig. 1. Along it we have enforced either Neumann (4) and Dirichlet (2)

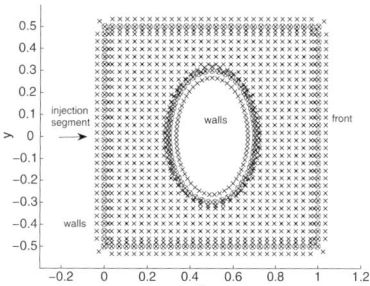

Fig. 1 Pointset 30×30 in the Dirichlet-injection case. Green/blue/red: PDE/Neumann BCs/Dirichlet BCs nodes. Crosses: RBF centers

Fig. 2 FEM solution

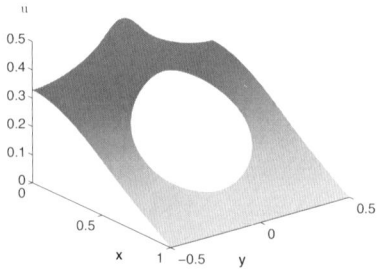

BCs. In the former case, the profile is

$$q_{IN}(y) = 0.9\left(1 - 3\left(\frac{y}{\varsigma}\right)^2 + 2\left(\frac{|y|}{\varsigma}\right)^3\right) \tag{27}$$

such that the exact solution is smooth at both ends $(0, \pm\varsigma)$ of the injection segment. In the injection-Dirichlet, the prescribed pressure $p_{IN}(y)$ is the exact solution of the problem with injection-Neumann BCs and profile like (27). Consequently, both versions of the problem have an identical solution. In lack of an analitical solution, a FEM approximation computed over 75,209 triangles has been used as reference. It is shown in Fig. 2.

4.2 Numerical Results

In order to trigger the Newton iterations, an initial guess of the solution must be provided to Algorithm 1. For simplicity, we have used as a guess the solution of the Laplace equation on the same domain and under identical BCs as the non-linear problem. However, other numerical experiments (not reported here) suggest that the scheme is reasonably robust with respect to the starting guess, especially in the Dirichlet-injection case.

Figure 3 shows the performance of the proposed method throughout the first seven iterations. Concretely, at iteration k, the error ε to the FEM solution, the non-linear residual R, and the linear residual (to the linearized equation) r are monitored over a fine mesh of 2,228 nodes scattered throughout the domain (and different from the collocation point set). The 'exact' solution values over this evaluation mesh have been linearly interpolated from the much finer FEM computational mesh. The goodness of the estimates ε, R, and r, is characterized by its root mean square (RMS) values, which is defined for a vector a of N elements as

$$RMS(a) = \sqrt{\frac{\sum_{i=1}^{i=N} a_i^2}{N}} \tag{28}$$

We first focus on the Dirichlet-injection case (solid line in Fig. 3). Starting from ≈ 0.01, the $RMS(\varepsilon)$ drops by two orders of magnitude in the two first Newton

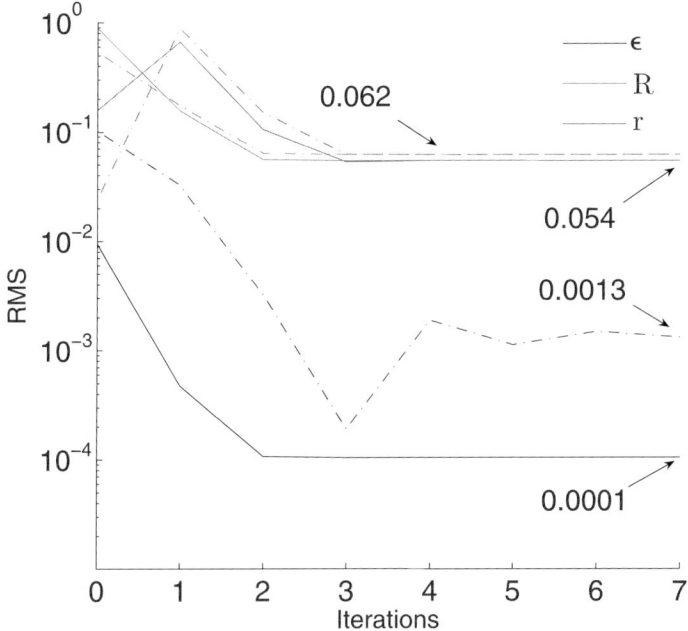

Fig. 3 Convergence of the operator-Newton scheme. Solid/broken line: Dirichlet/Neumann injection BCs. Iteration 0 corresponds to harmonic guess of solution

Fig. 4 A(x,y) it = 2
(Neumann injection)

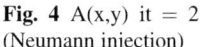

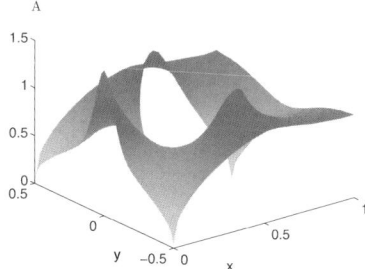

iterations (exhibiting a clearly superlinear convergence rate), and then stalls (although actually it keeps on dropping for a few more iterations, albeit at a negligible rate). Figures 6–9 demonstrate how the operator-Newton scheme accurately predicts the correction required to match the current error-notice that the algorithm itself of course ignores the true solution. Concerning R, it increases at first, and then also drops steadily until leveling off at about the $3rd-4th$ iteration. The fact that the *interpolation* non-linear residual R does not drop down to zero (or the error ε, for that purpose) is not surprising since the RBF method enforces the sequence of linear PDEs (and therefore the non-linear PDE) on a finite set of collocation nodes only.

Fig. 5 Initial guess (Neumann injection)

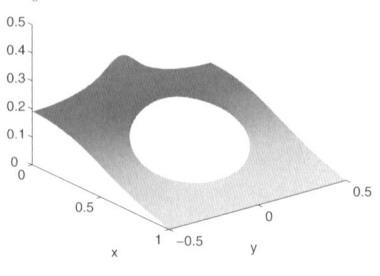

Fig. 6 Error it $= 0$

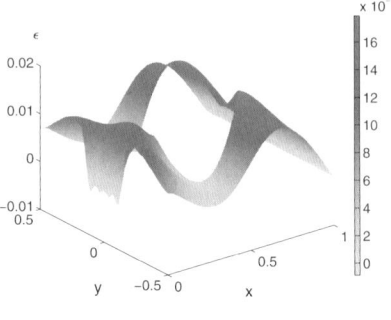

Fig. 7 Correction it $= 1$

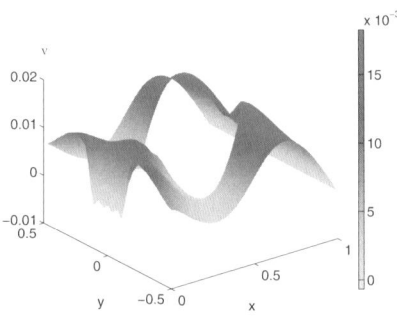

Fig. 8 Error it $= 1$

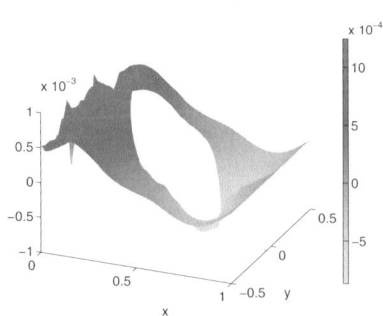

Indeed, the residual to the non-linear PDE averaged among the collocation nodes *does* drop to about 10^{-10} in some ten Newton iterations (not shown). However, the global accuracy depends on the collocation point set which models the domain and on the RBF interpolation space containing the numerical solution.

Fig. 9 Correction it $= 2$

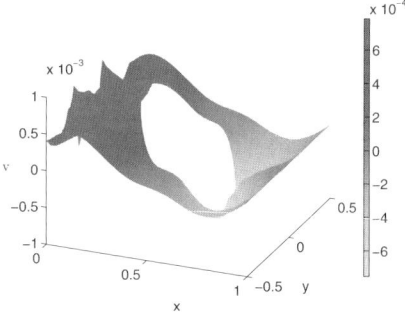

For the Neumann-injection case (broken line), the starting guess is shown in Fig. 5. Compared to the FEM solution in Fig. 2, the $0th$ error is apparent to the naked eye. Next, the $RMS(\varepsilon)$ decreases superlinearly during the first three iterations, but then 'bounces' and levels off, after some oscillations, at about 0.001. In both injection versions, the gain in accuracy from the guess is of the same order of magnitude. The performance of the numerical scheme in the Neumann case is understandably poorer, since derivative BCs are traditionally more difficult to deal with and because the accuracy of the RBF interpolation worsens roughly by an order of magnitude per derivative – which also fits the gap between the $RMSs(\varepsilon)$: 0.0001 and 0.0013. On the other hand, the asymptotic non-linear residual (chiefly made up of second derivatives) is very similar in both cases: 0.054 and 0.062. The quality of the approximation provided by the RBF-Newton scheme can be assessed by comparison with the accuracy of interpolation of the FEM solution with the same set of nodes and RBFs (without the exterior centers for PDEBC). Such interpolation values are: RMS(interpolation error) $= 5.9 \times 10^{-5}$, RMS(non-linear residual) $= 0.52$, and $MAX(|\text{nonlinear residual}|) = 4.73$ (the residual is based on derivating the interpolation solution).

Finally, it is apparent that the plots of the linear residual r merge with those of R. Moreover, the lines for R and r may cross but their collapsing together neatly signals the end of convergence of the error to the FEM solution. Since both estimates are available without knowledge of the exact solution, this phenomenon could furnish an efficient stopping criterion (without need to wait for the collocation non-linear residual to converge). The numerical equivalence of R and r means that the RBF scheme is no longer capable of capturing the features of the current linear PDE, because they take place *among* collocation nodes. Since the linear residual is defined as

$$r = R - A\frac{\partial^2 v}{\partial x^2} - B\frac{\partial^2 v}{\partial x \partial y} - C\frac{\partial^2 v}{\partial y^2} - \vec{D} \cdot \nabla v \qquad (29)$$

the fact that $r = R$ means that the RBF method is effectively 'seeing' a Laplace equation, whose solution is zero by virtue of the homogeneous BCs. Therefore, v vanishes and the convergence stagnates. Inversely, R is in turn affected by the linear residual r, which shows overshoots in intrincated or oversampled regions (like corners or where several RBF centers are too closely packed). Such spurious oscillations contaminate the non-linear residual R throughout the iterations until eventually masking it up. The final non-linear residual that remains in the pointset after convergence has stalled resembles noise, for it has zero average and is evenly distributed, except for the overshoots – see Fig. 12. The situation could be loosely summed up with the insight that the non-linear asymptotic residual settles at a value which is *not* the true R, but a value of it contaminated with the linear residual of the iterations prior to convergence.

Based on the above analysis, at least two improvement strategies, which turn out to be complementary, may be devised, namely: (a) reducing the linear residual at each linearized iteration, and (b) filter the spurious features out of the coefficients of the iterated PDEs. Regarding the first approach, it may be implemented by adaptively choosing the collocation nodes and the RBF centers, while keeping the size of the discretization support approximately constant, or by simply refining the point set (the effect of this on the injection-Neumann problem can be seen in the columns labeled I in Table 2. As for (b), we have tried out a simple idea related to the smoothing step of Algorithm 1, which will be explained in the following section.

5 Implicit Smoothing

5.1 Noise Removal in Interpolation Problems

Loosely speaking, the basic idea of denoising is to remove from a set of data those features whose characteristic length is below a given threshold. Convolution of the data with a low-pass filter is one such technique which allows a global interpretation of the denoising process: it attenuates (or supresses) the high frequencies present in the Fourier transform of the data. The usefulness of low-pass filters lies in the following two facts: the spectral components of the noise are typically higher than those of the data (which amounts to a shorter physical scale), and, if the highest frequency corresponding to a genuine structure in the signal can be established *a priori*, the filter cut-off frequency can be adjusted accordingly.

Since the action of a low-pass filter also removes the sharp features from the signal, convolution with a filter is also known as smoothing. The linear nature of the RBF representation enables a simple and unexpensive form of filtering through basic function substitution called *implicit smoothing*. We will just sketch the procedure and refer the reader to [2] for details. Consider the RBF expansion of a function u

$$u = \sum_{i=1}^{N} \lambda_i \phi(|\vec{x} - \vec{x}_i|) \tag{30}$$

By linearity, convolution with a linear filter ψ is

$$\tilde{u} := u \star \psi = \sum_{i=1}^{N} \lambda_i \left(\phi(|\vec{x} - \vec{x}_i|) \star \psi \right) \tag{31}$$

Therefore, if

$$\varphi := \phi \star \psi \tag{32}$$

is another RBF, denoising amounts to a change of basis in the interpolation space. In order to perform the filtering, one only has to interpolate the noisy data with the RBF ϕ, retain the expansion coefficients and replace ϕ by the smoothed up RBF φ. Both RBFs are implicitly related by the smoothing parameters (such as the cut-off frequency). The applicability of this technique is restricted to the triplets of viable functions (ϕ, ψ, φ) for which the relation (32) holds. A list of the possibilities currently available may be found in [2]. Among them, the scaled Matérn RBF is particularly simple, since

$$M_{d,\alpha,c} \star M_{d,\beta,c} = M_{d,\alpha+\beta,c} \tag{33}$$

The scaling parameter c can be interpreted in the physical space as the distance below which detail becomes severely blurred. For a gridded set of collocation points, it is natural to identify $c \approx h$ (with h the grid constant), because any feature below that length should be safely removable without loss of information, since the point set is not expected to capture it in the first place. For a point set made up of scattered data, determination of c is more elusive. We stress the heuristic nature of this approach, which performs surprisingly well in practical interpolation problems [6]. We will next try to take advantage of it in the PDE setting.

5.2 Application to PDEs

In order to illustrate the idea of smoothing up PDE coefficients, consider the following Poisson equation:

$$\nabla^2 u(x,y) = RHS, \ (x,y) \in \Omega \qquad u(x,y) = f(x,y), \ (x,y) \in \partial\Omega \tag{34}$$

The domain Ω is a box $[0,1] \times [-\frac{1}{2}, \frac{1}{2}]$, discretized into 30×30 gridded collocation nodes. $h = 1/29$ is the grid constant and an external layer of RBFs has been added at a distance h off the boundary for PDEBC. The RBF is $M_{2,5,c=h}$ and $f(x,y) = e^{-x^2}\cos(y)$. In order to assess the effect of noise on the RHS of (34), we have carried out three numerical experiments:

- RHS $= \nabla^2 f(x,y)$ (Clean RHS)
- RHS $= \nabla^2 f(x,y) + J\delta$ (Noisy RHS without smoothing)
- RHS $= \left(\nabla^2 f(x,y) + J\delta \right) \star M_{2,2,c=h}$ (Noisy RHS smoothed up)

Table 1 Poisson equation with a noisy RHS

	Clean RHS	Noisy RHS without smoothing	Noisy RHS smoothed up
RMS(ε)	4.5×10^{-6}	0.012	0.013
MAX(ε)	1.9×10^{-5}	0.036	0.038
RMS(r)	0.02	4.80	1.46
MAX(r)	0.11	18.54	6.86

Fig. 10 RHS
(after smoothing)
minus RHS (noisy source)

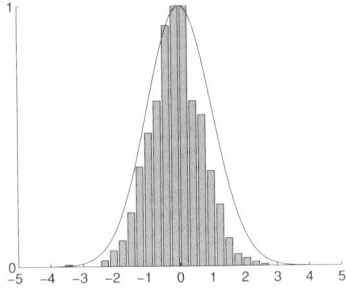

Fig. 11 RHS
(after smoothing)
minus RHS (clean source)

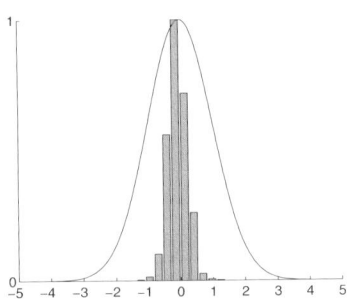

$J = 5$ is the amplitude of the noise and δ is a Gaussian distribution (of mean zero and unit variance and standard deviation). The results of the tests are shown in Table 1. It can be seen that the effect of the filter is small on the accuracy but evident on the residual. In the noisy and unsmoothed case, the residual is dominated by the RHS noise. The Matérn filter $M_{2,2,c=h}$ removes a fair amount of this noise from the residual without essentially affecting the signal, for the errors remain the same. The histogram in Fig. 10 shows the normalized distribution of the 'extracted' noise compared with the (also normalized) noise distribution. Figure 11 shows the 'remaining' noise in the RHS after smoothing. It is remarkable how the filter has eliminated all of the noise components except those of the smallest amplitude, even though the noise amplitude J is not considered among the smoothing parameters. The points of interest to the application of implicit smoothing to the operator-Newton method are however these two: (a) the residual to the elliptic PDE is easily masked up by the PDE coefficient noise, and (b) the convolution with a proper filter can remove an important part of the noise both from the coefficients and from the residual.

5.3 Operator-Newton Method with PDE Coefficient Smoothing

We have introduced implicit smoothing into the operator-Newton iterations in a diferent fashion than Algorithm 1. Instead of smoothing up the correction at every iteration, it is the set of coefficients to each linearized PDE that are smoothed up.

Regarding the implementation, we have adopted an approach in which the RBF involved in the operator-Newton iterations is independent from those involved in the smoothing process. At each iteration, the nodal values of each PDE coefficient are computed through Kansa's method with the RBF $M_{2,11,c=0.1}$, and captured on the same set of collocation nodes by the first scaled Matérn in the implicit smoothing triplet, the $M_{2,5,c=h}$. Then, they are interpolated on the same nodes by the 'smoothed up' RBF $M_{2,7,c=h}$, and this new vector of nodal values represents the smoothed up PDE coefficient.

As long as the collocation pointset remains fixed throughout the iterations, the smoothing of any quantity defined by a nodal vector reduces to a matrix-vector multiplication, where the (interpolation) matrix is computed at the beginning and stored. This makes of implicit smoothing a particularly efficient regularization technique.

Table 2 shows the effect of PDE coefficient smoothing on the error in the injection-Neumann BC version. For three different point sets based on 20×20, 30×30, and 40×40 grids, RMS(ε) is listed without (columns labeled I) and with smoothing (label S). In all cases, the iterated PDEs are solved with the RBF $M_{2,11,0.1}$, whereas the implicit smoothing is implemented with the triplet $M_{2,5,t}$ (capturing RBF), $M_{2,2,t}$ (filter), and $M_{2,7,t}$ (smoothed RBF).

Table 2 Effect of smoothing on Newton iterations

	$RMS(\varepsilon)$					
it	20×20		30×30		40×40	
0	0.10304		0.10195		0.10205	
1	0.02920		0.03266		0.03382	
2	0.00498	0.00519	0.00322	0.00438	0.00309	0.00358
3	0.00259	0.00253	0.00019	0.00187	0.00032	0.00329
4	0.00387	0.00252	0.00188	0.00172	0.00164	0.00245
5	0.00335	0.00276	0.00112	0.00088	0.00072	0.00208
6	0.00357	0.00267	0.00148	0.00120	0.00117	0.00193
7	0.00348	0.00266	0.00130	0.00097	0.00094	0.00127
8	0.00352	0.00266	0.00139	0.00103	0.00106	0.00153
9	0.00350	0.00264	0.00135	0.00095	0.00100	0.00076
10	0.00351	0.00264	0.00137	0.00095	0.00103	0.00123
11	0.00351	0.00263	0.00136	0.00091	0.00101	0.00047
12	0.00351	0.00263	0.00136	0.00089	0.00102	0.00101
13	0.00351	0.00264	0.00136	0.00087	0.00102	0.00032
14	0.00351	0.00265	0.00136	0.00086	0.00102	0.00085
15	0.00351	0.00266	0.00136	0.00085	0.00102	0.00026
	I	S	I	S	I	S

Fig. 12 Residual it = 2

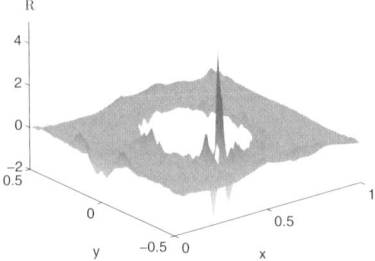

Fig. 13 Smoothed up residual
it = 2

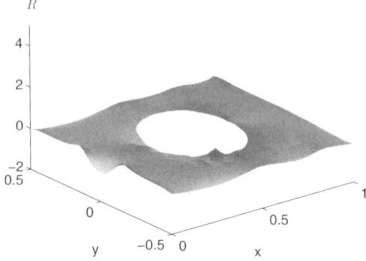

The smoothing length t (scaling parameter of the filter Matérn $M_{2,2,t}$) is set to the corresponding value of the grid constant, i.e. $1/19$, $1/29$, and $1/39$. Implicit smoothing is applied from the second Newton iteration only. The reason is that for the two first iterations a superlineal convergence rate is indeed observed and there-fore there should be no need for regularization. The effect of smoothing is different depending on the PDE coefficient. Coefficient $A(x,y)$, for instance, is fairly smooth (see Fig. 4) so that the application of smoothing only slightly 'files' the central peak. The effect on R, on the other hand, is quite remarkable as it can be seen in Figs. 12 and 13. Altogether, filtering the coefficients seems to lead to some improvement in the injection-Neumann case, despite the simplicity of the idea. However, in the Dirichlet-injection case (not shown), smoothing lacks any noticeable effect. It might be due to the fact that the error is already very close (twice as much) to the error of interpolating of the exact solution.

6 Conclusions

This paper builds up on our previous work on the Hele-Shaw flow which mod-els plastic injection molding [3–5]. This problem features a non-linear, unsteady, and free-boundary flow confined inside a potentially irregular 2D geometry, and therefore well suited to meshless discretizations. Here, we have focused on the numerical solution of the p-Laplace equation for the pressure. The numerical solu-tion of non-linear PDEs through asymmetric RBF collocation (Kansa's method) is still under-represented in the literature. Most authors have relied on general-purpose routines such as Powell's method or Levenberg-Marquardt's. We have preferred to investigate the operator-Newton algorithm proposed by G. Fasshauer, which recasts the non-linear PDE into a sequence of linear ones. Thus, the original problem remains inside the conceptual framework of Kansa's method, and can take advantage of its excellent performance in solving elliptic PDEs. Although the linearization of the p-Laplace equation is far from trivial, our numerical examples show good results, with the accuracy of solving the non-linear PDE coming very close to that of interpolating the exact solution after just a few operator-Newton iterations.

In order to improve the results without affecting efficiency, we are currently following two lines of research. The first one looks into an adaptive RBF approximation along the lines of the Greedy Algorithm in [16], and is still under investigation. The second one aims at regularization of the Newton iterates. A particularly unexpensive and RBF-friendly way of performing it is through implicit smoothing, which we have explored in this work with promising results.

Acknowledgements This work has been supported by the Spanish MECD grants MAT2005-05730 and FIS2007-62673 and by Madrid Autonomous Region grant S-0505. F. Bernal acknowledges E. Kansa, G. Fasshauer, R. Beatson and L. Ling for their most useful comments.

Appendix: The Scaled Matérn RBF and Its Derivatives

The scaled Matérn RBF in $\mathbf{R^d}$ is defined as

$$M_{d,\alpha,c}(r) = \frac{1}{c^d \pi^{d/2} 2^{(d+\alpha-2)/2} \Gamma(\alpha/2)} \left(\frac{r}{c}\right)^{(\alpha-d)/2} K_{(\alpha-d)/2}\left(\frac{r}{c}\right) \qquad (35)$$

where c is a scaling parameter and $K_v(z)$ is the modified Bessel function of the second kind. The parameter $v > 0$ controls the smoothness of the given Matérn function, such that $z^v K_v(z)$ is $\lfloor v - 1 \rfloor$ times differentiable.

The value at the origin is

$$M_{d,\alpha,c}(0) = \lim_{r \to 0} M_{d,\alpha,c}(r) = \frac{\Gamma(\frac{\alpha-d}{2})}{(2c\sqrt{\pi})^d \Gamma(\frac{\alpha}{2})} \qquad (36)$$

In order to simplify the notation, define

$$S_\mu(z) = z^\mu K_\mu(z) \qquad (37)$$

$$C_{d,\alpha} = \pi^{d/2} 2^{(d+\alpha-2)/2} \Gamma(\alpha/2) \qquad (38)$$

$$\mu = \frac{\alpha - d}{2} \qquad (39)$$

The derivatives of the scaled Matérn RBF are

$$M_{d,\alpha,c}(r) = \frac{1}{c^d C_{d,\alpha}} S_\mu\left(\frac{r}{c}\right) \tag{40}$$

$$\frac{\partial M_{d,\alpha,c}(r)}{\partial x} = -\frac{1}{c^{d+1} C_{d,\alpha}} \left(\frac{x - x_j}{c}\right) S_{\mu-1}\left(\frac{r}{c}\right) \tag{41}$$

$$\frac{\partial M_{d,\alpha,c}(r)}{\partial y} = -\frac{1}{c^{d+1} C_{d,\alpha}} \left(\frac{y - y_j}{c}\right) S_{\mu-1}\left(\frac{r}{c}\right) \tag{42}$$

$$\frac{\partial^2 M_{d,\alpha,c}(r)}{\partial x^2} = -\frac{1}{c^{d+2} C_{d,\alpha}} \left[S_{\mu-1}\left(\frac{r}{c}\right) - \left(\frac{x - x_j}{c}\right)^2 S_{\mu-2}\left(\frac{r}{c}\right)\right] \tag{43}$$

$$\frac{\partial^2 M_{d,\alpha,c}(r)}{\partial y^2} = -\frac{1}{c^{d+2} C_{d,\alpha}} \left[S_{\mu-1}\left(\frac{r}{c}\right) - \left(\frac{y - y_j}{c}\right)^2 S_{\mu-2}\left(\frac{r}{c}\right)\right] \tag{44}$$

$$\frac{\partial^2 M_{d,\alpha,c}(r)}{\partial x \partial y} = \frac{1}{c^{d+2} C_{d,\alpha}} \frac{(x - x_j)(y - y_j)}{c^2} S_{\mu-2}\left(\frac{r}{c}\right) \tag{45}$$

$$\nabla^2 M_{d,\alpha,c}(r) = \frac{1}{c^{d+2} C_{d,\alpha}} \left[\left(\frac{r}{c}\right)^2 S_{\mu-2}\left(\frac{r}{c}\right) - 2 S_{\mu-1}\left(\frac{r}{c}\right)\right] \tag{46}$$

References

1. G. Aronsson and U. Janfalk, *On Hele-Shaw flow of power-law fluids*, Eur. J. Appl. Math. **3**, 343–336 (1992).
2. R. K. Beatson and H.-Q. Bui, *Mollification Formulas and Implicit Smoothing*, Research report UCDMS 2003/19 (2003).
3. F. Bernal and M. Kindelan, *An RBF meshless method for injection molding modeling*, Lecture Notes in Computational Science and Engineering, Springer (2006).
4. F. Bernal and M. Kindelan, *RBF meshless modeling of non-Newtonian Hele-Shaw flow*, Eng. Anal. Boun. Elem. **31**, 863–874 (2007).
5. F. Bernal and M. Kindelan, *Meshless Simulation of Hele-Shaw Flow*, 14th European Conference on Mathematics for Industry ECMI-2006, to appear in Progress in Industrial Mathematics at ECMI 2006, Mathematics in Industry (2006).
6. J. C. Carr, R. K. Beatson, B. C. McCallum, W. R. Fright, T. J. McLennan, and T. J. Mitchell, *Smooth Surface Reconstruction from Noisy Range Data*. In GRAPHITE 2003, ACM Press, New York, 119–126 (2003).
7. A. H. D. Cheng, M. A. Golberg, E. J. Kansa, and G. Zammito, *Exponential convergence and h-c multiquadric collocation method for partial differential equations*. Numer. Meth. Part. D. Eq. **19**(5), 571–594 (2003).
8. G. E. Fasshauer, *Newton iteration with multiquadrics for the solution of nonlinear PDEs*, Comput. Math. Appl. **43**, 423–438 (2002).
9. G. E. Fasshauer, *Solving differential equations with radial basis functions: Multilevel methods and smoothing*, Adv. Comput. Math. **11**, 139–159 (1999).
10. G. E. Fasshauer, *On Smoothing for Multilevel Approximation with Radial Basis Functions*, Approximation Theory XI, Vol. II: Computational Aspects, C. K. Chui, and L. L. Schumaker (eds.), Vanderbilt University Press, Nashville, TN 55–62 (1999).
11. G. E. Fasshauer, E. C. Gartland, and J. W. Jerome, *Algorithms defined by Nash iteration: Some implementations via multilevel collocation and smoothing*, J. Comp. Appl. Math. **119**, 161–183 (2000).

12. A. I. Fedoseyev, M. J. Friedman, and E. J. Kansa, *Improved multiquadric method for elliptic partial differential equations via PDE collocation on the boundary*, Comput. Math. Appl. **43**, 439–455 (2002).
13. C. A. Hieber and S. F. Shen, *A finite-element/finite difference simulation of the injection-molding filling process*, J. Non-Newton. Fluid Mech. **7**, 1–32 (1979).
14. E. J. Kansa, *Multiquadrics - a scattered data approximation scheme with applications to computational fluid-dynamics. I. Surface approximations and partial derivative estimates*, Comput. Math. Appl. **19**, 127–145 (1990).
15. E. J. Kansa, *Multiquadrics - a scattered data approximation scheme with applications to computational fluid-dynamics. II. Solutions to parabolic, hyperbolic and elliptic partial differential equations*, Comput. Math. Appl. **19**, 147–161 (1990).
16. L. Ling, R. Opfer, and R. Schaback, *Results on meshless collocation techniques*, Eng. Anal. Bound. Elems. **30**(4), 247–253 (2006).
17. C. T. Mouat and R. K. Beatson, *RBF Collocation*, Research report UCDMS 2002/3 (2002).

Localized Radial Basis Functions with Partition of Unity Properties

Jiun-Shyan (JS) Chen(✉), Wei Hu, and Hsin-Yun Hu

Abstract In this paper we introduce a localization of radial basis function (RBF) under the general framework of partition of unity. A reproducing kernel that reproduces polynomials is used as the localizing function of RBF. It is shown that the proposed approach yields a similar convergence to that of the non localized RBF, while a better conditioned discrete system than that of the radial basis collocation method is achieved. Analyses of error and stability of the proposed method for solving boundary value problems are presented. Numerical examples are given to demonstrate the effectiveness of the proposed method.

Keywords: Radial basis functions · Local Radial Basis Functions · Partition of the Unity

1 Introduction

The radial basis function (RBF) was originally constructed for interpolation [14], and the multiquadrics RBF was shown to be related to the solution of the biharmonic potential problem and thus has a physical foundation [15]. RBF performs very well in interpolating highly irregular scattered data compared to many interpolation methods [5], and it has been introduced in high-dimensional interpolation in neural networks [13]. The work by [4] showed that RBFs are prewavelets (wavelets without orthogonality properties), and certain RBFs are effective projectors in multiresolution analysis particularly when the data structure is scattered [9]. RBF was first

J.-S. Chen and W. Hu
Department of Civil and Environmental Engineering, University of Califormia, Los Angeles, CA 90095-1593, USA, e-mail: jschen@seas.ucla.edu

H.-Y. Hu
Department of Mathematics, Tunghai University, Taichung 407, Taiwan, R.O.C.

A.J.M. Ferreira et al. (eds.) *Progress on Meshless Methods, Computational Methods in Applied Sciences.*
© Springer Science + Business Media B.V. 2009

applied to solving partial differential equations (PDEs) in [20,21] and the theoretical foundation of RBF method for solving PDEs has been well studied [12]. Applications of RBF for solving PDEs include, for example, singularity problems [18], Hamilton-Jacobi equations [6], fourth-order elliptic and parabolic problems [24], approximation in boundary element method for nonlinear elliptic PDEs [28], hyperbolic conservation laws [30], and smoothed multilevel approach [11]. When solving boundary value problems, RBF collocation method is shown to be more effective if boundary conditions are properly weighted [19].

While enjoying the exponential convergence, the RBF collocation method yields a fully dense discrete system and consequently ill-conditioned as the discrete dimension increases. This leads to a convergence problem in addition to the high computational cost in dealing with a full matrix, and it constitutes the major bottleneck in applying RBF to large scale computation. Several attempts have been made to resolve this difficulty. The ill-conditioning of the RBF collocation method has been reduced by introducing a compactly supported RBF truncated from polynomials that are strictly positive definite [31]. However, reasonable accuracy of these truncated polynomials can be achieved only when sufficiently large support is employed. Alternatively, several methods have been proposed to deal with solution of ill-conditioned discrete equations. The block partitioning method takes the advantage of better conditioning of each sub-block [22]. The multizone method for transient problem performs the transient solution of each smaller nonoverlapping zone that is better conditioned [32]. An adaptive algorithm [25] has been proposed to properly select suitable test and trial spaces iteratively.

In contrast to the global approximation in RBF collocation method based on strong form, meshfree local approximations have been introduced for solving partial differential equations based on weak form, for example, the moving least squares approximation [2, 23], the reproducing kernel approximation [7, 26], and the partition of unity [1, 10]. In these methods, the locality and smoothness of the approximation are defined in the kernel function with compact support, and basis functions are introduced either intrinsically or extrinsically to the kernel function to achieve certain order of completeness or to embed characteristic functions of the partial differential equation in the approximation. While these local approximations yield a well-conditioned banded discrete system, their solutions are typically with algebraic convergence.

In this work we introduce a combined reproducing kernel (RK) and radial basis function (RBF) approximation to achieve a local approximation that has similar convergence property as the RBF collocation method while producing a banded and better conditioned discrete system. The essential idea is to correct RBF with a compactly supported RK function that reproduces polynomials. Localizing RBF with polynomial reproducibility yields a convergence similar to RBF exponential convergence. Several numerical examples are analyzed to examine the performance of the proposed method.

2 Admissible Functions

2.1 Radial Basis Functions (RBF)

Let $\Omega \subset \Re^d$, $d \geq 1$, be a closed region with the boundary $\partial\Omega$, and let $\mathscr{S}$ be a set of N_s source points,

$$\mathscr{S} = \{\mathbf{x}_1, \mathbf{x}_2, \ldots, \mathbf{x}_{N_s}\} \subseteq \Omega \cup \partial\Omega. \tag{1}$$

For a function $u(\mathbf{x})$, the approximation, denoted by $v(\mathbf{x})$, is expressed as

$$v(\mathbf{x}) = \sum_{I=1}^{N_s} g_I(\mathbf{x}) a_I, \tag{2}$$

where a_I is the expansion coefficient, and $g_I(\mathbf{x})$ is the radial basis function, for example, the multiquadrics radial basis function (MQ RBF)

$$g_I(\mathbf{x}) = ((\mathbf{x} - \mathbf{x}_I)^2 + c^2)^{n-\frac{3}{2}}, \quad n = 1, 2, 3, \cdots \tag{3}$$

where $\mathbf{x}_I \in \mathscr{S}$, and the constant c in Eq. (3) is called the shape parameter of RBF. The convergence of RBF has been studied by [27], and it has been shown that there exist an exponential convergence in RBF

$$|u(\mathbf{x}) - v(\mathbf{x})| \leq C\theta^{c/h} \|u\|_t, \tag{4}$$

where $0 < \theta < 1$, and h is the radial distance defined as

$$h := h(\Omega, \mathscr{S}) = \sup_{\mathbf{x} \in \Omega} \min_{\mathbf{x}_I \in \mathscr{S}} \|\mathbf{x} - \mathbf{x}_I\|, \tag{5}$$

and $\| \cdot \|_t$ is induced from the following restrictive conditions

$$\int_{\Re^d} \frac{|\tilde{u}(\mathbf{s})|^2}{\tilde{g}_I(\mathbf{s})} \, d\mathbf{s} \leq \infty, \tag{6}$$

$$\int_{\Re^d} \|\mathbf{s}\|^z \tilde{g}_I(\mathbf{s}) \, d\mathbf{s} \leq \rho^z z!, \tag{7}$$

where $\tilde{u}(\mathbf{s})$ and $\tilde{g}_I(\mathbf{s})$ are Fourier transformations of $u(\mathbf{x})$ and $g_I(\mathbf{x})$, respectively, ρ is a positive constant, and z is a sufficiently large number.

The accuracy and rate of convergence of MQ RBF approximation is determined by the shape parameter c and the number of basis functions (the number of source points) N_s. The exponential convergence of RBF, however, is overshadowed by its global (nonlocal) approximation in solving PDEs, which yields a full matrix in the discrete system. Further, under the collocation framework in solving PDEs, the condition number of the discrete differential operator increases radically as the number of source points increases. It is also imperative to understand, as has been

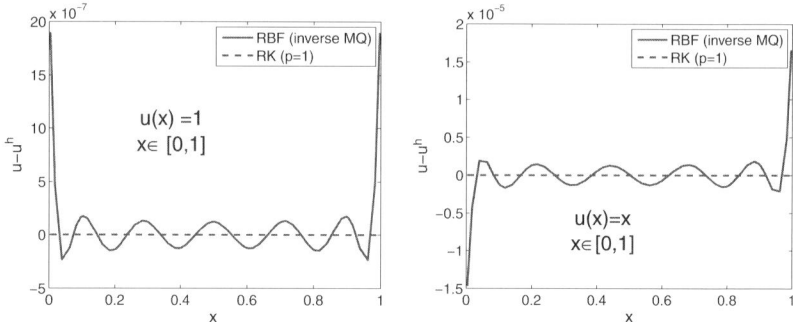

Fig. 1 Errors in the reproduction of one-dimensional constant and linear polynomial functions by RBF and RK functions

shown [3,29], that RBF is capable of reproducing constant and linear functions only in infinite domain:

$$\sum_{I \in Z} (\alpha + \beta hI) \, g_I(\mathbf{x}) = \alpha + \beta \mathbf{x}. \tag{8}$$

In other words, RBF cannot reproduce constant and linear functions in finite domain with finite number of source points. As shown in Fig. 1, RBF exhibits errors in reproducing constant and linear functions in a finite domain $[0,1]$, while reproducing kernel function [7,26] with linear basis $(p = 1)$ yields exact reproduction of constant and linear functions using 11 random points.

2.2 Reproducing Kernel (RK) Shape Functions

Let $\mathscr{T}$ be a set of N_p points in Ω and on boundary $\partial\Omega$,

$$\mathscr{T} = \{\mathbf{x}_1, \mathbf{x}_2, \dots, \mathbf{x}_{N_p}\}, \quad \mathbf{x}_I \in \bar{\Omega}, \quad I = 1, 2, \cdots, N_p. \tag{9}$$

The set is used to define a finite open covering $\mathscr{C} = \{\omega_I\}_{I=1}^{N_p}$ of Ω, $\Omega \subset \cup_{I=1}^{N_p} \omega_I$, as shown in Fig. 2. A class of functions $\{\phi_I(\mathbf{x})\}_{I=1}^{N_p}$ is called a partition of unity subordinated to the open covering $\mathscr{C}$ if it possesses the following property:

$$\sum_{I=1}^{N_p} \phi_I(\mathbf{x}) = 1, \quad \forall \mathbf{x} \in \Omega. \tag{10}$$

The partition of unity function can be constructed by

$$\phi_I(\mathbf{x}) = \varphi_a(\mathbf{x} - \mathbf{x}_I) b(\mathbf{x}). \tag{11}$$

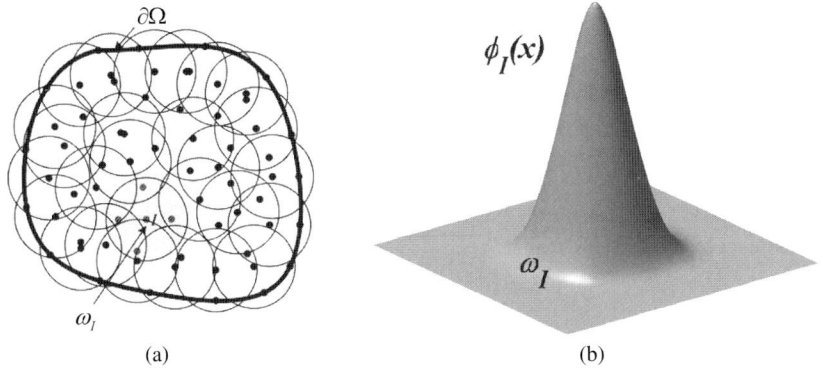

Fig. 2 (a) Discretization of domain by finite covers; (b) Reproducing kernel function

where $\varphi_a(\mathbf{x} - \mathbf{x}_I)$ kernel function defined on the finite cover ω_I centered at $\mathbf{x}_I$, for example, a cubic B-spline function, and a is the radius of the finite cover ω_I. Usually $\varphi_a(\mathbf{x} - \mathbf{x}_I) > 0$ for $\mathbf{x} \in \omega_I$ is selected. Here $b(\mathbf{x})$ is a function to be thought to meet partition of unity condition

$$\sum_{J=1}^{N_p} \varphi_a(\mathbf{x} - \mathbf{x}_J)b(\mathbf{x}) = 1. \tag{12}$$

By obtaining $b(\mathbf{x}) = 1/\sum_{J=1}^{N_p} \varphi_a(\mathbf{x} - \mathbf{x}_J)$, we have

$$\phi_I(\mathbf{x}) = \frac{\varphi_a(\mathbf{x} - \mathbf{x}_I)}{\sum_{J=1}^{N_p} \varphi_a(\mathbf{x} - \mathbf{x}_J)}. \tag{13}$$

The partition of unity function in Eq. (13) is the Shepard function. To achieve higher order completeness, a modification of Eq. (11) has been proposed

$$\phi_I(\mathbf{x}) = \left\{ \sum_{|\alpha| \leq p} (\mathbf{x} - \mathbf{x}_I)^\alpha b_\alpha(\mathbf{x}) \right\} \varphi_a(\mathbf{x} - \mathbf{x}_I). \tag{14}$$

Here we use the multi-dimensional notation $\alpha = (\alpha_1, \ldots, \alpha_d)$ with $d > 1$ being integer representing the dimension, the quantity $|\alpha| = \sum_{I=1}^{d} \alpha_I$ is the length of α, $\mathbf{x}^\alpha = x_1^{\alpha_1} \cdots x_d^{\alpha_d}$, $\{(\mathbf{x} - \mathbf{x}_I)^\alpha\}_{|\alpha| \leq p}$ is a set of monomial basis functions and $b_\alpha(\mathbf{x}), |\alpha| \leq p$, are the coefficients of the basis functions that vary with the location of approximation $\mathbf{x}$.

The coefficients $b_\alpha(\mathbf{x})$ is determined from the following reproducing conditions:

$$\sum_{I=1}^{N_p} \phi_I(\mathbf{x})\mathbf{x}_I^\alpha = \mathbf{x}^\alpha, \quad |\alpha| \leq p. \tag{15}$$

Equation (15) is equivalent to

$$\sum_{I=1}^{N_p} \phi_I(\mathbf{x})(\mathbf{x} - \mathbf{x}_I)^\alpha = \delta_{|\alpha|,0}, |\alpha| \leq p, \tag{16}$$

where δ_{ij} denotes the Kronecker delta function, or

$$\sum_{I=1}^{N_p} \phi_I(\mathbf{x})\mathbf{H}(\mathbf{x} - \mathbf{x}_I) = \mathbf{H}(\mathbf{0}), \tag{17}$$

$$\mathbf{H}^T(\mathbf{x} - \mathbf{x}_I) = [1, (x_1 - x_{1I}), \cdots, (x_d - x_{dI}), (x_1 - x_{1I})^2, \cdots, (x_d - x_{dI})^p]. \tag{18}$$

Substituting Eq. (14) into Eq. (17) to obtain $b_\alpha(\mathbf{x})$, and the reproducing kernel $\phi_I(\mathbf{x})$ in Eq. (14) reads

$$\phi_I(\mathbf{x}) = \mathbf{H}^T(\mathbf{0})\mathbf{M}^{-1}(\mathbf{x})\mathbf{H}(\mathbf{x} - \mathbf{x}_I)\varphi_a(\mathbf{x} - \mathbf{x}_I), \tag{19}$$

where

$$\mathbf{M}(\mathbf{x}) = \sum_{I=1}^{N_p} \mathbf{H}(\mathbf{x} - \mathbf{x}_I)\mathbf{H}^T(\mathbf{x} - \mathbf{x}_I)\varphi_a(\mathbf{x} - \mathbf{x}_I), \tag{20}$$

The function $\phi_I(\mathbf{x})$ is called the reproducing kernel approximation function. Conditions assuring the non-singularity of the matrix $\mathbf{M}(\mathbf{x})$ are discussed in [16], where rigorous convergence analysis and error estimates of the method are also provided.

3 Localized RBF with Polynomial Reproducibility

Different from the localization of RBF by Wendland [31] where RBF are localized by a truncated polynomials, here we introduce reproducing kernel with polynomial reproducibility as the localizing function of RBF under partition of unity framework [1].

Consider the following approximation

$$u^h(\mathbf{x}) = \sum_{I=1}^{N} \{\phi_I(\mathbf{x})(a_I + g_I(\mathbf{x})d_I)\} \tag{21}$$

where we let $N_p = N_s = N$. In general, different RBF or other basis functions can be employed in the approximation (21) to yield:

$$u^h(\mathbf{x}) = \sum_{I=1}^{N} \left\{ \phi_I(\mathbf{x})(a_I + \sum_{j=1}^{M} g_{Ij}(\mathbf{x}) \, d_{Ij}) \right\} \tag{22}$$

where g_{Ij} denotes RBF or other basis functions, and d_{Ij} is the corresponding coefficient. Rewrite the localized RBF approximation in Eq. (22) as

$$u^h(\mathbf{x}) = \sum_{I=1}^{N} \phi_I(\mathbf{x}) u_I^h(\mathbf{x}) \tag{23}$$

where

$$u_I^h(\mathbf{x}) \in V_I, \quad V_I = V_I(\omega_I) = \text{span}\{1, g_I^1, g_I^2, \cdots, g_I^M\}, \tag{24}$$

and

$$u^h(\mathbf{x}) \in V, \quad V = \cup_{I=1}^{N} V_I. \tag{25}$$

Assume that local approximation space V_I possesses the following property:

$$\|u - u_I^h\|_{\ell, \Omega \cap \omega_I} \le T(c, \ell, I, N, M) \|u\|_t, \quad \forall I, \ \ell \ge 0, \tag{26}$$

where $\|\cdot\|_{\ell, \Omega \cap \omega_I}$ denotes the Sobolev norm, $T(\cdot)$ is a term dependent on some parameters, and $\|\cdot\|_t$ is defined as in Sect. 2.1. Since $h = \mathscr{O}(N^{-1/d})$, thus $T(c, \ell, I, N, M) = \bar{T}(c, \ell, I, h, M)$. There exist an algebraic decay if monomial bases are chosen, and there exist an exponential decay [17, 18] if the Fourier functions or radial basis functions are used within the radius of convergence a. There exist a general error bound as follows [17, 18]

$$\|u(\mathbf{x}) - v(\mathbf{x})\|_{\ell, \Omega_a} \le C \eta_\ell^{c/h} \|u\|_t, \ \ell \ge 0, \tag{27}$$

where η_ℓ denotes a positive real number, $0 < \eta_0 < \eta_1 < \eta_2 < \cdots < 1$, and subdomain $\Omega_a \subset \Omega$. Denote

$$T_\ell = \max_I \bar{T}(c, \ell, I, h, M) = \mathscr{O}(\eta_\ell^{c/h}), \quad \ell = 0, 1, 2, \cdots . \tag{28}$$

Consider a quasi-uniform support size distribution. By using the condition of partition of unity Eq. (10), we have

$$\sum_{I=1}^{N} \phi_I(u - u_I^h) = u - \sum_{I=1}^{N} \phi_I u_I^h = u - u^h. \tag{29}$$

By using Eq. (26), there exist a global estimate [1] as follows

$$\|u - u^h\|_{0,\Omega}^2 = \|\sum_{I=1}^{N} \phi_I(u - u_I^h)\|_{0,\Omega}^2 = \int_{\Omega} |\sum_{I=1}^{N} \phi_I(u - u_I^h)|^2 d\Omega$$

$$\le \kappa \int_{\Omega} \sum_{I=1}^{N} |\phi_I(u - u_I^h)|^2 d\Omega = \kappa \int_{\Omega} \sum_{I=1}^{N} |\phi_I|^2 |u - u_I^h|^2 d\Omega$$

$$\leq \kappa \, C_\infty^2 \sum_{I=1}^{N} \int_{\Omega \cap \omega_I} |u - u_I^h|^2 d\Omega$$

$$= \kappa C_\infty^2 \sum_{I=1}^{N} \|u - u_I^h\|_{0,\Omega \cap \omega_I}^2 \leq C_\infty^2 \kappa^2 \, T_0^2 \, \|u\|_t^2, \tag{30}$$

where κ denotes the maximal number of covers for any $\mathbf{x} \in \Omega$, usually, $1 < p < \kappa << N$, and $|\phi_I|_\infty \leq C_\infty$.

By taking the square root of Eq. (30), and then using Eq. (27), we have

$$\|u - u^h\|_{0,\Omega} \leq \kappa C_\infty T_0 \|u\|_t \leq C \, \eta_0^{c/h} \|u\|_t. \tag{31}$$

where $C = \kappa C_\infty$ is a generic constant.

Note that for the boundary estimates, $\|u - u^h\|_{0,\partial\Omega}$ and $\|u_\nu - u_\nu^h\|_{0,\partial\Omega}$, where ν is the outer normal on boundary, they can be done in a similar manner. Those estimates can be obtained as a domain case in $d - 1$ dimensional space.

Let $D^\ell, \ell \geq 1$, be the ℓ-th order partial derivative operator. It follows that

$$\|u - u^h\|_{\ell,\Omega}^2 = \| \sum_{I=1}^{N} \phi_I (u - u_I^h)\|_{\ell,\Omega}^2 = \|D^\ell \sum_{I=1}^{N} \phi_I (u - u_I^h)\|_{0,\Omega}^2$$

$$\leq 2\| \sum_{I=1}^{N} (D^\ell \phi_I)(u - u_I^h)\|_{0,\Omega}^2 + 2\| \sum_{I=1}^{N} \phi_I D^\ell (u - u_I^h)\|_{0,\Omega}^2$$

$$\leq 2\kappa \int_\Omega \sum_{I=1}^{N} |D^\ell \phi_I|^2 |u - u_I^h|^2 d\Omega + 2\kappa \int_\Omega \sum_{I=1}^{N} |\phi_I|^2 |D^\ell (u - u_I^h)|^2 d\Omega$$

$$\leq 2\kappa \int_{\Omega \cap \omega_I} \sum_{I=1}^{N} |D^\ell \phi_I|^2 |u - u_I^h|^2 d\Omega + 2\kappa \int_{\Omega \cap \omega_I} \sum_{I=1}^{N} |\phi_I|^2 |D^\ell (u - u_I^h)|^2 d\Omega$$

$$\leq 2\kappa C^2 a^{-2\ell} \sum_{I=1}^{N} \|u - u_I^h\|_{0,\Omega \cap \omega_I}^2 + 2\kappa C_\infty^2 \sum_{I=1}^{N} \|u - u_I^h\|_{\ell,\Omega \cap \omega_I}^2,$$

$$\leq 2C^2 a^{-2\ell} \kappa^2 \, T_0^2 \, \|u\|_t^2 + 2C_\infty^2 \kappa^2 \, T_\ell^2 \, \|u\|_t^2, \tag{32}$$

where T_ℓ is defined as in Eq. (28), and $|D^\ell \phi_I(\mathbf{x})|_\infty \leq Ca^{-\ell}$ has been used. Moreover, we have

$$\|u - u^h\|_{\ell,\Omega} \leq (C_1 \kappa a^{-\ell} \eta_0^{c/h} + C_2 \kappa \eta_\ell^{c/h}) \|u\|_t. \tag{33}$$

In addition, the inverse inequalities are presented here for the study of convergence and stability in the next section. Let the approximation in Eq. (22) be written as

$$v := u^h(\mathbf{x}) = \sum_{I=1}^{N} \phi_I(\mathbf{x}) a_I + \sum_{I=1}^{N} \sum_{j=1}^{M} \psi_{Ij}(\mathbf{x}) d_{Ij} =: v_1 + v_2, \tag{34}$$

where

$$\psi_{Ij}(\mathbf{x}) = \phi_I(\mathbf{x})g_{Ij}(\mathbf{x}). \tag{35}$$

We have the local inverse estimates as follows (see [8] for details)

$$\|v_1\|_{\ell,\omega_I}^2 \leq C_1 a^{-\ell d} p^{2\ell d} \|v_1\|_{0,\omega_I}^2, \quad \text{for } \ell \geq 1,$$
$$\|v_2\|_{\ell,\omega_I}^2 \leq (C_2 a^{-3\ell d/2} p^{2\ell d} + C_3\, a^{-3\ell d/2} \mu^{\ell d}) \|v_2\|_{0,\omega_I}^2, \quad \text{for } \ell \geq 1, \tag{36}$$

where a is the maximal radius of finite cover ω_I, p is the reproducing degree, d is the space dimension, μ is the maximal number of the radial basis function within cover ω_I, and C_i is a generic constant. Furthermore, we obtain a global inverse estimate

$$\|v\|_{\ell,\Omega}^2 = \|v_1 + v_2\|_{\ell,\Omega}^2 \leq 2\|v_1\|_{\ell,\Omega}^2 + 2\|v_2\|_{\ell,\Omega}^2$$
$$\leq C_4 \left\{ \sum_{I=1}^N \|v_1\|_{\ell,\omega_I}^2 \right\} + C_5 \left\{ \sum_{I=1}^N \|v_2\|_{\ell,\omega_I}^2 \right\}. \tag{37}$$

It follows that

$$\|v\|_{\ell,\Omega} \leq (C_6\, \kappa^{1/2} a^{-\ell d/2} p^{\ell d} + C_7\, \kappa^{1/2} a^{-3\ell d/4} p^{\ell d} + C_8\, \kappa^{1/2} a^{-3\ell d/4} \mu^{\ell d/2}) \|v\|_{0,\Omega}, \tag{38}$$

where $1 < p < \kappa, \mu << N$, and C_i is a generic constant.

4 Weighted Local Radial Basis Collocation Method

4.1 Weighted Collocation Method

Consider the following general form of a boundary value problem:

$$\begin{aligned} \mathbf{L}\,\mathbf{u} &= \mathbf{f}, && \text{in} \quad \Omega, \\ \mathbf{B}^h \mathbf{u} &= \mathbf{h}, && \text{on} \quad \partial\Omega^h, \\ \mathbf{B}^g \mathbf{u} &= \mathbf{g}, && \text{on} \quad \partial\Omega^g, \end{aligned} \tag{39}$$

where Ω is the problem domain, $\partial\Omega^h$ is the Neumann boundary, $\partial\Omega^g$ is the Dirichlet boundary, $\partial\Omega^h \cup \partial\Omega^g = \partial\Omega$, $\mathbf{L}$ is the differential operator in Ω, $\mathbf{B}^h$ is the differential operator on $\partial\Omega^h$, and $\mathbf{B}^g$ is the operator on $\partial\Omega^g$.

We consider the weighted collocation method [19] to seek the solution of problem Eq. (39). For simplicity, let the component u_k of unknown $\mathbf{u}$ be approximated by

$$v_k = \sum_{I=1}^N \phi_I(\mathbf{x})a_{Ik} + \sum_{I=1}^N \psi_I(\mathbf{x})b_{Ik}, \quad \forall v_k \in V, \tag{40}$$

where v_k is the component of $\mathbf{v}$. Thus, we have

$$\mathbf{v} = \begin{bmatrix} v_1 \\ v_2 \\ \vdots \\ v_d \end{bmatrix} = \mathbf{\Psi}^T(\mathbf{x})\mathbf{c}, \tag{41}$$

where $\mathbf{v} \in U$, and $U = V \times V \times \cdots \times V$. For example, in two dimension, $d = 2$, it follows that

$$\mathbf{\Psi}^T(\mathbf{x}) = [\mathbf{\Psi}_1(\mathbf{x}), \mathbf{\Psi}_2(\mathbf{x}), \cdots, \mathbf{\Psi}_N(\mathbf{x})], \quad \mathbf{\Psi}_I(\mathbf{x}) = \begin{bmatrix} \phi_I & 0 & \psi_I & 0 \\ 0 & \phi_I & 0 & \psi_I \end{bmatrix} \tag{42}$$

$$\mathbf{c}^T = [\mathbf{c}_1, \mathbf{c}_2, \cdots, \mathbf{c}_N], \quad \mathbf{c}_I^T = [a_{I1}, a_{I2}, b_{I1}, b_{I2}].$$

The optimal solution $\mathbf{u}^h \in U$ satisfies the following discrete variational problem

$$\hat{E}(\mathbf{u}^h) = \min_{\mathbf{v} \in U} \hat{E}(\mathbf{v}), \tag{43}$$

where

$$\hat{E}(\mathbf{v}) = \frac{1}{2} \left\{ \widehat{\int}_\Omega (\mathbf{Lv} - \mathbf{f})^2 d\Omega + \alpha^h \widehat{\int}_{\partial\Omega^h} (\mathbf{B}^h\mathbf{v} - \mathbf{h})^2 dL + \alpha^g \widehat{\int}_{\partial\Omega^g} (\mathbf{B}^g\mathbf{v} - \mathbf{g})^2 dL \right\},$$

and $\widehat{\int}$ denotes the numerical integral. To balance errors in domain Ω and on boundaries $\partial\Omega^h$ and $\partial\Omega^g$, the boundary terms are weighted by α^h and α^g [19]. The linear system resulting from the weighted least squares functional is equivalent to solving the following overdetermined system (with the number of collocation points greater than the number of source points) by least squares approach:

$$\begin{bmatrix} \mathbf{A}^1 \\ \sqrt{\alpha^h}\,\mathbf{A}^2 \\ \sqrt{\alpha^g}\,\mathbf{A}^3 \end{bmatrix} \mathbf{a} = \begin{bmatrix} \mathbf{b}^1 \\ \sqrt{\alpha^h}\,\mathbf{b}^2 \\ \sqrt{\alpha^g}\,\mathbf{b}^3 \end{bmatrix}. \tag{44}$$

where $\mathbf{A}^1, \mathbf{A}^2, \mathbf{A}^3$ are the matrices associated with the operators $\mathbf{L}$, $\mathbf{B}^h$, and $\mathbf{B}^g$ evaluated at the collocation points, respectively,

$$\mathbf{A}^1 = \begin{bmatrix} \mathbf{L}(\mathbf{\Psi}^T(\mathbf{p}_1)) \\ \mathbf{L}(\mathbf{\Psi}^T(\mathbf{p}_2)) \\ \vdots \\ \mathbf{L}(\mathbf{\Psi}^T(\mathbf{p}_{N_p})) \end{bmatrix}, \quad \mathbf{A}^2 = \begin{bmatrix} \mathbf{B}^h(\mathbf{\Psi}^T(\mathbf{q}_1)) \\ \mathbf{B}^h(\mathbf{\Psi}^T(\mathbf{q}_2)) \\ \vdots \\ \mathbf{B}^h(\mathbf{\Psi}^T(\mathbf{q}_{N_q})) \end{bmatrix}, \quad \mathbf{A}^3 = \begin{bmatrix} \mathbf{B}^g(\mathbf{\Psi}^T(\mathbf{r}_1)) \\ \mathbf{B}^g(\mathbf{\Psi}^T(\mathbf{r}_2)) \\ \vdots \\ \mathbf{B}^g(\mathbf{\Psi}^T(\mathbf{r}_{N_r})) \end{bmatrix}, \tag{45}$$

where

$$[\mathbf{p}_1, \mathbf{p}_2, \cdots, \mathbf{p}_{N_p}] \subseteq \Omega, \; [\mathbf{q}_1, \mathbf{q}_2, \cdots, \mathbf{q}_{N_q}] \subseteq \partial\Omega^h, \; [\mathbf{r}_1, \mathbf{r}_2, \cdots, \mathbf{r}_{N_r}] \subseteq \partial\Omega^g, \tag{46}$$

and the $\mathbf{b}^1, \mathbf{b}^2, \mathbf{b}^3$ are induced vectors associated with the functions $\mathbf{f}, \mathbf{h}$ and $\mathbf{g}$, respectively,

$$
\mathbf{b}^1 = \begin{bmatrix} \mathbf{f}(\mathbf{p}_1) \\ \mathbf{f}(\mathbf{p}_2) \\ \vdots \\ \mathbf{f}(\mathbf{p}_{N_p}) \end{bmatrix}, \quad
\mathbf{b}^2 = \begin{bmatrix} \mathbf{h}(\mathbf{q}_1) \\ \mathbf{h}(\mathbf{q}_2) \\ \vdots \\ \mathbf{h}(\mathbf{q}_{N_q}) \end{bmatrix}, \quad
\mathbf{b}^3 = \begin{bmatrix} \mathbf{g}(\mathbf{r}_1) \\ \mathbf{g}(\mathbf{r}_2) \\ \vdots \\ \mathbf{g}(\mathbf{r}_{N_r}) \end{bmatrix}.
\tag{47}
$$

Note that the α^h and α^g are weights associated with the Neumann boundary $\partial\Omega^h$ and the Dirichlet boundary $\partial\Omega^g$, respectively. As derived from error analysis in [19], balanced errors in domain and boundary terms can be achieved if the weights are selected as follows:

$$
\sqrt{\alpha^h} \approx \mathscr{O}(1), \sqrt{\alpha^g} \approx \mathscr{O}(N), \qquad \text{for Poisson problem,} \tag{48}
$$
$$
\sqrt{\alpha^h} \approx \mathscr{O}(1), \sqrt{\alpha^g} \approx \mathscr{O}(N\max\{\lambda,\mu\}), \qquad \text{for elasticity problem,} \tag{49}
$$

where λ and μ are Lame's constants. The scheme in (44) is called the *weighted collocation method*.

4.2 Convergence and Stability

The discrete variational problem in Eq. (43) is equivalent to the following formulation

$$
\hat{b}(\mathbf{u}^h, \mathbf{v}) = \hat{f}(\mathbf{v}), \tag{50}
$$

where $\hat{b}(\mathbf{u}, \mathbf{v})$ denotes discrete form of the following bilinear and linear forms

$$
b(\mathbf{u}, \mathbf{v}) = \int_{\Omega} \mathbf{L}\mathbf{u} \cdot \mathbf{L}\mathbf{v}\, d\Omega + \alpha^h \int_{\partial\Omega^h} \mathbf{B}^h \mathbf{u} \cdot \mathbf{B}^h \mathbf{v}\, dL + \alpha^g \int_{\partial\Omega^g} \mathbf{B}^g \mathbf{u} \cdot \mathbf{B}^g \mathbf{v}\, dL, \tag{51}
$$
$$
f(\mathbf{v}) = \int_{\Omega} \mathbf{f} \cdot \mathbf{L}\mathbf{v}\, d\Omega + \alpha^h \int_{\partial\Omega^h} \mathbf{h} \cdot \mathbf{B}^h \mathbf{v}\, dL + \alpha^g \int_{\partial\Omega^g} \mathbf{g} \cdot \mathbf{B}^g \mathbf{v}\, dL. \tag{52}
$$

Define a norm associated with (51)

$$
|||\mathbf{v}||| = \{\|\mathbf{L}\mathbf{v}\|_{0,\Omega}^2 + \|\mathbf{v}\|_{1,\Omega}^2 + \alpha^h \|\mathbf{B}^h \mathbf{v}\|_{0,\partial\Omega^h}^2 + \alpha^g \|\mathbf{B}^g \mathbf{v}\|_{0,\partial\Omega^g}^2\}^{\frac{1}{2}}. \tag{53}
$$

Following the analysis given in [17, 18], we can prove

$$
\hat{b}(\mathbf{u}, \mathbf{v}) \le C_a |||\mathbf{u}||| \times |||\mathbf{v}|||, \quad \mathbf{v} \in U, \tag{54}
$$
$$
C_b |||\mathbf{v}|||^2 \le \hat{b}(\mathbf{v}, \mathbf{v}), \quad \mathbf{v} \in U \tag{55}
$$

where C_a and C_b are positive constants. The solution of the collocation method (43) has error bound as follows

$$\||\mathbf{u} - \mathbf{u}^h\|| = C \min_{\mathbf{v} \in U} \||\mathbf{u} - \mathbf{v}\||. \tag{56}$$

Suppose that $\mathbf{L}$ is a second order elliptic operator [19], we obtain the estimate

$$\||\mathbf{u} - \mathbf{u}^h\|| \le C_1 \rho \, \|\mathbf{u} - \mathbf{v}\|_{2,\Omega} + C_2 \sqrt{\alpha^h} \|\mathbf{u} - \mathbf{v}\|_{0,\partial\Omega^h} + C_3 \sqrt{\alpha^g} \|\mathbf{u}_v - \mathbf{v}_v\|_{0,\partial\Omega^g}, \tag{57}$$

where $\rho = 1$ for Poisson problem and $\rho = \max\{\lambda, \mu\}$ for elasticity problem, C_i are constants independent of parameters κ, a, c and h. For the first term, the estimates given in Section 3.2 for the case of $\ell = 2$ can be applied to both for the Poisson and elasticity problems. For the remaining boundary terms, error estimates can be done in a similar manner.

In the following we shall investigate the condition number of local RB collocation method. The discrete form of the weighted collocation method Eq. (44) can be expressed as

$$\mathbf{F}\mathbf{y} = \mathbf{r}, \tag{58}$$

where

$$\mathbf{F} = \mathbf{W}^{1/2}\mathbf{A}, \quad \mathbf{r} = \mathbf{W}^{1/2}\mathbf{b}, \tag{59}$$

where $\mathbf{W}$ is associated with the weights α^h and α^g. The condition number of the matrix $\mathbf{F}$ is defined as

$$\text{Cond}(\mathbf{F}) = \frac{\sigma_{max}(\mathbf{F})}{\sigma_{min}(\mathbf{F})} = \frac{\sqrt{\lambda_{max}(\mathbf{F}^T\mathbf{F})}}{\sqrt{\lambda_{min}(\mathbf{F}^T\mathbf{F})}}, \tag{60}$$

where $\sigma_{max}(\cdot)$ and $\sigma_{min}(\cdot)$ denote the maximal and minimal of singular values in the singular value decomposition for matrix $\mathbf{F}$, and $\lambda_{max}(\cdot)$ and $\lambda_{min}(\cdot)$ denote the maximal and minimal eigenvalues for matrix $\mathbf{F}^T\mathbf{F}$. The matrix is resulting from the discrete norm $\|\cdot\|_E$ defined as

$$\overline{\|\mathbf{v}\|}_E^2 = \overline{\|\mathbf{L}\mathbf{v}\|}_{0,\Omega}^2 + \alpha^h \overline{\|\mathbf{B}^h\mathbf{v}\|}_{0,\partial\Omega^h}^2 + \alpha^g \overline{\|\mathbf{B}^g\mathbf{v}\|}_{0,\partial\Omega^g}^2 \tag{61}$$

$$= \hat{b}(\mathbf{v}, \mathbf{v}) = \int_\Omega (\mathbf{L}\mathbf{v})^2 \, d\Omega + \alpha^h \int_{\partial\Omega^h} (\mathbf{B}^h\mathbf{v})^2 \, dL + \alpha^g \int_{\partial\Omega^g} (\mathbf{B}^g\mathbf{v})^2 \, dL.$$

Equation (61) can be written in a quadratic form

$$\overline{\|\mathbf{v}\|}_E^2 = \mathbf{y}^T\mathbf{F}^T\mathbf{F}\mathbf{y} =: \mathbf{y}^T\mathbf{G}\mathbf{y}. \tag{62}$$

Since the matrix $\mathbf{G}$ is a symmetric and positive definite matrix, we then have the following inequalities

$$\lambda_{min}(\mathbf{G})\mathbf{y}^T\mathbf{y} \le \mathbf{y}^T\mathbf{G}\mathbf{y} \le \lambda_{max}(\mathbf{G})\mathbf{y}^T\mathbf{y}, \tag{63}$$

and

$$| \overline{\|\mathbf{v}\|}_E^2 - \|\mathbf{v}\|_E^2 | = |\hat{b}(\mathbf{v},\mathbf{v}) - b(\mathbf{v},\mathbf{v})| \leq E(\hbar,\kappa,a,p,\mu), \tag{64}$$

where $\|\cdot\|_E$ denotes the norm for the bilinear form (51), and $E(\cdot)$ denotes an error term of approximation which is dependent on some parameters, and $\hbar$ is the maximal spacing of the collocation points. To ensure that an optimal solution $\mathbf{u}^h$ exists, the term $E(\cdot)$ should be of order $o(1)$, where $o(1) << 1$, see [17, 18].

It follows from Eqs. (62) and (64) that

$$\|\mathbf{v}\|_E^2 - E(\hbar,\kappa,a,p,\mu) \leq \mathbf{y}^T \mathbf{G} \mathbf{y} \leq \|\mathbf{v}\|_E^2 + E(\hbar,\kappa,a,p,\mu). \tag{65}$$

and by Rayleight-Ritz theorem

$$\lambda_{min}(\mathbf{G}) = \min \frac{\mathbf{y}^T \mathbf{G} \mathbf{y}}{\mathbf{y}^T \mathbf{y}} \geq \min \frac{\|\mathbf{v}\|_E^2 - E(\hbar,\kappa,a,p,\mu)}{\mathbf{y}^T \mathbf{y}} \tag{66}$$

$$\lambda_{max}(\mathbf{G}) = \max \frac{\mathbf{y}^T \mathbf{G} \mathbf{y}}{\mathbf{y}^T \mathbf{y}} \leq \max \frac{\|\mathbf{v}\|_E^2 + E(\hbar,\kappa,a,p,\mu)}{\mathbf{y}^T \mathbf{y}}.$$

Moreover, we have

$$c_1 \|\mathbf{v}\|_{2,\Omega}^2 \leq \|\mathbf{v}\|_E^2 \leq c_2 \|\mathbf{v}\|_{2,\Omega}^2, \quad C_1 \|\mathbf{v}\|_{2,\Omega}^2 \leq \mathbf{y}^T \mathbf{G} \mathbf{y} \leq C_2 \|\mathbf{v}\|_{2,\Omega}^2. \tag{67}$$

The lower and upper bounds for the eigenvalues are given as follows

$$\lambda_{max}(\mathbf{G}) \leq \max \frac{C_2 \|\mathbf{v}\|_{2,\Omega}^2}{\mathbf{y}^T \mathbf{y}}$$

$$\leq \left(C_3 \frac{\kappa a^{-2d} p^{4d}}{\mathbf{y}^T \mathbf{y}} + C_4 \frac{\kappa a^{-3d} p^{4d}}{\mathbf{y}^T \mathbf{y}} + C_5 \frac{\kappa a^{-3d} \mu^{2d}}{\mathbf{y}^T \mathbf{y}} \right) \|\mathbf{v}\|_{0,\Omega}^2, \tag{68}$$

and

$$\lambda_{min}(\mathbf{G}) \geq \min \frac{C_1 \|\mathbf{v}\|_{2,\Omega}^2}{\mathbf{y}^T \mathbf{y}} \geq \frac{C_0 \|\mathbf{v}\|_{0,\Omega}^2}{\mathbf{y}^T \mathbf{y}}, \tag{69}$$

in which inverse inequality (38) is used. Consequently, the condition number is bounded by

$$\text{Cond}(\mathbf{F}) = \left\{ \frac{\lambda_{max}(\mathbf{G})}{\lambda_{min}(\mathbf{G})} \right\}^{1/2}$$

$$\leq C_6 \sqrt{\kappa} a^{-d} p^{2d} + C_7 \sqrt{\kappa} a^{-3d/2} p^{2d} + C_8 \sqrt{\kappa} a^{-3d/2} \mu^d, \tag{70}$$

where C_i are generic constants. The first term in Eq. (70) is resulting from RK shape function, whereas the second and third terms are resulting from the RBF multiplied by RK shape function.

In two-dimensional case, $d = 2$, under the collocation framework, the condition number of the MQ RBF is $\mathcal{O}(N^4)$, for reproducing kernel of degree p is $\mathcal{O}(\sqrt{\kappa}a^{-2}p^4)$, and for the proposed local RBF is

$$\text{Cond}(\mathbf{F}) = \mathcal{O}(\sqrt{\kappa}a^{-3}p^4) + \mathcal{O}(\sqrt{\kappa}a^{-3}\mu^2), \tag{71}$$

where $1 < p < \kappa, \mu << N$, the κ, μ are the parameters defined as before. Since

$$a = (p+1)h, \quad h = \mathcal{O}(N^{-1/d}) = \mathcal{O}(1/\sqrt{N}), \tag{72}$$

and p, κ, μ are fixed in computation, we obtain the bounds in condition numbers as follows

$$
\begin{array}{lll}
\text{RBF} & : & \text{Cond} \approx \mathcal{O}(N^4) \approx \mathcal{O}(h^{-8}), \\
\text{RK} & : & \text{Cond} \approx \mathcal{O}(\sqrt{\kappa}a^{-2}p^4) \approx \mathcal{O}(h^{-2}), \\
\text{proposed local BRF} & : & \text{Cond} \approx \mathcal{O}(\sqrt{\kappa}a^{-3}p^4) + \mathcal{O}(\sqrt{\kappa}a^{-3}\mu^2) \approx \mathcal{O}(h^{-3}).
\end{array}
\tag{73}
$$

We see that there exists a significant reduction in condition number in the proposed localized RBF compared to the standard RBF.

5 Numerical Experiments

In the following study, MQ RBF, Wendland function $g_{5,3}$ [31], pure RK function with quadratic basis ($p = 2$) and cubic basis ($p = 3$), and the proposed localized RBF (MQ RBF localized with RK function) are compared. For RK function to converge well under collocation, sufficiently smooth kernel function is used. A fifth order B-spline is employed as the kernel in RK function in this study. In all examples, the number of collocation points is selected to be four times the number of source points (discrete points), unless specified otherwise, c denotes the shape parameter in RBF, h represents the radial distance, a is the size of finite cover in the kernel function, and p is the order of monomial basis functions in RK approximation.

5.1 Two-Dimensional Poisson's Problem

Consider the following Poisson problem

$$
\begin{aligned}
\Delta u(x,y) &= (x^2 + y^2)e^{xy}, &&\text{in} \quad \Omega = (0,1) \times (0,1), \\
u(x,y) &= e^{xy}, &&\text{on} \quad \partial\Omega.
\end{aligned}
\tag{74}
$$

The condition number and convergence in L^2 and H^1 error norms with uniform discretization are shown in Figs. 3–5, respectively. The results show that L-RBF achieves a much better conditioning compared to RBF while possesses a comparable

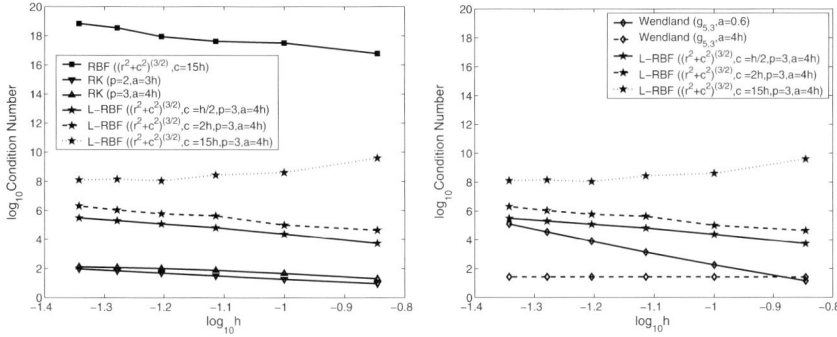

Fig. 3 Condition numbers change as refinement in 2D Poisson problem

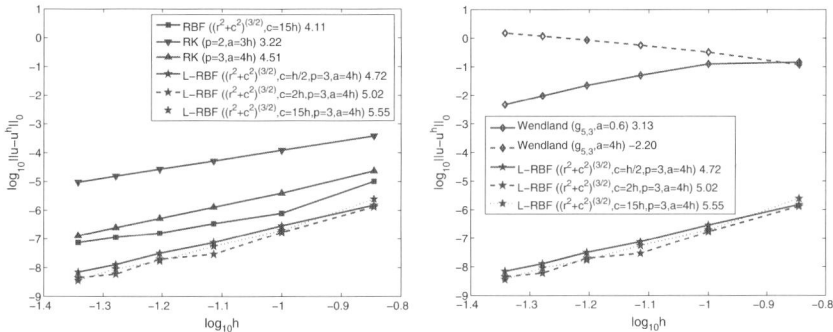

Fig. 4 Convergence of L^2 in error norm for 2D Poisson problem

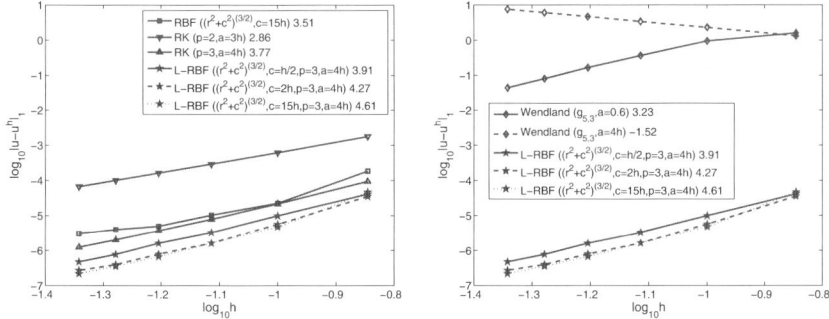

Fig. 5 Convergence of H^1 in error norm for 2D Poisson problem

accuracy and convergence rates compared to the RBF solution. Although the Wendland function also yields a better conditioning compared to the RBF functions, the condition number increases rapidly as the model is refined when a fixed large support ($a = 0.6$) is used. Further, the accuracy of Wendland function compares poorly with the proposed L-RBF. In fact, the use of small support ($a = 4h$) in Wendland

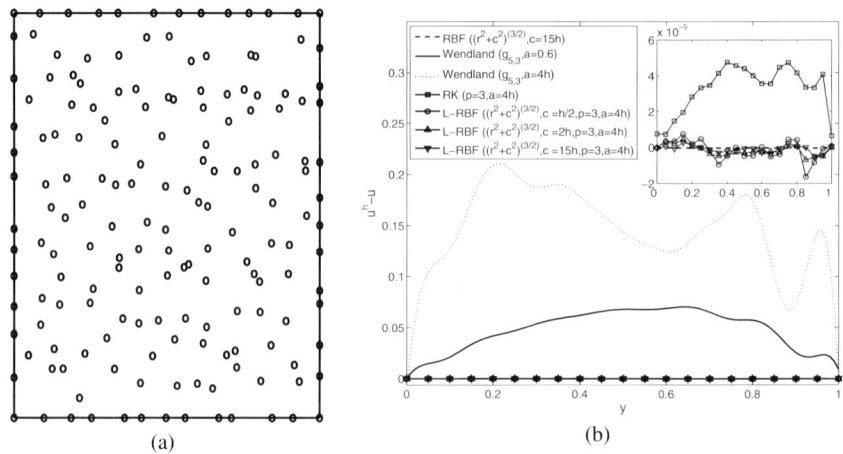

Fig. 6 (**a**) Randomly distributed source points; (**b**) Error distribution along center line ($x = 0.5$) in 2D Poisson problem

function does not yield a convergence in both error norms. For Wendland function to converge, a relatively large support size ($a = 0.6$ in this example) needs to be used, but this in turn causes the conditioning problem according to Fig. 3. The RK functions offer the best conditioning in the discrete system, but they converge slightly slower than L-RBF functions. This example demonstrates that L-RBF is the best approach to achieve both good convergence and well-conditioned banded discrete system.

The rate of convergence in the $\log(\text{error norm}) - \log(h)$ plot presented here requires a careful examination as we keep c/h constant in the refinement. For RBF and L-RBF, we have the error in the form $\|u - u^h\|_\ell \leq C\eta^{c/h}$. This yields $\log\|u - u^h\|_\ell = \log C + c/h \log \eta = \log C + c/h \log_h \eta \log h$. By keeping c/h constant, the slope of the $\log - \log$ plots, $c/h \log_h \eta$, decrease as h decreases since $0 < \eta < 1$ and c/h is kept constant.

One irregular model with randomly distributed source points shown in Fig. 6(a) is analyzed, and Fig. 6(b) shows the error distribution along the center line ($x = 0.5$), where h denotes the average radial distance. Although the solution of L-RBF is not as good as that of the global RBF approximation, L-RBF achieves the better solution accuracy compared to that of RK functions, and much better than that of Wendland function.

5.2 Elastic Cylinder Problem

An infinitely long elastic cylinder subjected to an internal pressure is shown in Fig. 7(a). The corresponding governing equation is

$$\sigma_{ij,j} = 0, \quad i = 1, 2 \quad \text{in } \Omega, \tag{75}$$

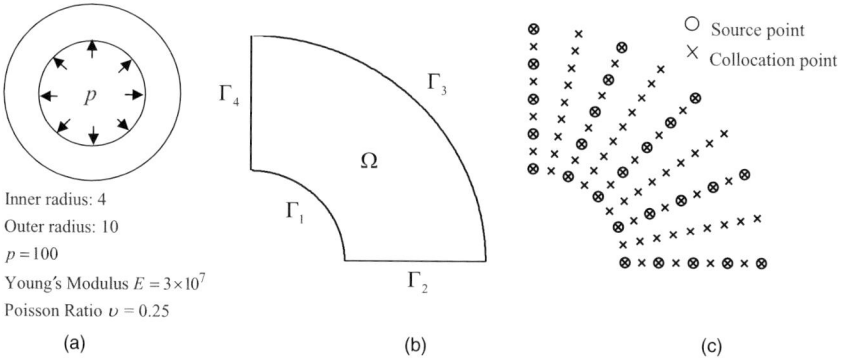

Fig. 7 (**a**) Problem statement, (**b**) Quarter model, (**c**) Distribution of source points and collocation points

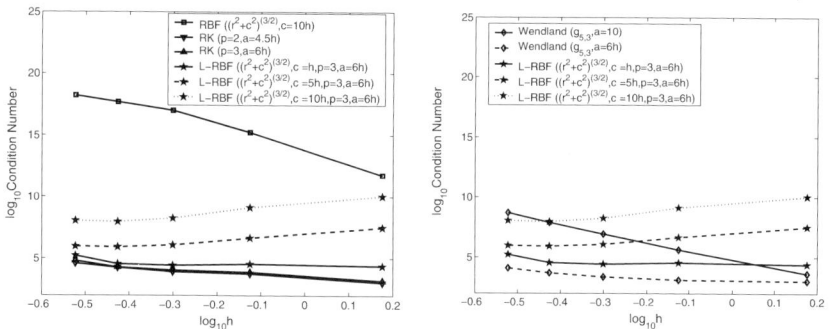

Fig. 8 Condition numbers change as refinement in cylinder problem

with boundary conditions for the quarter model as shown in Fig. 7(b) as:

$$
\begin{array}{lll}
h_1 = -Pn_1, \ h_2 = -Pn_2, & \text{on} & \Gamma_1, \\
h_1 = 0, u_2 = 0, & \text{on} & \Gamma_2, \\
h_1 = 0, h_2 = 0, & \text{on} & \Gamma_3, \\
u_1 = 0, h_2 = 0, & \text{on} & \Gamma_4,
\end{array} \tag{76}
$$

where $\sigma_{ij} = C_{ijk\ell} u_{(k,\ell)}$ is the stress, $C_{ijk\ell}$ is the elasticity tensor, $h_i = \sigma_{ij} n_j$ is the surface traction, and P is the pressure.

The condition number and convergence in L^2 and H^1 error norms are shown in Figs. 8–10, respectively. Although Wendland function improves the conditioning of the discrete system compared to that of RBF, the condition number increases as the model is refined when large support is used. The L-RBF approach, on the other hand, yields a very well-conditioned system in all discretizations. Further, L-RBF functions offer exceptional convergence rates compared to RK functions and RBF, whereas Wendland functions converge poorly.

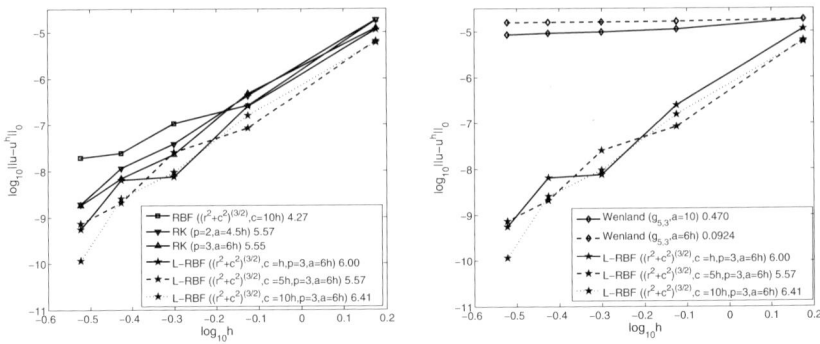

Fig. 9 Convergence of L^2 in error norm for cylinder problem

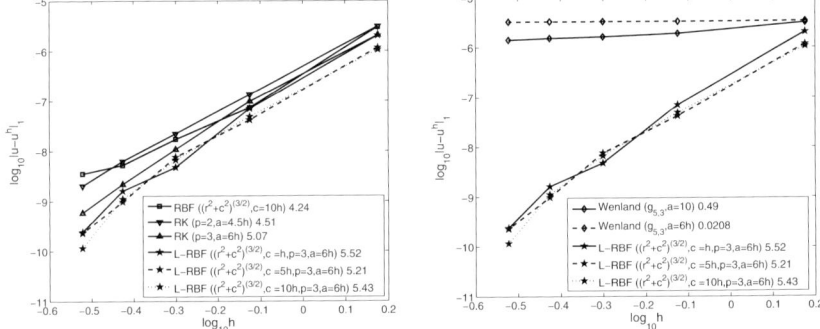

Fig. 10 Convergence of H^1 in error norm for cylinder problem

6 Conclusions

We formulated a localized RBF by introducing a reproducing kernel as the localizing function. The proposed method introduces combination of a local reproducing kernel approximation and RBF function localized with reproducing kernel that possesses polynomial reproducibility. The error analysis shows that if the error of reproducing kernel is sufficiently small, the proposed method maintains the exponential convergence of RBF, while significantly improving the conditioning of the discrete system and yielding a banded matrix. It is also shown that for Wendland function to converge, relatively large support has to be employed, but this in turn results in ill-conditioning in the discrete system.

Acknowledgements We are grateful for the anonymous reviewer for their valuable comments. The support of this work by Lawrence Livermore National Laboratory (USA) to the first and second authors and the support by National Science Council (Taiwan, R.O.C.) to the third author are greatly acknowledged.

References

1. I. Babuska and J. M. Melenk, *The partition of unity method*, Int. J. Numer. Meth. Eng., Vol. 40, pp. 727–758, 1997.

2. T. Belytschko, Y. Y. Lu, and L. Gu, *Element-free Galerkin methods*, Int. J. Numer. Meth. Eng., Vol. 37, pp. 229–256, 1994.

3. M. D. Buhmann, *radial basis functions*, Cambridge Monographs on Applied and Computational Methematics, Cambridge University Press, Cambridge, 18–19, 2003.

4. M. D. Buhmann and C. A. Micchelli, *Multiquadric interpolation improved advanced in the theory and applications of radial basis functions*, Comput. Math. Appl., Vol. 43, pp. 21–25, 1992.

5. R. E. Carlson and T. A. Foley, *Interpolation of track data with radial basis methods*, Comp. Math. Appl., Vol. 24, pp. 27–34, 1992.

6. T. Cecil, J. Qian, and S. Osher, *Numerical methods for high dimensional Hamilton-Jacobi equations using radial basis functions*, J. Comput. Physics., Vol. 196, pp. 327–347, 2004.

7. J. S. Chen, C. Pan, C. T. Wu, and W. K. Liu, *Reproducing kernel particle methods for large deformation analysis of nonlinear structures*, Comput. Method. Appl. Mech. Eng., Vol. 139, pp. 195–227, 1996.

8. J. S. Chen, W. Hu, and H. Y. Hu, *Reproducing kernel enhanced local radial basis collocation method*, International Journal for Numerical Methods in Engineering, Vol. 75(5), pp. 600–627, 2008.

9. C. K. Chui, J. Stoeckler, and J. D. Ward, *Analytic wavelets generated by radial functions*, Adv. Comput. Math., Vol. 5, pp. 95–123, 1996.

10. C. A. M. Duarte and J. T. Oden, *Hp clouds – an hp meshless method*, Numer. Meth. Part. D. E., Vol. 12, pp. 673–705, 1996.

11. G. F. Fasshauer, *Solving differential equations with radial basis functions: multilevel methods and smoothing*, Adv. Comput. Math., Vol. 11, pp. 139–159, 1999.

12. R. Franke and R. Schaback, *Solving partial differential equations by collocation using radial functions*, Appl. Math. Comput., Vol. 93, pp. 73–82, 1998.

13. F. Girosi, *On some extensions of radial basis functions and their applications in artificial intelligence*, Comp. Math. Appl., Vol. 24, pp. 61–80, 1992.

14. R. L. Hardy, *Multiquadric equations of topography and other irregular surfaces*, J. Geophys. Res., Vol. 176, pp. 1905–1915, 1971.

15. R. L. Hardy, *Theory and applications of the multiquadric-biharmonic method: 20 years of discovery*, Comp. Math. Appl., Vol. 19(8/9), pp. 163–208, 1990.

16. W. Han and X. Meng, *Error analysis of the reproducing kernel particle method*, Comput. Method Appl. Mech. Eng., Vol. 190, pp. 6157–6181, 2001.

17. H. Y. Hu and Z. C. Li, *Collocation methods for Poisson's equation*, Comput. Method Appl. Mech. Eng., Vol. 195, pp. 4139–4160, 2006.

18. H. Y. Hu, Z. C. Li, and A. H. D. Cheng, *Radial basis collocation method for elliptic equations*. Comput. Math. Appli., Vol. 50, pp. 289–320, 2005.

19. H. Y. Hu, J. S. Chen, and W. Hu, *Weighted radial basis collocation method for boundary value problems*, Int. J. Numer. Meth. Eng., Vol. 69, pp. 2736–2757, 2007.

20. E. J. Kansa, *Multiqudrics - a scattered data approximation scheme with applications to computational fluid-dynamics - I Surface approximations and partial derivatives*. Comput. Math. Appli., Vol. 19, pp. 127–145, 1992.

21. E. J. Kansa, *Multiqudrics - a scattered data approximation scheme with applications to computational fluid-dynamics - II Solutions to parabolic, hyperbolic and elliptic partial differential equations*, Comput. Math. Appli., Vol. 19, pp. 147–161, 1992.

22. E. J. Kansa and Y. C. Hon, *Circumventing the ill-conditioning problem with multiquadric radial basis functions: applications to elliptic partial differential equations*, Comput. Math. Appli., Vol. 39, pp. 123–137, 2000.

23. P. Lancaster and K. Salkauskas, *Surfaces generated by moving least squares methods*, Math. Comput., Vol. 37, pp. 141–158, 1981.

24. J. Li, *Mixed methods for forth-order elliptic and parabolic problems using radial basis functions*, Adv. Comput. Math., Vol. 23, pp. 21–30, 2005.

25. L. Ling, R. Opfer, and R. Schaback, *Results on meshless collocation techniques*, Eng., Anal. Boundary Elements, Vol. 30, pp. 247–253, 2006.

26. W. K. Liu, S. Jun, and Y. F. Zhang, *Reproducing kernel particle methods*, Int. J. Numer. Meth. Fluids., Vol. 20, pp. 1081–1106, 1995.

27. W. R. Madych, *Miscellaneous error bounds for multiquadric and related interpolatory*, Comput. Math. Appli., Vol. 24, No.12, pp. 121–138, 1992.

28. R. Pollandt, *Solving nonlinear equations of mechanics with the boundary element method and radial basis functions*, Int. J. Numer. Meth. Eng., Vol. 40, pp. 61–73, 1997.

29. M. J. D. Powell, *Univariate multiquadric interpolation: some recent results*, Curves and Surfaces, Academic, New York, pp. 371–382, 1991.

30. T. Sonar, *Optimal recovery using thin-plate splines in finite volume methods for the numerical solution of hyperbolic conservation laws*, IMA J. Numer. Anal., Vol. 16, pp. 549–581, 1996.

31. H. Wendland, *Piecewise polynomial, positive definite and compactly supported radial basis functions of minimal degree*, Adv. Comput. Math., Vol. 4, pp. 389–396, 1995.

32. S. M. Wong, Y. C. Hon, T. S. Li, S. L. Chung and E. J. Kansa, *Multizone decomposition for simulation of time-dependent problems using the multiquadric scheme*, Comput. Math. Appli., Vol. 37, pp. 23–43, 1999.

Preconditioning of Radial Basis Function Interpolation Systems via Accelerated Iterated Approximate Moving Least Squares Approximation

Gregory E. Fasshauer(✉) and Jack G. Zhang

Abstract The standard approach to the solution of the radial basis function interpolation problem has been recognized as an ill-conditioned problem for many years. This is especially true when infinitely smooth basic functions such as multiquadrics or Gaussians are used with extreme values of their associated shape parameters. Various approaches have been described to deal with this phenomenon. These techniques include applying specialized preconditioners to the system matrix, changing the basis of the approximation space or using techniques from complex analysis. In this paper we present a preconditioning technique based on residual iteration of an approximate moving least squares quasi-interpolant that can be interpreted as a change of basis. In the limit our algorithm will produce the perfectly conditioned cardinal basis of the underlying radial basis function approximation space. Although our method is motivated by radial basis function interpolation problems, it can also be adapted for similar problems when the solution of a linear system is involved such as collocation methods for solving differential equations.

Keywords: Preconditioning methods · Radial Basis Functions · Accelerated Iterated least squares

1 Motivation

The solution of ill-conditioned problems has been a major challenge throughout the history of numerical analysis affecting computational accuracy, complexity and stability. One should avoid dealing with ill-conditioning whenever possible. However,

G.E. Fasshauer
Department of Applied Mathematics, Illinois Institute of Technology, Chicago, IL 60616, USA,
e-mail: fasshauer@iit.edu

J.G. Zhang
Department of Mathematics and Statistics, University of New Mexico, Albuquerque, NM 87131,
USA, e-mail: jackunm@math.unm.edu

A.J.M. Ferreira et al. (eds.) *Progress on Meshless Methods, Computational Methods in Applied Sciences.*
© Springer Science + Business Media B.V. 2009

there are cases when finding a theoretically good solution may initially lead to an ill-conditioned situation. *Radial basis function* (RBF) methods have various nice features such as a certain insensitivity to the dimension of the problem domain and ease of implementation (see e.g. [7, 16, 31]). Therefore they have recently gained a great deal of attention and have been implemented in various applications such as multidimensional scattered data interpolation and the numerical solution of differential equations by collocation (see, e.g. [6, 13, 15]). In their standard formulation RBF methods often involve the solution of linear systems whose system matrices usually are full and severely ill-conditioned – especially if certain popular radial basic functions such as Gaussians or inverse multiquadrics [3, 4, 21] are used.

In this paper we extend our earlier work on iterated approximate moving least squares (IAMLS) approximation [18] by providing an accelerated version of the residual iteration algorithm that serves as a preconditioner for RBF systems and can be interpreted as a change of basis procedure. In the limit our algorithm will produce the perfectly conditioned cardinal basis of the underlying radial basis function approximation space. Although our method is motivated by radial basis function interpolation problems, it can also be adapted to similar problems involving the solution of a linear system such as collocation methods for solving differential equations.

The paper is organized as follows. In the remainder of this section we will illustrate how ill-conditioning may happen for RBF interpolation systems by describing the RBF interpolation problem and its solution approach. In Sect. 2 we provide a quick review of some standard preconditioning techniques, and relate our method to polynomial preconditioners. In Sect. 3 we go into the details of our iterated approximate MLS preconditioner including the acceleration procedure. Sect. 4 contains the numerical algorithm we have implemented in MATLAB®. We end the paper with a presentation and discussion of some numerical experiments in Sect. 5.

1.1 Conditioning of RBF Interpolation

Given $\{(x_j, f_j): j = 1, 2, \ldots, N\}$, with datasites $x_j \in \mathbb{R}^s$ and values $f_j = f(x_j) \in \mathbb{R}$ assumed to come from some unknown smooth function f, we are to construct a continuous function $\mathscr{P}_f$ such that

$$\mathscr{P}_f(x_j) = f_j, \quad \text{for } j = 1, 2, \ldots, N. \tag{1}$$

Due to the *Mairhuber-Curtis* result for multi-dimensional interpolation problems (details may be found in, e.g. [16] or [31]) the function $\mathscr{P}_f$ is usually assumed to be a linear expansion of shifts of a real-valued radial basic function ϕ, that is

$$\mathscr{P}_f(\cdot) := \sum_{j=1}^{N} c_j \phi(\cdot - x_j). \tag{2}$$

Thus, condition (1) leads to the following linear system

$$\mathbf{Ac} = \mathbf{f}, \tag{3}$$

with the interpolation matrix $\mathbf{A}_{ij} := \phi(x_i - x_j)$ for $i, j = 1, \ldots, N$, the coefficient vector $\mathbf{c} := [c_1, \ldots, c_N]^T$, and the right-hand side $\mathbf{f} := [f_1, \ldots, f_N]^T$. If ϕ is taken to be an s-variate *strictly positive definite* function, then $\mathbf{A}$ is guaranteed to be invertible and thus (3) has a unique solution $\mathbf{c} = \mathbf{A}^{-1}\mathbf{f}$.

The definition of ϕ will sometimes involve a shape or scaling factor ε, e.g., the Gaussian basic function is given by $\phi(\cdot) = e^{-\varepsilon^2\|\cdot\|^2}$, where $\|\cdot\|$ is usually the Euclidean norm. In this paper, we restrict our attention only to fixed constant $\varepsilon > 0$. An important issue that researchers have been working on for some time is to optimize the RBF interpolant with respect to the shape parameter ε of a certain chosen radial basis (see e.g. [19, 22, 23, 30]). That is, to find ε_{opt} so that

$$\varepsilon_{opt} = \operatorname*{argmin}_{\varepsilon > 0}\{\|f - \mathscr{P}_f\|\}.$$

Such an optimization usually involves both a theoretical and a numerical component. Of these at least the latter is related to the conditioning of the problem. This may be demonstrated by the following simple typical examples. The left part of Fig. 1 shows an error plot for 1D interpolation problems with Gaussian RBFs (as defined above), and the right part contains analogous results for the inverse multiquadrics $\phi(\cdot) := \frac{1}{(1+\varepsilon^2\|\cdot\|^2)}$. Both sets of results were obtained using the test function $f(x) = x(1-x)$ on $[0, 1]$ with 101 data points $x_j = 0.01(j-1)$, $j = 1, \ldots, 101$. The solution $\mathbf{c} = \mathbf{A}^{-1}\mathbf{f}$ was computed by the default built-in solver in MATLAB®.

The main trend of the curves shows that the optimal ε value seems to fall in an interval where the accuracy behavior of the interpolant is extremely unstable. If the same experiment is performed with a slightly different data point resolution or different ε resolution, the main trend of the resulting curves remains similar, but the oscillatory segments change unpredictably and nowhere match each other. This sawtooth instability has not yet been well understood although it has been recognized

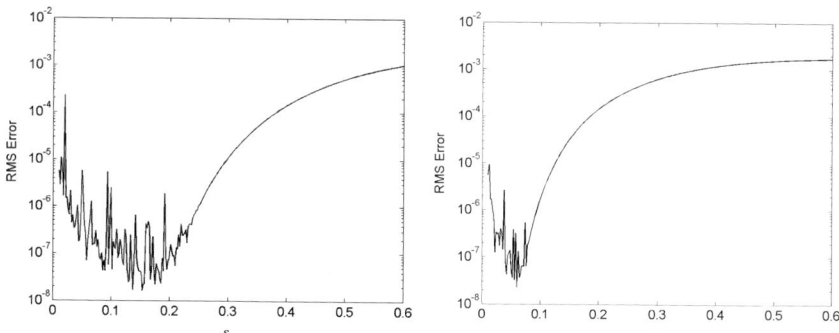

Fig. 1 Errors for different ε values

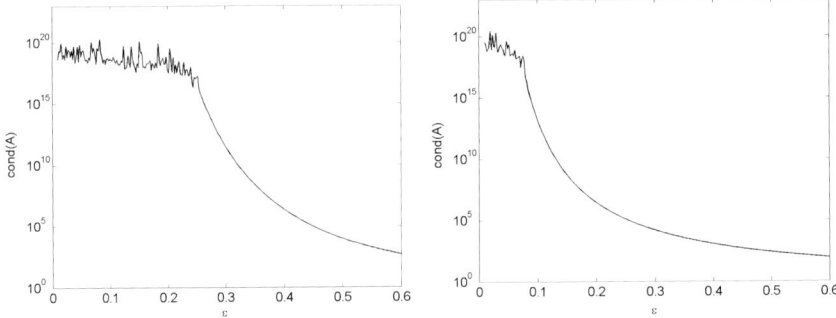

Fig. 2 cond(A) by MATLAB®

that RBF interpolants in 1-D in general converge to the Lagrange interpolating polynomial which in turn gives rise to the *Runge phenomenon* (see [11, 22]). However,
it is reasonable to believe that there may be some kind of connection between this
instability and the conditioning of the associated interpolation matrix $\mathbf{A}$. When $\mathbf{A}$ is
ill-conditioned, i.e., its condition number $\kappa(\mathbf{A})$ is large, then most of standard linear
system solvers may become unreliable because the solution $\mathbf{c}$ found by these solvers
loses a significant amount of accuracy. The next two figures (Fig. 2) are estimated
condition number curves corresponding to the set of results recorded in Fig. 1.

As we have seen in the examples above, the smallest error in RBF interpolation
is associated with the choice of a basic function that has a rather flat shape (i.e., ε
is small), and therefore the corresponding interpolation matrix is dense and close to
singular (due to almost parallel rows or columns of the interpolation matrix). This
does not necessarily mean that we have to give up on the solution space spanned by
such a basis because it is conceivable that there exists a "better" basis for the same
linear space. Various techniques have been proposed to deal with this (seemingly)
ill-conditioned problem. One obvious strategy is to apply specialized preconditioners to the system matrix. A number of papers exist on this subject starting from
work of Dyn and co-workers in the mid 1980s (see, e.g. [13, 14]) or the more recent
papers [2, 3, 6, 25, 26]. Another approach is to introduce a new – hopefully better –
basis of the approximation space (see, e.g. [4]). Complex analysis techniques in
the form of a Contour-Padé algorithm were suggested in [21], and numerical linear techniques based on the QR or singular value decompositions have also been
proposed [9, 20, 24, 27].

Our approach to dealing with the ill-conditioned standard RBF basis is via a
preconditioning algorithms based on our earlier work on iterated *approximate moving least-squares* (AMLS) approximation [18]. We will show that iteration on the
AMLS residuals can effectively reduce the condition number of the linear system of the RBF interpolant. In fact, it reflects a change of basis which generates
approximate cardinal functions. A partial theoretical justification for this change-of-
basis approach (at least for RBFs with finite smoothness) is provided by the recent
paper [10] where the authors show the stability of the radial basis function *space* –
even if the standard *basis* may be ill-conditioned. As a result of our algorithm the

eigenvalues of the preconditioned system are tightly clustered around unity, and it is known that such well conditioned linear system allow most Krylov solvers to converge quickly and also yield accurate and reliable solutions.

2 A Short Review of Preconditioning Techniques

A classical approach to overcome the difficulties associated with the ill-conditioning of systems of linear algebraic equations is to find an appropriate preconditioning matrix $\mathbf{P}$ so that $\kappa(\mathbf{PA}) \ll \kappa(\mathbf{A})$ or $\kappa(\mathbf{AP}) \ll \kappa(\mathbf{A})$. The ideal preconditioner is given by $\mathbf{A}^{-1}$ itself. Of course, use of $\mathbf{P} = \mathbf{A}^{-1}$ is impractical.

As indicated by the two different notations used above, there are different ways to apply the preconditioning. One is to left-multiply $\mathbf{P}$ to the original linear system, i.e., to consider

$$\mathbf{PAc} = \mathbf{Pf}. \tag{4}$$

Then, the solution is given as $\mathbf{c} = (\mathbf{PA})^{-1}(\mathbf{Pf})$ which is theoretically equal to the solution $\mathbf{c} = \mathbf{A}^{-1}\mathbf{f}$ given by (3) but expected to be numerically more accurate since $\mathbf{PA}$ is better conditioned than $\mathbf{A}$. However, an undesired phenomenon may occur. The relative residual $\dfrac{\|\mathbf{PAc} - \mathbf{Pf}\|}{\|\mathbf{Pf}\|}$ may be small partially because $\|\mathbf{P}\|$ is very large in magnitude. Thus, the absolute residual $\|\mathbf{Ac} - \mathbf{f}\|$ may not be guaranteed to be small. Moreover, if we use the coefficient vector $\mathbf{c}$ thus obtained to construct the approximant $\mathscr{P}_f$ and then evaluate it at a new set of points $y_i \in \mathbb{R}^s$, $i = 1, \ldots, M$, then the resulting values of $\mathscr{P}_f(y_i)$ are often inaccurate. This phenomenon was observed in some of our numerical examples.

A second preconditioning strategy is to change the original linear system to

$$\mathbf{APc} = \mathbf{f}, \tag{5}$$

that is

$$\mathbf{c} = (\mathbf{AP})^{-1}\mathbf{f}. \tag{6}$$

Use of the right-preconditioned system (5) is equivalent to reformulating the RBF interpolant $\mathscr{P}_f$ in (2) as

$$\mathscr{P}_f(\cdot) := \sum_{j=1}^{N} c_j \gamma_j(\cdot), \tag{7}$$

where the set $\{\gamma_j(\cdot)\}$ represents a new basis (since $\mathbf{P}$ is usually non-singular) of the space spanned by the original basis set $\{\phi(\cdot - x_j)\}$. This change of basis is provided by the transformation

$$\Gamma(\cdot) := \begin{bmatrix} \gamma_j(\cdot) \\ \vdots \\ \gamma_N(\cdot) \end{bmatrix} = \mathbf{P}^T \begin{bmatrix} \phi(\cdot - x_1) \\ \vdots \\ \phi(\cdot - x_N) \end{bmatrix} =: \mathbf{P}^T \Phi(\cdot). \tag{8}$$

As noted earlier, when $\mathbf{P} = \mathbf{A}^{-1}$, $\Gamma(\cdot)$ becomes the cardinal basis of $\mathrm{span}\{\Phi(\cdot)\}$, i.e., $\gamma_j(x_i) = \delta_{ij}$. In that case the two basis sets are related as

$$\Phi(\cdot)^T = \Gamma(\cdot)^T \mathbf{A} \text{ or } \Phi(\cdot)^T \mathbf{A}^{-1} = \Gamma(\cdot)^T, \quad (\text{since } \mathbf{P} = \mathbf{A}^{-1}).$$

Note that the notation used here may appear to be a less natural one. However, we feel compelled to use it since we are working with right preconditioning defined via (4).

The evaluation of the resulting interpolant $\mathscr{P}_f$ formulated in (7) can be put in the following matrix-vector notation. Define an evaluation matrix

$$\mathbf{B}_{ij} := \phi(y_i - x_j), \ i = 1, \dots, M, \ j = 1, \dots, N.$$

Then the evaluation vector

$$\mathbf{y} = \left[\mathscr{P}_f(y_1), \dots, \mathscr{P}_f(y_M) \right]^T$$

is given as

$$\mathbf{y} = \mathbf{BPc}. \tag{9}$$

Note that the two linear systems defined in (4) and (5) are different in general, meaning that their coefficient vectors are not equal. However, we use the same notation $\mathbf{c}$ for convenience. In the next section we start our discussion of the construction of the preconditioner $\mathbf{P}$.

As an additional reason for using the right-preconditioning scheme we mention a classic preconditioning technique known as *polynomial preconditioning*. This technique is related to the method we are going to describe. According to Benzi [5] the idea to precondition a linear system goes back to Cesari in 1937 [8]. In fact, Cesari used a low degree polynomial $p(\mathbf{A})$ in $\mathbf{A}$. However, Benzi also states that polynomial preconditioners for Krylov subspace methods came into vogue in the late 1970s but are currently out of favor because of their limited effectiveness and robustness, especially for nonsymmetric problems. In the classic formulation only polynomial preconditioners of low degree (2–16) were suggested for practical use (see [1]). As we will see later, our method can stably lift the polynomial to a much higher degree.

3 Preconditioning by Iterated AMLS

Iterated approximate MLS approximation is based on the concept of approximate approximation first suggested by Maz'ya in the early 1990s [28]. The iterated AMLS approximation starts with the definition of a quasi-interpolant. Then, a sequence of approximants is constructed by adding residuals computed on the data sites to the previous approximant [18].

3.1 Iterated AMLS

The formulation of this approach can be summarized as

$$\mathscr{Q}_f^{(n)}(\cdot) := \left[\Phi(\cdot)^T \sum_{k=0}^{n} (\mathbf{I} - \mathbf{A})^k \right] \mathbf{f}, \quad n = 0, 1, 2, \ldots, \tag{10}$$

where $\Phi(\cdot)$ is the vector of original basis functions as defined in (8). Denote

$$\Gamma(\cdot)^T = \Phi(\cdot)^T \sum_{k=0}^{n} (\mathbf{I} - \mathbf{A})^k.$$

Clearly, $\Gamma(\cdot)$ is also a vector of functions. Moreover, its entries are linear combinations of $\phi(\cdot - x_j)$ so that it corresponds to a change of basis for $\mathrm{span}\{\phi(\cdot - x_j)\}$. This follows since it can be shown that the transformation matrix $\sum_{k=0}^{n} (\mathbf{I} - \mathbf{A})^k$ has full rank.

It is known that the truncated Neumann series $\sum_{k=0}^{n} (\mathbf{I} - \mathbf{A})^k$ converges to $\mathbf{A}^{-1}$ if and only if $\|\mathbf{I} - \mathbf{A}\| < 1$. Thus $\mathscr{Q}_f^{(n)} \to \mathscr{P}_f$ and $\Gamma(\cdot)$ converges to a cardinal basis as $n \to \infty$. If we denote the truncated Neumann series by $\mathbf{P}^{(n)}$ then $\mathbf{P}^{(n)}\mathbf{A} = \mathbf{A}\mathbf{P}^{(n)} \to \mathbf{I}$ as $n \to \infty$ (the equality holds because $\mathbf{P}^{(n)}$ is a polynomial of $\mathbf{A}$).

We summarize some of the main properties of the iterated AMLS method, while more details are presented in [18]. Iterated AMLS can be used to compute

- An approximate inverse of $\mathbf{A}$

$$\mathbf{A}^{-1} \approx \mathbf{P}^{(n)}$$

- Approximate expansion coefficients for the standard RBF interpolation problem (3)

$$\mathbf{c} \approx \mathbf{P}^{(n)}\mathbf{f}$$

- An approximate cardinal basis $\Gamma(\cdot)$ by

$$\Gamma(\cdot)^T = \Phi(\cdot)^T \mathbf{P}^{(n)}$$

In all of these formulations

$$\mathbf{P}^{(n)} = \sum_{k=0}^{n} (\mathbf{I} - \mathbf{A})^k \tag{11}$$

and $\mathbf{A}$ denotes the original interpolation matrix with entries $\phi(x_i - x_j)$. Note that $\mathbf{A}$ is symmetric.

We formalize and prove the last of these statements.

Theorem 1. *The n-th iterated quasi-interpolant can be written as*

$$\mathscr{Q}_f^{(n)}(\cdot) = \Phi(\cdot)^T \sum_{k=0}^{n} (\mathbf{I} - \mathbf{A})^k \mathbf{f} =: \Gamma(\cdot)^T \mathbf{f},$$

i.e., $\{\gamma_1(\cdot),\ldots,\gamma_N(\cdot)\}$ *provides a new — approximately cardinal — basis for* $\mathrm{span}\{\phi(\cdot-x_1),\ldots,\phi(\cdot-x_N)\}$.

Proof. We use induction on n. By definition we have

$$\mathscr{Q}_f^{(n+1)}(\cdot) = \mathscr{Q}_f^{(n)}(\cdot) + \sum_{j=1}^{N}\left[f(x_j) - \mathscr{Q}_f^{(n)}(x_j)\right]\phi(\cdot-x_j).$$

Next, using the induction hypothesis yields

$$\mathscr{Q}_f^{(n+1)}(\cdot) = \Phi(\cdot)^T\sum_{k=0}^{n}(\mathbf{I}-\mathbf{A})^k\mathbf{f} + \sum_{j=1}^{N}\left[f(x_j) - \Phi(x_j)^T\sum_{k=0}^{n}(\mathbf{I}-\mathbf{A})^k\mathbf{f}\right]\phi(\cdot-x_j)$$

$$= \Phi(\cdot)^T\sum_{k=0}^{n}(\mathbf{I}-\mathbf{A})^k\mathbf{f} + \Phi(\cdot)^T\left[\mathbf{I} - \mathbf{A}\sum_{k=0}^{n}(\mathbf{I}-\mathbf{A})^k\right]\mathbf{f}.$$

If we simplify further we obtain

$$\mathscr{Q}_f^{(n+1)}(\cdot) = \Phi(\cdot)^T\left[\mathbf{I} - \sum_{k=0}^{n}(\mathbf{I}-\mathbf{A})^{k+1}\right]\mathbf{f}$$

$$= \Phi(\cdot)^T\left[\sum_{k=0}^{n+1}(\mathbf{I}-\mathbf{A})^k\right]\mathbf{f} = \Gamma(\cdot)^T\mathbf{f}$$

which completes the proof.   □

Clearly, $\mathbf{P}^{(n)}$ can be used as a preconditioner for the interpolation problem discussed in previous section.

3.2 Accelerating Convergence of the Iterations

As described earlier, preconditioning by iterated AMLS requires both the coefficient vector $\mathbf{c}$ and the preconditioning matrix $\mathbf{P}^{(n)}$ to be explicitly computed. This requires expensive matrix-matrix multiplication. So, it is desirable to find a computational algorithm that can reduce the operation count and thus increase numerical accuracy. Writing out (10) for $n = 0$ and $n = 1$, we can see

$$\mathscr{Q}_f^{(0)}(\cdot) = \Phi(\cdot)^T\mathbf{f}, \tag{12}$$

$$\mathscr{Q}_f^{(1)}(\cdot) = \Phi(\cdot)^T\left(\mathbf{I} + (\mathbf{I}-\mathbf{A})\right)\mathbf{f} = \Phi(\cdot)^T\left(2\mathbf{I} - \mathbf{A}\right)\mathbf{f}, \tag{13}$$

where the matrix $\mathbf{A}_{ij} = \phi(x_i - x_j)$ arises from the evaluation of the approximant $\mathscr{Q}_f^{(0)}$ at the data sites as required for the residual calculation.

The iterative process (10) can be accelerated via the following scheme. Take

$$\tilde{\mathscr{Q}}_f^{(0)} := \mathscr{Q}_f^{(n)} = \left[\Phi(\cdot)^T \sum_{k=0}^{n} (\mathbf{I} - \mathbf{A})^k \right] \mathbf{f}, \tag{14}$$

and perform one iteration following the pattern (12–13). Then we have

$$\tilde{\mathscr{Q}}_f^{(1)}(\cdot) = \left[\Phi(\cdot)^T \sum_{k=0}^{n} (\mathbf{I} - \mathbf{A})^k \right] \left[2\mathbf{I} - \mathbf{A} \sum_{k=0}^{n} (\mathbf{I} - \mathbf{A})^k \right] \mathbf{f}$$

$$= \Phi(\cdot)^T \left[\sum_{k=0}^{2n+1} (\mathbf{I} - \mathbf{A})^k \right] \mathbf{f}$$

$$= \mathscr{Q}_f^{(2n+1)}(\cdot). \tag{15}$$

Surely, the acceleration (14) and (15) can be performed continuously and consecutively starting as early as from the beginning of the original iteration.

This is formalized in

Theorem 2. *Acceleration of the iterated approximate MLS method (10) is achieved with*

$$\tilde{\mathscr{Q}}_f^{(n)}(\cdot) = \Phi(\cdot)^T \left[\sum_{k=0}^{2^n - 1} (\mathbf{I} - \mathbf{A})^k \right] \mathbf{f}, \quad n = 0, 1, 2, \ldots. \tag{16}$$

Proof. As in Theorem 1 we get

$$\mathscr{Q}_f^{(n+1)}(\cdot) = \Phi(\cdot)^T \sum_{k=0}^{n} (\mathbf{I} - \mathbf{A})^k \mathbf{f} + \Phi(\cdot)^T \left[\mathbf{I} - \mathbf{A} \sum_{k=0}^{n} (\mathbf{I} - \mathbf{A})^k \right] \mathbf{f}.$$

According to the acceleration strategy explained above we now replace $\Phi(\cdot)^T$ by its iterated version $\Phi^{(n)}(\cdot)^T$. That yields

$$\tilde{\mathscr{Q}}_f^{(n+1)}(\cdot) = \Phi(\cdot)^T \sum_{k=0}^{n} (\mathbf{I} - \mathbf{A})^k \mathbf{f} + \Phi^{(n)}(\cdot)^T \left[\mathbf{I} - \mathbf{A} \sum_{k=0}^{n} (\mathbf{I} - \mathbf{A})^k \right] \mathbf{f}$$

$$= \Phi^{(n)}(\cdot)^T \left[2\mathbf{I} - \mathbf{A} \sum_{k=0}^{n} (\mathbf{I} - \mathbf{A})^k \right] \mathbf{f}$$

$$= \Phi(\cdot)^T \sum_{k=0}^{n} (\mathbf{I} - \mathbf{A})^k \left[2\mathbf{I} - \mathbf{A} \sum_{k=0}^{n} (\mathbf{I} - \mathbf{A})^k \right] \mathbf{f}$$

$$= \Phi(\cdot)^T \left[\sum_{k=0}^{2n+1} (\mathbf{I} - \mathbf{A})^k \right] \mathbf{f} = \Phi^{(2n+1)}(\cdot)^T \mathbf{f} = \mathscr{Q}_f^{(2n+1)}(\cdot).$$

We are done by observing that the upper index of summation satisfies $\tilde{a}_{n+1} = 2\tilde{a}_n + 1$, $\tilde{a}_0 = 0$, i.e., $\tilde{a}_n = 2^n - 1$. □

Clearly, $\left\{ \tilde{\mathcal{Q}}_f^{(n)} \right\}$ inherits all convergence properties of $\left\{ \mathcal{Q}_f^{(n)} \right\}$ but with a faster speed of convergence.

Now, we update the notation for the preconditioner (11) to the accelerated version

$$\mathbf{P}^{(n)} := \sum_{k=0}^{2^n - 1} (\mathbf{I} - \mathbf{A})^k . \tag{17}$$

In the next section we will describe how this reduction of operations may be carried out and moreover how matrix-matrix multiplications can actually be avoided during the iterations.

4 Computational Algorithm

According to the general preconditioning strategies outlined in Section 1 we have:

1. For the system (4), i.e., $\mathbf{PAc} = \mathbf{Pf}$ we can proceed as follows:

   - $\mathbf{P}^{(0)} = \mathbf{I}$
   - For $k = 1, 2, \ldots, n$, $\mathbf{P}^{(k)} = \left(2\mathbf{I} - \mathbf{P}^{(k-1)}\mathbf{A} \right) \mathbf{P}^{(k-1)}$
   - Use a standard linear solver to compute $\mathbf{c} = \left(\mathbf{P}^{(n)}\mathbf{A} \right)^{-1} \left(\mathbf{P}^{(n)}\mathbf{f} \right)$
   - Evaluate $\mathbf{y} = \mathbf{Bc}$

2. For the system (5), i.e., $\mathbf{APc} = \mathbf{f}$ we can proceed as:

   - $\mathbf{P}^{(0)} = \mathbf{I}$
   - For $k = 1, 2, \ldots, n$, $\mathbf{P}^{(k)} = \mathbf{P}^{(k-1)} \left(2\mathbf{I} - \mathbf{AP}^{(k-1)} \right)$
   - Use a standard linear solver to compute $\mathbf{c} = \left(\mathbf{AP}^{(n)} \right)^{-1} \mathbf{f}$
   - Evaluate $\mathbf{y} = \left(\mathbf{BP}^{(n)} \right) \mathbf{f}$

For the reasons stated in Sect. 1.1 we use only the second preconditioning strategy. Note that we use $\left(\mathbf{BP}^{(n)} \right)$ to indicate that this quantity will be computed first since it will give a better accuracy based on our experimental observation. The specific computational algorithm is listed below. Note that this algorithm needs only one matrix diagonal decomposition. No further matrix-matrix multiplications are needed during the iterations.

Algorithm 1

- *Perform the eigen-decomposition $\mathbf{A} = \mathbf{X}\mathbf{\Lambda}^{(0)}\mathbf{X}^{-1}$ and initialize $\mathbf{P}^{(0)} = \mathbf{I}$*
- *For $k = 1, 2, \ldots, n$, $\mathbf{P}^{(k)} = \mathbf{P}^{(k-1)} \left(\mathbf{I} - \mathbf{\Lambda}\mathbf{P}^{(k-1)} \right)$*
- *Update the preconditioner $\mathbf{P}^{(n)} \leftarrow \mathbf{X}\mathbf{P}^{(n)}\mathbf{X}^{-1}$*

- *Compute* $\mathbf{c} = \left(\mathbf{AP}^{(n)}\right)^{-1}\mathbf{f}$
- *Evaluate* $\mathbf{y} = \left(\mathbf{BP}^{(n)}\right)\mathbf{c}$

Note that the diagonalization of $\mathbf{A}$ provides theoretical equivalences for ordering or arranging the computation in Algorithm 1. For example, it is not necessary to actually compute $\left(\mathbf{AP}^{(n)}\right)^{-1}$ since $\left(\mathbf{AP}^{(n)}\right)$ was already given in the form of a diagonal decomposition. Thus its inverse may be easily obtained via its diagonal decomposition. However, these different arrangements may yield different computational accuracies and it is not clear which arrangement is best.

Finally, it should be clear that this preconditioning method is not necessarily restricted to RBF interpolation. Indeed, it could be applied to generic linear systems as long as the system matrix satisfies the convergence requirements stated in [18].

5 Numerical Experiments and Discussion

5.1 The Basic Functions Used in Our Experiments

The experiments presented in this section use shifts of normalized radial functions such as Laguerre-Gaussians and generalized inverse multiquadrics defined on $[0,1]^2$. The following is proved in [32]:

Theorem 3.

1. Let

$$\psi(t) = \frac{1}{\sqrt{\pi^s}} e^{-t} L_d^{s/2}(t),$$

where $L_d^{s/2}(\cdot)$ are generalized Laguerre polynomials of order $s/2$ and degree d. This will yield the family of Laguerre-Gaussians.

2. Let

$$\psi(t) = \frac{1}{\pi^{s/2}} \frac{1}{(1+t)^{2d+s}} \sum_{j=0}^{d} \frac{(-1)^j (2d+s-j-1)!(1+t)^j}{(d-j)!\, j!\, \Gamma(d+s/2-j)},$$

which gives rise to generalized inverse multiquadrics.

In either case the function $\phi(x) = \psi\left(\|x\|^2\right)$ is strictly positive definite in $\mathbb{R}^s$ and satisfies the continuous moment conditions

$$\int_{\mathbb{R}^s} x^\alpha \phi(x)dx = \delta_{\alpha,0}, \quad 0 \le |\alpha| \le 2d+1$$

of order $2d+1$.

The specific examples of Laguerre-Gaussians and generalized inverse multiquadrics for space dimension $s = 1,2,3$ and degree $d = 0,1,2$ used in some of our numerical experiments are listed in Tables 1 and 2.

Table 1 Examples of Laguerre-Gaussians: $\phi(x) = e^{-\|x\|^2} \times$ table entry

$s\backslash d$	0	1	2
1	$\dfrac{1}{\sqrt{\pi}}$	$\dfrac{1}{\sqrt{\pi}}\left(\dfrac{3}{2} - \|x\|^2\right)$	$\dfrac{1}{\sqrt{\pi}}\left(\dfrac{15}{8} - \dfrac{5}{2}\|x\|^2 + \dfrac{1}{2}\|x\|^4\right)$
2	$\dfrac{1}{\pi}$	$\dfrac{1}{\pi}\left(2 - \|x\|^2\right)$	$\dfrac{1}{\pi}\left(3 - 3\|x\|^2 + \dfrac{1}{2}\|x\|^4\right)$
3	$\dfrac{1}{\pi^{3/2}}$	$\dfrac{1}{\pi^{3/2}}\left(\dfrac{5}{2} - \|x\|^2\right)$	$\dfrac{1}{\pi^{3/2}}\left(\dfrac{35}{8} - \dfrac{7}{2}\|x\|^2 + \dfrac{1}{2}\|x\|^4\right)$

Table 2 Examples of generalized inverse multiquadrics

$s\backslash d$	0	1	2
1	$\dfrac{1}{\pi}\dfrac{1}{1+\|x\|^2}$	$\dfrac{1}{\pi}\dfrac{(3-\|x\|^2)}{(1+\|x\|^2)^3}$	$\dfrac{1}{\pi}\dfrac{(5-10\|x\|^2+\|x\|^4)}{(1+\|x\|^2)^5}$
2	$\dfrac{1}{\pi}\dfrac{1}{(1+\|x\|^2)^2}$	$\dfrac{2}{\pi}\dfrac{(2-\|x\|^2)}{(1+\|x\|^2)^4}$	$\dfrac{3}{\pi}\dfrac{(3-6\|x\|^2+\|x\|^4)}{(1+\|x\|^2)^6}$
3	$\dfrac{4}{\pi^2}\dfrac{1}{(1+\|x\|^2)^3}$	$\dfrac{4}{\pi^2}\dfrac{(5-3\|x\|^2)}{(1+\|x\|^2)^5}$	$\dfrac{8}{\pi^2}\dfrac{(7-14\|x\|^2+3\|x\|^4)}{(1+\|x\|^2)^7}$

In our experiments we combine the basic functions with a shape scaling factor ε which has a strong influence on the condition number of **A** (as already illustrated in Sect. 1.1 and Fig. 1). Also, a multivariate spacing factor h is used in our experiments which is taken to be the average of the data point spacing, i.e., for 2D experiments with N points in $[0,1]^2$ we have $h = \frac{1}{\sqrt{N}-1}$. As a result we end up with, e.g., a scaled Gaussian basis function of the form

$$\phi(\cdot - x_j) = \frac{\varepsilon^2}{\pi} e^{-\varepsilon^2 \|(\cdot - x_j)/h\|^2}.$$

The use of h in the definition of the basis functions usually appears in the context of *stationary* (with h) and *non-stationary* (without h) approximation. When the domain of the problem is taken to be the unit cube $[0, 1]^s$, the standard approximate MLS formulation must be in stationary form which will then lead to a convergence scenario with a so-called *saturation error* [28, 29]. A similar phenomenon is also observed in standard RBF interpolation [16, 31]. Although standard RBF interpolation can be formulated in the non-stationary setting, it is observed that as the number of data points gets denser, the interpolation matrix becomes more ill-conditioned and therefore the solution becomes increasingly inaccurate and unstable [16]. In light of such numerical difficulties, the use of h reduces the effect on the conditioning of the interpolation matrix caused by the number of data points since then $\phi(x_i - x_j)$ is (approximately, if the x_j are not evenly spaced) invariant. Hence, conditioning of the interpolation matrix will *roughly* only depend on the shape parameter

ε as long as the distribution of data points is reasonably even. We use the word "*roughly*" because condition numbers for far apart numbers of data points are still significantly different based on our experiments even with fixed ε and in the stationary setting. Certainly, h may be combined with ε and associated with the data points x_j (see [12]). In such a case distinguishing the two scaling factors becomes trivial.

5.2 Effects of the Preconditioner

As mentioned earlier, the standard (left) preconditioning method defined in (4) is often unreliable. Thus, we only present results for the right preconditioning method defined in (5) and carried out by Algorithm 1.

Figure 3 demonstrates how the condition number of the system matrix changes during the preconditioning iterations. The right plot is a zoom-in of the left one. As described earlier, the preconditioner $\mathbf{P}^{(n)}$ is a truncated Neumann series which can be viewed as a polynomial preconditioner. This technique is simple and easy to implement. However, when the interpolation matrix $\mathbf{A}$ is rather ill-conditioned (which is often true), direct computation with products of $\mathbf{A}$ will rapidly lose its accuracy. Hence, the polynomial preconditioner is likely to become useless (cf. the earlier insights reported in the literature [1]). We reach a similar conclusion based on our experiments. Direct computation of $\mathbf{AP}^{(n)}$ is extremely inaccurate and unstable especially in the beginning of the iterations (i.e., with low-degree polynomial preconditioners). In a striking contrast to this, the accelerated computation is stable and yields a satisfactory condition number drop. However, as n increases (corresponding to polynomial degrees of $2^n - 1$), the accelerated computation may still gather enough numerical error so that the convergence of the Neumann series (i.e., convergence of the iterated AMLS algorithm) are destroyed.

A more comprehensive series of condition number drops is presented in Fig. 4. The left plot uses Laguerre-Gaussians (with $\varepsilon = 0.2, 0.3, 0.4, d = 0, 1, 2$) and the right one uses the generalized inverse multiquadrics (with $\varepsilon = 0.008, 0.08, 0.25, d = 0, 1, 2$). It can be observed that corresponding functions behave similarly.

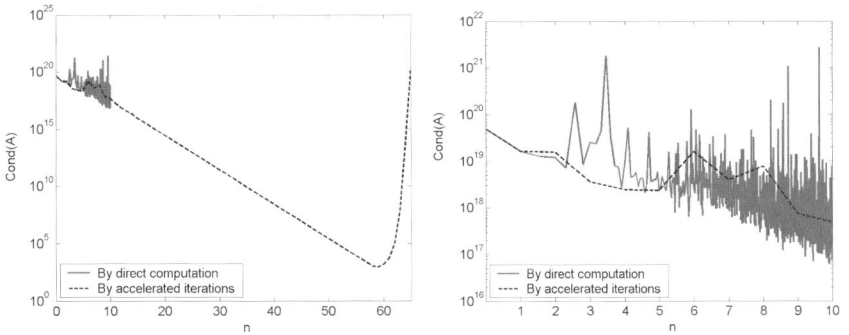

Fig. 3 Condition numbers with accelerated (black/dashed) and with standard polynomial (red/solid) preconditioning for $N = 289$ Halton points in $[0,1]^2$

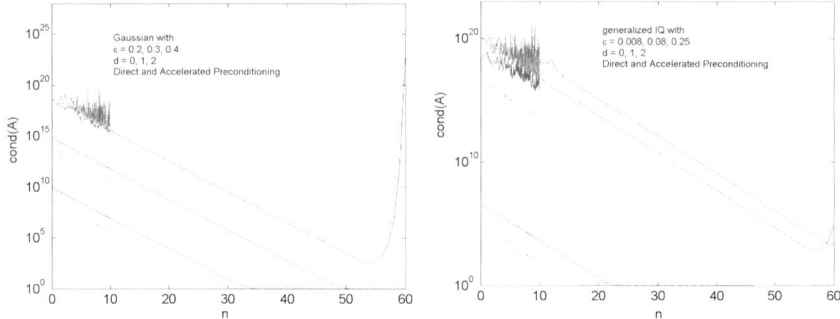

Fig. 4 Condition numbers with accelerated preconditioning, $N = 289$ Halton points

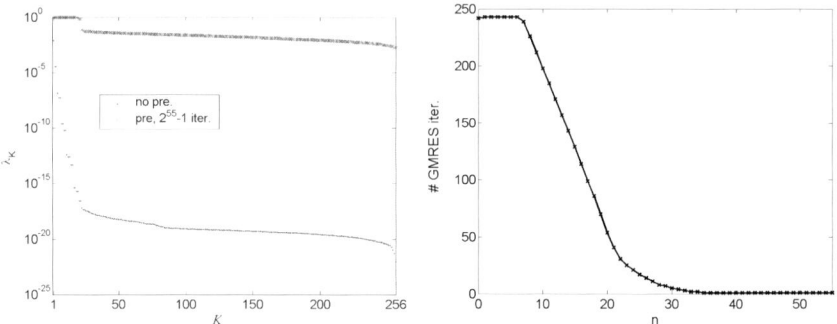

Fig. 5 Eigenvalue distribution and GMRES convergence, $N = 256$ uniform points

Generalized inverse multiquadrics and Laguerre-Gaussians seem to behave similarly with generalized inverse multiquadrics being slightly better conditioned overall.

Figure 5 shows an example of the eigenvalue clustering achieved by the accelerated preconditioning (left plot), and an example of the improvements in GMRES convergence (right figure) for the solution of a test problem based on data sampled from the 2D modified Franke function g defined on $[0, 1]^2$ via

$$
\begin{aligned}
f(x_1, x_2) = \frac{3}{4} &\left[\exp\left(-\frac{(9x_1 - 2)^2}{4} - \frac{(9x_2 - 2)^2}{4} \right) \right. \\
&\left. + \exp\left(-\frac{(9x_1 + 1)^2}{49} - \frac{(9x_2 + 1)^2}{10} \right) \right] \\
&+ \frac{1}{2} \exp\left(-\frac{(9x_1 - 7)^2}{4} - (9x_2 - 3)^2 \right) \\
&- \frac{1}{5} \exp\left(-(9x_1 - 4)^2 - (9x_2 - 7)^2 \right)
\end{aligned}
$$
$$
g(x_1, x_2) = 15 f(x_1, x_2) \exp\left(\frac{-1}{1 - 4(x_1 - 1/2)^2} \right) \exp\left(\frac{-1}{1 - 4(x_2 - 1/2)^2} \right).
$$

Table 3 Condition number comparison with [3]

Cond. no.	289		1,089		4,225	
	No pre	Pre	No pre	Pre	No pre	Pre
MQ [3]	1.506(8)	5.742(1)	2.154(9)	2.995(3)	3.734(10)	4.369(4)
TPS [3]	4.005(6)	3.330(0)	2.753(8)	1.411(2)	2.605(9)	2.025(3)
Gauss	8.796(9)	1.000(0)	6.849(10)	1.000(0)	7.632(10)	1.000(0)
IQ	1.186(8)	1.000(0)	4.284(8)	1.000(0)	1.082(9)	1.000(0)

Table 4 GMRES iteration counts as compared with [3]

No.	289		1,089		4,225	
GMRES iter.	No pre	Pre	No pre	Pre	No pre	Pre
MQ [3]	145	8	>150	15	>150	28
TPS [3]	103	5	145	6	>150	9
Gauss	>150	2	>150	2	>150	2
IQ	>150	2	>150	2	>150	2

In Table 3 we list a set of comparisons to results of the local cardinal basis method presented in [3]. Our experiments listed in Tables 3 and 4 employ 2D Halton points, a value of $n = 40$ for the accelerated preconditioner and shape parameters of $\varepsilon = 0.4$ for the Gaussian and $\varepsilon = 0.2$ for the inverse quadratic basis. Results for higher-order Laguerre-Gaussians and higher-order generalized inverse multiquadrics are similar and therefore omitted.

5.3 A Stopping Criterion

Now it is time to ask the question how to determine the number of iterations (or polynomial degree) n (or $2^n - 1$) used with the preconditioner. It is clear that, as the iteration goes on, the preconditioner $\mathbf{P}^{(n)}$ changes from $\mathbf{I}$ to $\mathbf{A}^{-1}$, that is, $\kappa(\mathbf{P}^{(0)}) = 1$ and $\kappa(\mathbf{P}^{(\infty)}) = \kappa(\mathbf{A})$, while $\kappa(\mathbf{AP}^{(0)}) = \kappa(\mathbf{A})$ and $\kappa(\mathbf{AP}^{(\infty)}) = 1$. Thus, considering only solution of the linear system, we would like to have $\kappa(\mathbf{AP}^{(n)})$ as small as possible.

However, since $\mathbf{P}^{(n)}$ will also be used for the evaluation of the interpolant $\mathscr{P}_f$ at the point set $\{y_j\}$ it is desired that $\kappa(\mathbf{P}^{(n)})$ should be kept small so as to obtain evaluation accuracy. Hence, we suggest that the iteration stops when

$$\kappa\left(\mathbf{P}^{(n)}\right) = \kappa\left(\mathbf{AP}^{(n)}\right). \tag{18}$$

Let σ_{max} and σ_{min} be the largest and smallest singular values of $\mathbf{A}$. Recall that ϕ is strictly positive definite, that is, $\mathbf{A}$ is positive definite and symmetric and $\|\mathbf{I} - \mathbf{A}\|_2 < 1$. Thus, $0 < \sigma_{min} < \sigma_{max} < 1$. It can be verified via singular value decomposition for $\mathbf{A}$ that

$$\kappa\left(\mathbf{P}^{(n)}\right) = \frac{\sum_{k=0}^{2^n-1}(1-\sigma_{min})^k}{\sum_{k=0}^{2^n-1}(1-\sigma_{max})^k} = \frac{1-(1-\sigma_{min})^{2^n}}{1-(1-\sigma_{max})^{2^n}}\frac{\sigma_{max}}{\sigma_{min}}, \tag{19}$$

and

$$\kappa\left(\mathbf{A}\mathbf{P}^{(n)}\right) = \frac{1-(1-\sigma_{max})^{2^n}}{1-(1-\sigma_{min})^{2^n}}. \tag{20}$$

Thus, the ideal number of iterations can be estimated by solving the nonlinear equation

$$\frac{1-(1-\sigma_{max})^{2^n}}{1-(1-\sigma_{min})^{2^n}} = \sqrt{\frac{\sigma_{max}}{\sigma_{min}}} \tag{21}$$

for n. Note that for a symmetric positive definite $\mathbf{A}$ its eigen-decomposition in Algorithm 1 is identical to its singular value decomposition. Thus, there is no extra computation needed for finding σ_{max} and σ_{min}. If Shepard's method is used, then $\mathbf{A}$ is just a product of a symmetric positive definite matrix and a diagonal matrix (performing row or column scaling). Thus, with a little adaption the formulation that we have discussed will still be applicable.

In Fig. 6 and Table 5 we present a set of error comparisons obtained with this optimally stopped preconditioning algorithm. Both the original system and the iteratively preconditioned system (computed by Algorithm 1 with the suggested optimal number of iterations) are solved by a MATLAB® GMRES method with

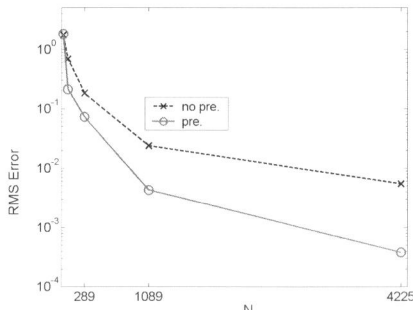

Fig. 6 Error drop comparison with automatic stopping criterion. N uniform points in 2D

Table 5 "Optimal" number of preconditioning iterations, n, for the results shown in Fig. 6

N	289			1,089			4,225		
$\kappa(\mathbf{A})$	1.4e + 11			2.1e + 12			4.8e + 12		
	RMSerr	GMRES	n	RMSerr	GRMES	n	RMSerr	GMRES	n
No pre	7.2e − 2	>10		2.4e − 2	>10		5.5e − 3	>10	
pre	1.8e − 1	>10	20	4.3e − 3	6	20	3.8e − 4	1	21

default settings, that is, $\mathbf{c} = \text{gmres}(\mathbf{A}, \mathbf{b})$ in MATLAB® syntax for $\mathbf{Ac} = \mathbf{b}$, and $\mathbf{c} = \text{gmres}(\mathbf{AP}, \mathbf{b})$ for $\mathbf{APc} = \mathbf{b}$. When $N = 4{,}225$ the condition number $\kappa(\mathbf{A}) \approx 10^{12}$, i.e., $\kappa(\mathbf{AP}) \approx \kappa(\mathbf{P}) \approx 10^6$, and the preconditioning algorithm terminates after $n = 21$ iterations. With our preconditioning the GMRES method converges within the default maximum number of iterations for a default tolerance while it does not converge without preconditioning. Note that the error drop without preconditioning is also reasonably stable although it is larger than that achieved with preconditioning. This happens because that lack of exactness at data sites is not necessarily reflected in the global accuracy of the solution.

5.4 Concluding Remarks

We have demonstrated that the proposed accelerated preconditioning method is effective and easy to implement. The diagonalization performed in the acceleration Algorithm 1 improves the speed of computation without contributing any extra numerical inaccuracy. The accuracy of $\mathbf{P}^{(n)}$ as a preconditioner is actually immaterial as long as $\kappa\left(\mathbf{AP}^{(n)}\right) \ll \kappa(\mathbf{A})$. However, since the accuracy of the evaluation (via $\mathbf{BP}^{(n)}$ or $\mathbf{BP}^{(n)}\mathbf{c}$) also depends highly on $\kappa\left(\mathbf{P}^{(n)}\right)$ it is clear that for extremely ill-conditioned problems (with $\kappa(\mathbf{A}) > 10^{20}$) this preconditioning method will not work very well.

Based on the numerical experiments we performed in MATLAB®, our preconditioning method works efficiently and accurately when $\kappa\left(\mathbf{AP}^{(n)}\right)$ is in the order of $10^{12} \sim 10^{14}$. Thus it has certain advantages over most of the standard MATLAB® solvers.

When $\kappa(\mathbf{A})$ exceeds 10^{20}, $\kappa\left(\mathbf{AP}^{(n)}\right)$ can still be significantly reduced, but then $\mathbf{P}^{(n)}$ becomes very ill-conditioned. Also, in such a case, a non-linear solver with higher precision is required to solve (21) for n.

Finally, recall that our preconditioning process starts with a well-formulated quasi-interpolant. Thus, the method can also give good performance in situations where interpolation is not required or preferred, such as, for example, optimized smooth approximation of noisy data (see [17]).

References

1. S. F. Ashby, T. A. Manteuffel, and J. S. Otto, "A comparison of adaptive Chebyshev and least squares polynomial preconditioning for Hermitian positive definite linear systems", *SIAM J. Sci. Statist. Comput.* Vol. 13, pp. 1–29, 1992.
2. B. J. C. Baxter, "Preconditioned conjugate gradients, radial basis functions, and Toeplitz matrices", *Comput. Math. Appl.* Vol. **43**, pp. 305–318, 2002.

3. R. K. Beatson, J. B. Cherrie, and C. T. Mouat, "Fast fitting of radial basis functions: methods based on preconditioned GMRES iteration", *Adv. Comput. Math.* Vol. **11**, pp. 253–270, 1999.

4. R. K. Beatson, W. A. Light, and S. Billings, "Fast solution of the radial basis function interpolation equations: domain decomposition methods", *SIAM J. Sci. Comput.* Vol. **22**, pp. 1717–1740, 2000.

5. M. Benzi, "Preconditioning techniques for large linear systems: a survey", *J. Comput. Phys.* Vol. **182**, pp. 418–477, 2002.

6. D. Brown, L. Ling, E. Kansa, and J. Levesley, "On approximate cardinal preconditioning methods for solving PDEs with radial basis functions", *Eng. Anal. Bound. Elem.* Vol. **29**, pp. 343–353, 2005.

7. M. D. Buhmann, "Radial Basis Functions", Cambridge University Press, New York, 2003.

8. L. Cesari, "Sulla risoluzione dei sistemi di equazioni lineari per approssimazioni successive", *Ricerca Sci., Roma* Vol. **2** 8_1, pp. 512–522, 1937.

9. C. S. Chen, H. A. Cho and M. A. Golberg, "Some comments on the ill-conditioning of the method of fundamental solutions", *Eng. Anal. Bound. Elem.* Vol. **30**, pp. 405–410, 2006.

10. S. De Marchi and R. Schaback, "Stability of Kernel-Based Interpolation", preprint, 2007.

11. T. A. Driscoll and B. Fornberg, "Interpolation in the limit of increasingly flat radial basis functions", *Comput. Math. Appl.*, Vol. **43**, pp. 413–422, 2002.

12. T. A. Driscoll and A. R. H. Heryudono, "Adaptive residual subsampling methods for radial basis function interpolation and collocation problems", *Comput. Math. Appl.*, Vol. **53**, pp. 927–939, 2007.

13. N. Dyn, "Interpolation of scattered data by radial functions", in *Topics in Multivariate Approximation*, C. K. Chui, L. L. Schumaker, and F. Utreras (eds.), Academic New York, pp. 47–61, 1987.

14. N. Dyn, D. Levin, and S. Rippa, "Numerical procedures for surface fitting of scattered data by radial functions", *SIAM J. Sci. Statist. Comput.* Vol. **7**, pp. 639–659, 1986.

15. G. E. Fasshauer, "Solving partial differential equations by collocation with radial basis functions", in *Surface Fitting and Multiresolution Methods*, A. Le Mehaute, C. Rabut, and L. L. Schumaker (eds.), Vanderbilt University Press, Nashville, TN, pp. 131–138, 1997.

16. G. E. Fasshauer, "Meshfree approximation methods with MATLAB", *Interdisciplinary Mathematical Sciences*, Vol. **6**, World Scientific Publishers, Singapore, 2007.

17. G. E. Fasshauer and J. G. Zhang, "Scattered data approximation of noisy data via iterated moving least squares", in *Curve and Surface Fitting: Avignon 2006*, T. Lyche, J. L. Merrien and L. L. Schumaker (eds.), Nashboro Press, Brentwood, TN, pp. 150–159, 2007.

18. G. E. Fasshauer and J. G. Zhang, "Iterated approximate moving least squares approximation", in *Advances in Meshfree Techniques*, V. M. A. Leitao, C. Alves and C. A. Duarte (eds.), Springer, New York, pp. 221–240, 2007.

19. G. E. Fasshauer and J. G. Zhang, "On choosing "optimal" shape parameters for RBF approximation", *Numer. Algorithms* Vol. **45**, pp. 345–368, 2007.

20. B. Fornberg and C. Piret, "A stable algorithm for flat radial basis functions on a sphere", *SIAM J. Sci. Comp.* Vol. **30**, pp. 60–80, 2007.

21. B. Fornberg and G. Wright, "Stable computation of multiquadric interpolants for all values of the shape parameter", *Comput. Math. Appl.* Vol. **47**, pp. 497–523, 2004.

22. B. Fornberg and J. Zuev, "The Runge phenomenon and spatially variable shape parameters in RBF interpolation", *Comput. Math. Appl.* Vol. **54**, pp. 379–398, 2007.

23. E. J. Kansa and R. E. Carlson. "Improved accuracy of multiquadric interpolation using variable shape parameters", *Comput. Math. Applic.* Vol. **24**, pp. 99–120, 1992.

24. C.-F. Lee, L. Ling and R. Schaback, "On convergent numerical algorithms for unsymmetric collocation", *Adv. Comp. Math*, to appear.

25. L. Ling and E. J. Kansa, "Preconditioning for radial basis functions with domain decomposition methods", *Math. Comput. Model.* Vol. **40**, pp. 1413–1427, 2004.

26. L. Ling and E. J. Kansa, "A least-squares preconditioner for radial basis functions collocation methods", *Adv. Comp. Math.* Vol. **23**, pp. 31–54, 2005.

27. L. Ling and R. Schaback, "Stable and convergent unsymmetric meshless collocation methods", *SIAM J. Numer. Anal.*, to appear.

28. V. Maz'ya, "A new approximation method and its applications to the calculation of volume potentials. Boundary point method", in *DFG-Kolloquium des DFG-Forschungsschwerpunktes "Randelementmethoden"*, 1991.
29. V. Maz'ya and G. Schmidt, "On quasi-interpolation with non-uniformly distributed centers on domains and manifolds", *J. Approx. Theory*, Vol. **110**, pp. 125–145, 2001.
30. S. Rippa, "Algorithm for selecting a good value for the parameter c in radial basis function interpolation", *Adv. Comput. Math.* Vol. **11**, pp. 193–210, 1999.
31. H. Wendland, Scattered Data Approximation, Cambridge University Press, Cambridge, 2005.
32. J. G. Zhang, "Iterated Approximate Moving Least-Squares: Theory and Applications", *Ph.D. Dissertation, Illinois Institute of Technology*, 2007.

Arbitrary Precision Computations of Variations of Kansa's Method

Leevan Ling

Abstract In this paper, we are interested in some convergent formulations for the unsymmetric collocation method or the so-called Kansa's method. The rates of convergence of two variations of Kansa's method are examined and verified in arbitrary–precision computations.

Keywords: Radial basis function · Kansa's method · convergence · error bounds · high precision computation

1 Introduction

In this paper, we are interested in the radial basis function (RBF) method for solving partial differential equations (PDE) in strong form.

We consider PDE in the general form of

$$Lu = f, \ L : \mathscr{U} \to \mathscr{F}, \tag{1}$$

where L is a linear operator between two normed linear spaces $\mathscr{U}$ and $\mathscr{F}$. The PDE (1) can be solved by a sufficiently smooth solution $u^* \in \mathscr{U}$ that generates the data $f := Lu^* \in \mathscr{F}$.

The unsymmetric RBF collocation method was first proposed by Kansa [5, 6]. In the original formulation, the trial and test spaces were closely related. Let Φ: $\mathbb{R}^d \times \mathbb{R}^d \to \mathbb{R}$ be a symmetric positive definite kernel and $X_n = (x_1, \ldots, x_n) \subset \mathbb{R}^d$ be a set of *centers*. Usually, these centers are irregularly placed in the domain $\overline{\Omega}$ in which the PDE is defined. In order to obtain a numerical approximation u, we

L. Ling
Department of Mathematics, Hong Kong Baptist University, Kowloon Tong, Hong Kong,
e-mail: lling@hkbu.edu.hk

A.J.M. Ferreira et al. (eds.) *Progress on Meshless Methods, Computational Methods in Applied Sciences.*

assume that there exists an approximation to u^* in the *trial space* given by the span of $\{u_1, \ldots, u_n\}$ such that

$$u := \sum_{j=1}^{n} \alpha_j u_j \in \mathcal{U}_n := \text{span}\{u_1, \ldots, u_n\}. \tag{2}$$

The trial function in (2) is given by $u_j := \Phi(\cdot, x_j)$ for $1 \le j \le n$.

Next, we let $Y_n = X_n$ be the set of *collocation* or *test* points. We write the set of collocation conditions at Y_n as an scalar equations

$$\lambda_i[u] := Lu(y_i) = g_i =: f(y_i) \in \mathbb{R}, \qquad \text{for all } 1 \le i \le n. \tag{3}$$

The test functional λ_i ($1 \le i \le n$) is defined by applying the differential (or boundary) operator followed by a point evaluation at $y_i \in Y_n$. In general, a single set $\Lambda_n := \{\lambda_i\}_{i=1}^n$ of functionals contains several types of differential or boundary operators. The resulting Kansa's unsymmetric collocation matrix K has the ij-entries

$$K_{ij} := \lambda_i[u_j] = \lambda_i^{(1)} \Phi(\cdot, x_j), \quad 1 \le i, j \le n,$$

where the superscript (1) of $\lambda_i^{(1)}$ indicates the *first* variable of Φ on which the functional operates. Unknown coefficients are obtained by solving an exactly determined matrix system

$$K\alpha = g.$$

Although the method is relatively easy for implementation, there are some open questions that need to be answered. From the theoretical point of view, near-flat RBFs forms a terrible base of an excellent approximation space. Therefore, the Kansa's matrix system suffers serious ill-conditioning problem. Moreover, the Kansa's formulation in some specially constructed situations [3] may result in singular systems. Hence, it has neither error bounds nor convergence proofs. In order to carry out some mathematical analysis on the Kansa's approach, it is necessary to make further assumptions and modify the formulation.

In this paper, we answer the latter question. In some of our earlier works [2,8,9], Kansa's method was modified in such a way that solvability is guaranteed. Later, we proposed in [10,11] another variant of the method so that error bounds become possible. The solvability and convergence theorems of the modified formulations are reviewed here. The goal of this paper is to numerically compare among a few formulations of the unsymmetric meshless collocation methods. We use the arbitrary–precision computations capability of *Mathematica*© to numerically verify the proven theory: the numerical solution of some modified unsymmetric collocation methods converge faster than the interpolant with respect to the PDE residual norm.

2 Modified Kansa's Methods

The following theorem addresses that solvability is guaranteed if the trial functions or equivalently the RBF centers are correctly chosen.

Theorem 1 ([8]). *Assume the kernel* Φ *to be smooth enough to guarantee that the functions* $u_\lambda := \lambda^y \Phi(y, \cdot)$ *for* $\lambda \in \Lambda$ *are continuous. Furthermore, let the m functionals* $\lambda_1, \ldots, \lambda_m$ *of* Λ_m *be linearly independent over* $\mathscr{U}$. *Then the set of functions* $\{u_\lambda\}$ *for* $\lambda \in \Lambda_m$ *constructed above is linearly independent, and hence the unsymmetric collocation matrix is nonsingular for properly chosen trial centers.*

Suppose (3) is well-posed in a Hilbert space $\mathscr{H}$. Assume that

$$\|u\|_\Lambda := \sup_{\lambda \in \Lambda} |\lambda(u)| \leq \|u\|_\mathscr{U} \text{ for all } u \in \mathscr{U} \tag{4}$$

is a norm on $\mathscr{U}$ and is weaker than the norm in $\mathscr{U}$. Moreover, let $\mathscr{U}_\varepsilon$ be a subspace of $\mathscr{U}$ such that for all $u \in \mathscr{U}$ there is some approximation $v_{u,\varepsilon} \in \mathscr{U}_\varepsilon$ with

$$\|u - v_{u,\varepsilon}\|_\Lambda \leq \varepsilon \|u\|_\mathscr{U} \text{ for all } u \in \mathscr{U} \tag{5}$$

for any small number $\varepsilon > 0$. We would like to construct a function $v_\varepsilon \in \mathscr{U}_\varepsilon$ that solves

$$v_\varepsilon = \arg \min_{v \in \mathscr{U}_\varepsilon} \|v - u^*\|_\Lambda. \tag{6}$$

We do not know whether the minimum is attained and how it can be obtained, but due to (5) we can assume that there is a function $v_\varepsilon^* \in \mathscr{U}_\varepsilon$ with

$$\|v_\varepsilon^* - u^*\|_\Lambda \leq 2\|v_{u^*,\varepsilon} - u^*\|_\Lambda \leq 2\varepsilon \|u^*\|_\mathscr{U} \tag{7}$$

which is sufficient for our purpose.

Theorem 2 ([9–11]). *Let* $\mathscr{U}$ *be a normed linear space with norm* $\|.\|_\mathscr{U}$, *dual space* $\mathscr{U}^*$ *and dual unit sphere* $\mathscr{U}_1^* := \{\lambda \in \mathscr{U}^* : \|\lambda\|_{\mathscr{U}^*} = 1\}$. *Let a test set* $\Lambda \subset \mathscr{U}_1^*$ *be given such that* $\|.\|_\Lambda$ *is defined on* $\mathscr{U}$ *with (4). Assume further that the general interpolation problem (3) is well-posed. Let* $\{\mathscr{U}_\varepsilon\}_\varepsilon$ *be a scale of subspaces of* $\mathscr{U}$ *for* $\varepsilon \to 0$ *such that for all* $u \in \mathscr{U}$ *there is a* $v_{u,\varepsilon} \in \mathscr{U}_\varepsilon$ *with (5). For all* $\varepsilon \to 0$, *take a function* v_ε^* *satisfying (7). Then the error bound (7) holds and there is convergence* $\|v_\varepsilon^* - u^*\|_\Lambda \to 0$. $\square$

Trial spaces generated by spans of RBFs with sufficiently dense centers form a sequence of subspaces getting dense in $\mathscr{U}$. We know the RBF interpolant satisfies (5) for all $\varepsilon > 0$. Hence, the convergence rate of the unsymmetric collocation method given in (6) is faster than the rate convergence of the interpolant (in the trial space $\mathscr{U}_\varepsilon$ to the exact solution in $\mathscr{U}$) with respect to the PDE residual norm $\|.\|_\Lambda$.

Theoretically, we need a sufficient dense set of test points. Given a set of *properly chosen* trial points in X_n, we can assume that there is a test points Y_m with $n < m$.

The resulting overdetermined matrix, also denoted by K, has the ij-entries

$$K_{ij} := \lambda_i[u_j] = \lambda_i^{(1)}\Phi(\cdot, x_j), \ 1 \le i \le m, \ 1 \le j \le n.$$

Theorem 2 suggests that the overdetermined system $K\alpha = g$ should be solved by linear optimization with respect to the PDE residual norm. We denote this formulation as *PDE-LO*. An adaptive algorithm for the linear optimization process can be found in [10].

Since the implementation of linear optimization is nontrival, we prefer to use a least–squares solver instead. We denote the formulation *PDE-LS* when the overdetermined system $K\alpha = g$ is solved by the least–squares minimization.

2.1 An Example with Arbitrary-Precision

We demonstrate some convergence results of the modified Kansa's formulations in arbitrary–precision computation. Here, RBF interpolations are tested against RBF-PDE with the original Kansa's method and its variations: PDE-LO and PDE-LS. The test problem is a two-dimensional Poisson equation found in [1,4]:

$$\nabla^2 u(x, y) = f(x, y) \quad (x, y) \in [0, 1]^2,$$

with Dirichlet type boundary conditions. The right-hand function f and the boundary data are generated by the exact solution

$$u^*(x, y) = \sin\frac{\pi x}{6}\sin\frac{7\pi x}{4}\sin\frac{3\pi y}{4}\sin\frac{5\pi y}{4}.$$

For each tested mesh norm h, the trial points $X_{n(h)}$ for all methods are formed by equally spaced points on grid with stepsize h. For the original Kansa's method, the set of trial points is identical to the set of test points for each tested case. The interpolation problem is set up using the same set of points. Interpolation conditions are generated by the exact solution u^*. For the two modified Kansa's variations, we need more test points than trial points–the set of test points is generated using stepsize $\frac{h}{2}$. The multiquadric kernel is used in all tests and its shape parameter is fixed at $c = 1$. This is, for each tested h, the trial spaces are identical for all methods.

All matrix systems are solved in *Mathematica*© with arbitrary-precision. Exactly determined systems are solved by a LU-algorithm. The overdetermined systems are solved by the adaptive linear optimization algorithm in [10] and the QR-algorithm, respectively, for PDE-LO and PDE-LS. Therefore, ill-conditioning of the matrix system has no effect in our study of convergence rate.

Maximum error are collected on uniformly spaced with mesh norm h between $h = \frac{1}{5}$ to $\frac{1}{25}$. The error profiles against various h are shown in Fig. 1. All tested formulations and the interpolation problem clearly demonstrate exponential rates of convergence. As h decreases, the accuracies among all methods become similar.

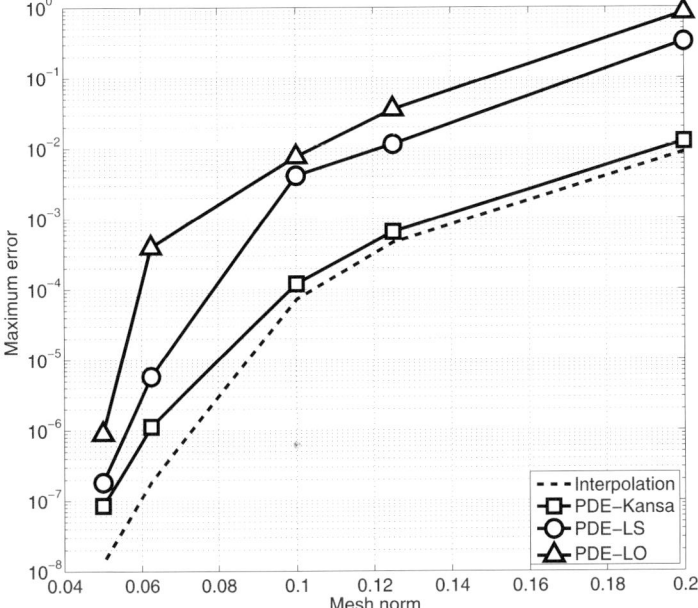

Fig. 1 Arbitrary–precision computations: error profiles for various unsymmetric meshless formulations with a fixed shape parameter $c = 1$

Although the original Kansa's method may give smaller error than the proposed variations for the tested h, its convergence rate seems to be slower than the others as $h \to 0$.

To see a clear demonstration of the convergence rates of smaller h, the error profiles are extrapolated using the exponential convergence formula given in [4]

$$\varepsilon = \mathcal{O}\left(\exp(ac^{3/2}) \lambda^{\sqrt{c}/h} \right),$$

and are shown in Fig. 2. The corresponding λ values are:

PDE-LO	0.417956
PDE-LS	0.364557
Kansa	0.482061
Interpolation	0.418077

We observe that, with respect to the maximum norm, the PDE-LO convergence at a similar rate as that of the interpolant to the exact solution. Among all, PDE-LS is not only easy to implement, but also has the faster convergence rate. More numerical examples can be found in [7]. It is shown that, when then PDE-optimal shape parameter is employed, the modified Kansa's variations PDE-LS and PDE-LO clearly show faster convergence rates than the Kansa's method and the interpolation problem.

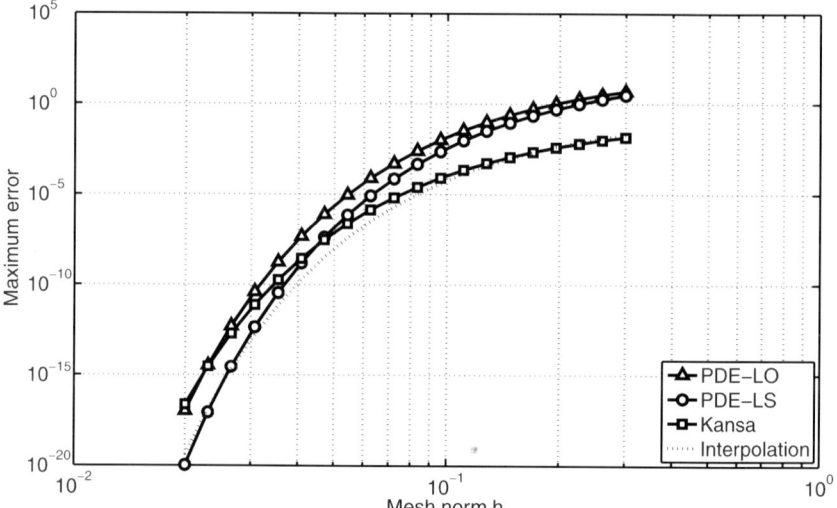

Fig. 2 Extrapolation using data in Fig. 1

3 Conclusion

In this paper, a new class of convergent formulations for solving PDEs via an unsymmetric radial basis function collocation method is examined. Some results obtained by arbitrary–precision computations are given. Numerical efficiency is greatly improved by replacing the linear optimization by least–squares. Despite the lack of a full convergence theory, the least–squares approach demonstrates a faster than interpolation rate of convergence.

For practical applications, an adaptive algorithm can be found in [8, 10]. The algorithm provides a subspace selection process to ensure the (original and modified) Kansa's matrix systems will not crash the linear solves. We refer the readers to the original articles for details.

References

1. A. H.-D. Cheng, M. A. Golberg, E. J. Kansa, and G. Zammito. Exponential convergence and *h-c* multiquadric collocation method for partial differential equations. *Numer. Meth. Partial Diff. Eq.*, 19(5):571–594, 2003.
2. Y. C. Hon and R. Schaback. Solvability of partial differential equations by meshless kernel methods. To appear in *Adv. Comput. Math.*, 2006.
3. Y. C. Hon and R. Schaback. On unsymmetric collocation by radial basis functions. *Appl. Math. Comput.*, 119(2–3):177–186, 2001.
4. C.-S. Huang, C.-F. Lee, and A. H.-D. Cheng. Error estimate, optimal shape factor, and high precision computation of multiquadric collocation method. *Eng. Anal. Bound. Elem.*, 31(7):614–623, July 2007.

5. E. J. Kansa. Multiquadrics—a scattered data approximation scheme with applications to computational fluid-dynamics. I. Surface approximations and partial derivative estimates. *Comput. Math. Appl.*, 19(8–9):127–145, 1990.
6. E. J. Kansa. Multiquadrics—a scattered data approximation scheme with applications to computational fluid-dynamics. II. Solutions to parabolic, hyperbolic and elliptic partial differential equations. *Comput. Math. Appl.*, 19(8–9):147–161, 1990.
7. C.-F. Lee, L. Ling, and R. Schaback. On convergent numerical algorithms for unsymmetric collocation. To appear in *Adv. Comput. Math.*, 2007.
8. L. Ling, R. Opfer, and R. Schaback. Results on meshless collocation techniques. *Eng. Anal. Bound. Elem.*, 30(4):247–253, April 2006.
9. L. Ling and R. Schaback. On adaptive unsymmetric meshless collocation. In S. N. Atluri and A. J. B. Tadeu, editors, *Proceedings of the 2004 International Conference on Computational & Experimental Engineering and Sciences*, volume CD-ROM, page paper # 270, Forsyth, USA, 2004. Advances in Computational & Experimental Engineering & Sciences, Tech Science Press.
10. L. Ling and R. Schaback. Stable and convergent unsymmetric meshless collocation methods. To appear in *SIAM J. Numer. Anal.*, 2007.
11. R. Schaback. Convergence of unsymmetric kernel-based meshless collocation methods. *SIAM J. Numer. Anal.*, 45(1):333–351, 2007.

A Meshless Approach for the Analysis of Orthotropic Shells Using a Higher-Order Theory and an Optimization Technique

Carla Maria da Cunha Roque, António Joaquim Mendes Ferreira(⊠), and Renato Manuel Natal Jorge

Abstract A higher order shear deformation theory is used for modeling orthotropic shells, with a meshless method based on global multiquadric radial basis functions. An optimization technique is presented for a quasi-independent choice of the shape parameter in the multiquadric function. The results obtained are compared with an analytical solution and show excellent accuracy.

Keywords: Meshless Methods · Radial Basis Functions · Composite Materials · Higher-order Theories

1 Introduction

The HOST9 higher order shell deformation theory presented by Khare et al. [1] is in this paper reduced to seven degrees of freedom, and produces accurate results in the modeling of shells. The theory is briefly presented in the present paper in Section 4 and readers should refer to reference [1] for more details. In this paper we compare the analytical solution for several shells as presented in [1] with the solution obtained with the meshless multiquadric radial basis functions method.

This collocation method, in a general way, can use a variety of radial basis functions as interpolation functions, such as those in Eqs. (1–4). Radial basis functions depend on a distance r between points in a grid (usually an euclidean distance, but not necessarily), and some may depend on a shape parameter, c. A plot of these functions is presented in Fig. 1. The radial basis function used in this paper is the multiquadric function. The multiquadric function depends on a shape parameter, c, that works as a fine tuning, producing a smooth curved surface or a hat shaped

A.J.M. Ferreira, C.M. da Cunha Roque, and R.M.N. Jorge

Departamento de Engenharia Mecânica e Gestão Industrial, Faculdade de Engenharia da Universidade do Porto, Rua Dr. Roberto Frias, 4200-465 Porto, Portugal, e-mail: ferreira@fe.up.pt

A.J.M. Ferreira et al. (eds.) *Progress on Meshless Methods, Computational Methods in Applied Sciences.*

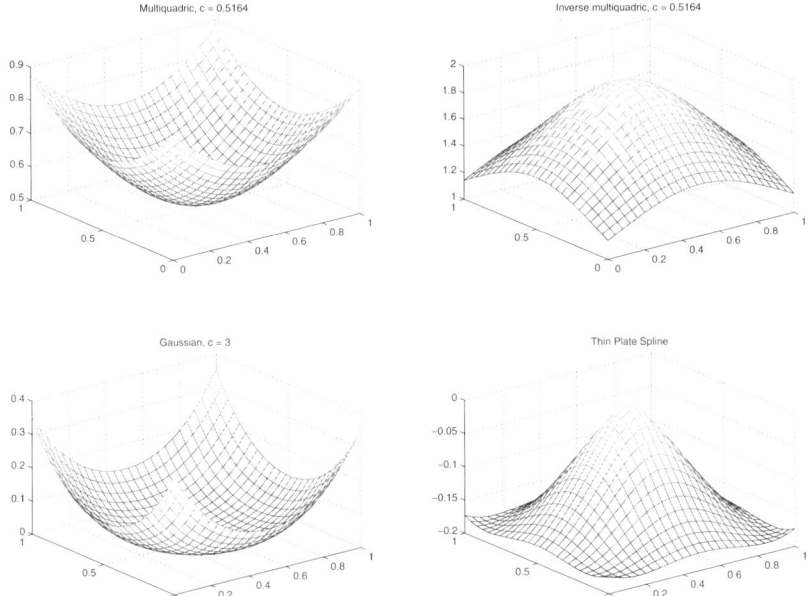

Fig. 1 Examples of radial basis functions

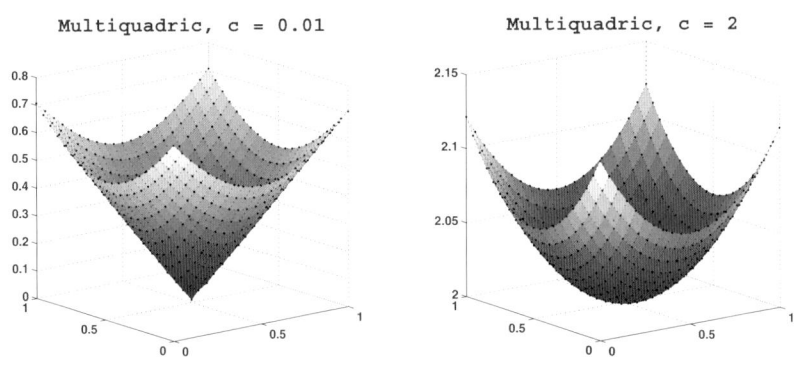

Fig. 2 The multiquadric function, with different shape parameters, c

function, as seen in Fig. 2.

$$g(r,c) = (c^2 + r^2)^\beta; \; \beta > 0 \qquad \text{multiquadric} \qquad (1)$$

$$g(r,c) = (c^2 + r^2)^\beta; \; \beta < 0 \qquad \text{inverse multiquadric} \qquad (2)$$

$$g(r,c) = e^{-cr^2}; \; c > 0 \qquad \text{gaussian} \qquad (3)$$

$$g(r) = r^2 \log r; \qquad \text{thin plate spline} \qquad (4)$$

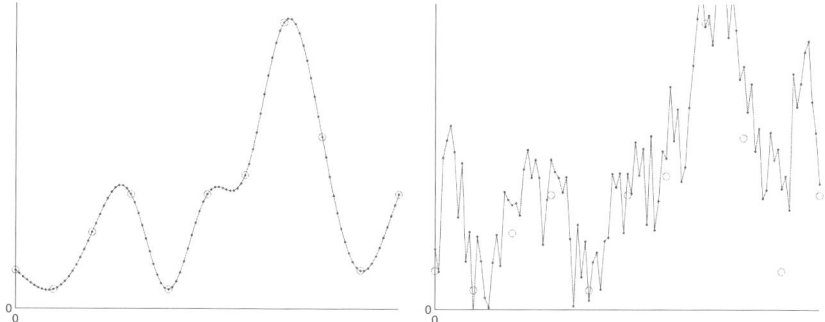

Fig. 3 Example of the influence of the multiquadric shape parameter, c in the interpolation of a data set. ○ – data set. *Left*: $c = 0.1$. *Right*: $c = 2$

Table 1 Some proposals for the choice of a shape parameter for the multiquadric function

Reference	Shape parameter, c
[9]	$c = 0.815d$
[10]	$c = 1.25D/\sqrt{N}$
[3]	$c^2 = c^2 \min(c^2_{max}/c^2_{min})^{(j-1)/(N-1)}$
[11]	$c = 2/\sqrt{N}$
	d, D – distances; j, N – number of points in grid

The meshless multiquadric method is well known for solving systems of partial differential equations with excellent accuracy [2, 3]. However, it has the problem of the choice of an adequate shape parameter for the multiquadric interpolation function.

For a verification of this dependence, we consider the interpolation of a data set. In Fig. 3, a data set is interpolated with the multiquadric function, with different shape parameters ($c = 0.1$ and $c = 2$, respectively). The parameter $c = 0.1$ produces a good interpolation, whereas $c = 2$ produces a bad interpolation of the data set. This simple example shows how the shape parameter can influence the quality of an interpolation, or in the case of shells in bending, the solution of a system of partial differential equations.

Several proposals for the choice of an adequate shape parameter can be found in the papers of Hardy [9], Franke [10], Kansa [2], Fasshauer [11]. All these proposals are somehow related with the number of points in the grid and the distance between those points (Table 1). However, according to Rippa [4], the shape parameter should depend on many other factors, such as:

- Number of grid points
- Distribution of points
- Interpolation function, g

- Conditioning number of matrix
- Computer precision

We then propose the use of a statistical technique based on Rippa's algorithm [4] that overcomes the choice of the shape parameter to the simple indication of a user-defined interval. The examples in this paper show an improvement of the static deformation results when using this optimization technique in the modeling of orthotopic shells.

2 The Multiquadric Method

Consider the generic boundary value problem with a domain Ω with boundary $\partial\Omega$, and some linear differential operators L and B:

$$Lu(x) = f(x), \ x \in \Omega \subset \mathbb{R}^n; \quad Bu|_{\partial\Omega} = q \tag{5}$$

The function $\mathbf{u(x)}$ is approximated by:

$$\mathbf{u} \simeq \bar{\mathbf{u}} = \sum_{j=1}^{N} \alpha_j g_j \tag{6}$$

were α_j are parameters to be determined after the collocation method is applied. We consider a global collocation method where the linear operators L and B acting in domain $\Omega \backslash \partial\Omega$ and boundary $\partial\Omega$ define a set of global equations in the form

$$\begin{pmatrix} \mathbf{L}_{ii} & \mathbf{L}_{ib} \\ \mathbf{B}_{bi} & \mathbf{B}_{bb} \end{pmatrix} \begin{pmatrix} \alpha_i \\ \alpha_b \end{pmatrix} = \begin{pmatrix} f_i \\ q_b \end{pmatrix}; \quad \text{or} \quad [\mathscr{L}][\alpha] = [\lambda] \tag{7}$$

where i and b denote domain and boundary nodes, respectively; f_i and q_b are some external conditions in domain and boundary (in plates in bending these can be external forces).

The function g represents a radial basis function. In the present case, we choose the multiquadric function, defined as:

$$g(r,c) = (\|x - x_j\|^2 + c^2)^{\frac{1}{2}} \tag{8}$$

were r is the euclidean distance between two nodes and c is a shape parameter that improves the function surface so that convergence gets faster [2, 3]. Other radial basis functions could be used (gaussians, splines, etc.). However, multiquadrics proved to be excellent for global, smooth, boundary-value problems, like shells in bending [5, 6].

3 An Optimization Technique

An optimal shape parameter c can be obtained for an interpolation problem $A\alpha = f$, $A = g(||\mathbf{x}_j - \mathbf{x}_i, c||)$, by the leave-one-out cross validation technique in regression analysis. The problem can be formulated as finding c in order to minimize a cost function given by the norm of an error vector $E(c)$ with components

$$E_i(c) = f_i - \sum_{j=1, j\neq i}^{N} \alpha_j^{(i)} g(||\mathbf{x}_j - \mathbf{x}_i, c||) \tag{9}$$

Here $\sum_{j=1, j\neq i}^{N} \alpha_j^{(i)} g(||\mathbf{x}_j - \mathbf{x}_i, c||)$ is the function value predicted at the i-th data point using RBF interpolation based on a set of data that excludes the i-th point.

A more efficient algorithm, from a computational point of view, is given by the following formula [4, 7]:

$$E_i(c) = \frac{\alpha_i}{A_{i,i}^{-1}} \tag{10}$$

where α_i is the i-th coefficient for the full interpolation problem and $A_{i,i}^{-1}$ is the i-th diagonal element of the inverse of the corresponding interpolation matrix A. In the case of our boundary value problem, the error to be minimized is a residual error, of the form [8]:

$$E_i(c) = \lambda_i - \sum_{j=1, j\neq i}^{N} \alpha_j^{(i)} \mathscr{L}g(||\mathbf{x}_j - \mathbf{x}_i, c||) \tag{11}$$

Now the generalization of the cross-validation algorithm is straightforward. Our BVP is given by Eq. (7). We can use the following formula that is analogous to Eq. (10):

$$E_i(c) = \frac{\alpha_i}{\mathscr{L}_{i,i}^{-1}} \tag{12}$$

where α_i is the i-th coefficient for the full collocation problem Eq. (7) and $\mathscr{L}_{i,i}^{-1}$ is the i-th diagonal element of the inverse of the corresponding collocation matrix $\mathscr{L}$. Having the cost function, we use the MATLAB function fminbnd to find a local minimum.

4 Equilibrium Equations and Boundary Conditions

In this paper we consider a shell with curvature radius R_x, R_y and thickness h (Fig. 4) The equilibrium equations are derived considering following the displacement field (u, v, w):

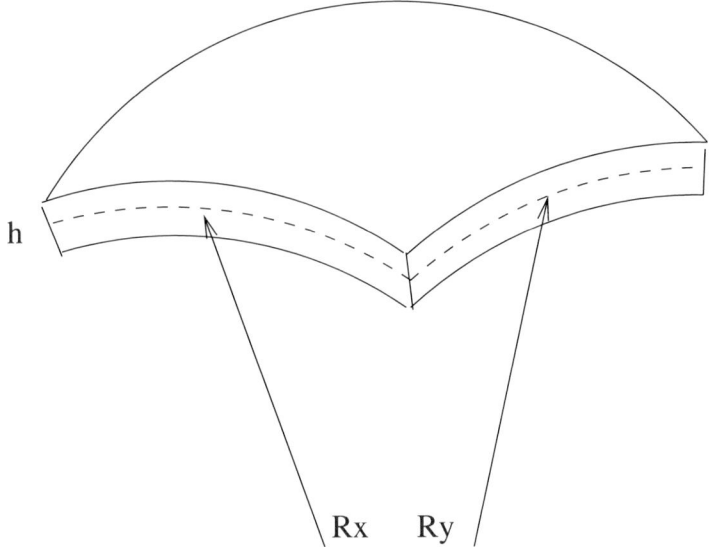

Fig. 4 Curved shell: radius of curvature R_x, R_y and thickness h

$$u(x,y,z) = u_0(x,y) + z\phi_x(x,y) + z^3\phi_x^*(x,y) \tag{13}$$

$$v(x,y,z) = v_0(x,y) + z\phi_y(x,y) + z^3\phi_y^*(x,y) \tag{14}$$

$$w(x,y,z) = w_0(x,y) \tag{15}$$

For small displacements and assuming $h/R_x, h/R_y \ll 1$, the strain-displacements relations are given by:

$$\varepsilon_{xx} = \frac{\partial u}{\partial x} + \frac{w}{R_x} \tag{16}$$

$$\varepsilon_{yy} = \frac{\partial v}{\partial y} + \frac{w}{R_y} \tag{17}$$

$$\gamma_{xy} = \frac{\partial u}{\partial y} + \frac{\partial v}{\partial x} \tag{18}$$

$$\gamma_{yz} = \frac{\partial v}{\partial z} + \frac{\partial w}{\partial y} - \frac{v}{R_y} \tag{19}$$

$$\gamma_{xz} = \frac{\partial u}{\partial z} + \frac{\partial w}{\partial x} - \frac{u}{R_x} \tag{20}$$

Replacing the expressions for the displacement field (u,v,w) in Eqs. (16–20), the strain-displacements relations are now expressed by:

$$
\begin{Bmatrix} \varepsilon_{xx} \\ \varepsilon_{yy} \\ \gamma_{xy} \\ \gamma_{yz} \\ \gamma_{xz} \end{Bmatrix} = \begin{Bmatrix} \varepsilon_{xx}^{(0)} \\ \varepsilon_{yy}^{(0)} \\ \gamma_{xy}^{(0)} \\ \gamma_{yz}^{(0)} \\ \gamma_{xz}^{(0)} \end{Bmatrix} + z \begin{Bmatrix} \varepsilon_{xx}^{(1)} \\ \varepsilon_{yy}^{(1)} \\ \gamma_{xy}^{(1)} \\ \gamma_{yz}^{(1)} \\ \gamma_{xz}^{(1)} \end{Bmatrix} + z^2 \begin{Bmatrix} \varepsilon_{xx}^{(2)} \\ \varepsilon_{yy}^{(2)} \\ \gamma_{xy}^{(2)} \\ \gamma_{yz}^{(2)} \\ \gamma_{xz}^{(2)} \end{Bmatrix} + z^3 \begin{Bmatrix} \varepsilon_{xx}^{(3)} \\ \varepsilon_{yy}^{(3)} \\ \gamma_{xy}^{(3)} \\ \gamma_{yz}^{(3)} \\ \gamma_{xz}^{(3)} \end{Bmatrix} \tag{21}
$$

$$
= \begin{Bmatrix} \frac{\partial u_0}{\partial x} + \frac{w_0}{R_x} \\ \frac{\partial v_0}{\partial y} + \frac{w_0}{R_y} \\ \frac{\partial u_0}{\partial y} + \frac{\partial v_0}{\partial x} \\ \phi_y + \frac{\partial w_0}{\partial y} - \frac{v_0}{R_x} \\ \phi_x + \frac{\partial w_0}{\partial x} - \frac{u_0}{R_y} \end{Bmatrix} + z \begin{Bmatrix} \frac{\partial \phi_x}{\partial x} \\ \frac{\partial \phi_y}{\partial y} \\ \frac{\partial \phi_x}{\partial y} + \frac{\partial \phi_y}{\partial x} \\ -\frac{\phi_y}{R_y} \\ -\frac{\phi_x}{R_x} \end{Bmatrix} + z^2 \begin{Bmatrix} 0 \\ 0 \\ 0 \\ 3\phi_y^* \\ 3\phi_x^* \end{Bmatrix} + z^3 \begin{Bmatrix} \frac{\partial \phi_x^*}{\partial x} \\ \frac{\partial \phi_y^*}{\partial y} \\ \frac{\partial \phi_x^*}{\partial y} + \frac{\partial \phi_y^*}{\partial x} \\ -\frac{\phi_y^*}{R_y} \\ -\frac{\phi_x^*}{R_x} \end{Bmatrix} \tag{22}
$$

By neglecting transverse normal stress, σ_z, the stress-strain relations in the local (material) cartesian system can be obtained as:

$$
\begin{Bmatrix} \sigma_1 \\ \sigma_2 \\ \tau_{12} \\ \tau_{23} \\ \tau_{31} \end{Bmatrix} = \begin{bmatrix} Q_{11} & Q_{12} & 0 & 0 & 0 \\ Q_{12} & Q_{22} & 0 & 0 & 0 \\ 0 & 0 & Q_{33} & 0 & 0 \\ 0 & 0 & 0 & Q_{44} & 0 \\ 0 & 0 & 0 & 0 & Q_{55} \end{bmatrix} \begin{Bmatrix} \varepsilon_1 \\ \varepsilon_2 \\ \gamma_{12} \\ \gamma_{23} \\ \gamma_{31} \end{Bmatrix} \tag{23}
$$

where subscripts 1 and 2 are respectively the fiber and the normal to fiber inplane directions, and 3 is the direction normal to the plate. The reduced stiffness components, Q_{ij} are given by

$$
Q_{11} = \frac{E_1}{1 - v_{12}v_{21}}; \quad Q_{22} = \frac{E_2}{1 - v_{12}v_{21}}; \quad Q_{12} = v_{21}Q_{11}; \tag{24}
$$

$$
Q_{33} = G_{12}; \quad\quad Q_{44} = G_{23}; \quad\quad Q_{55} = G_{31} \quad\quad v_{21} = v_{12}\frac{E_2}{E_1} \tag{25}
$$

in which E_1, E_2, v_{12}, G_{12}, G_{23} and G_{31} are material properties of the lamina. The material properties can then be transformed to global artesian axes by transformation [12]:

$$
\begin{Bmatrix} \sigma_{xx} \\ \sigma_{yy} \\ \tau_{xy} \\ \tau_{yz} \\ \tau_{zx} \end{Bmatrix} = \begin{bmatrix} \bar{Q}_{11} & \bar{Q}_{12} & \bar{Q}_{16} & 0 & 0 \\ \bar{Q}_{12} & \bar{Q}_{22} & \bar{Q}_{26} & 0 & 0 \\ \bar{Q}_{16} & \bar{Q}_{26} & \bar{Q}_{66} & 0 & 0 \\ 0 & 0 & 0 & \bar{Q}_{44} & \bar{Q}_{45} \\ 0 & 0 & 0 & \bar{Q}_{45} & \bar{Q}_{55} \end{bmatrix} \begin{Bmatrix} \varepsilon_{xx} \\ \varepsilon_{yy} \\ \gamma_{xy} \\ \gamma_{yz} \\ \gamma_{zx} \end{Bmatrix} \tag{26}
$$

The equilibrium equations are then obtained using the principle of virtual work,

$$\delta\Pi = \int_x \int_y \int_z (\sigma_{xx}\delta\varepsilon_{xx} + \sigma_{yy}\delta\varepsilon_{yy} + \tau_{xy}\delta\gamma_{xy} + \tau_{xz}\delta\gamma_{xz} + \tau_{yz}\delta\gamma_{yz}$$
$$- q\delta w_0)dx\,dy\,dz = 0 \tag{27}$$

producing seven equations of equilibrium:

$$\delta u_0 : \frac{\partial N_{xx}}{\partial x} + \frac{\partial N_{xy}}{\partial y} + \frac{Q_{xz}}{R_x} = 0 \tag{28}$$

$$\delta v_0 : \frac{\partial N_{yy}}{\partial y} + \frac{\partial N'_{xy}}{\partial x} + \frac{Q_{yz}}{R_y} = 0 \tag{29}$$

$$\delta w_0 : \frac{\partial Q_{xz}}{\partial x} + \frac{\partial Q_{yz}}{\partial y} - \frac{N_{xx}}{R_x} - \frac{N_{yy}}{R_y} = q \tag{30}$$

$$\delta\phi_x : \frac{\partial M_{xx}}{\partial x} + \frac{\partial M_{xy}}{\partial y} - Q_{xz} + \frac{K_{xz}}{R_x} = 0 \tag{31}$$

$$\delta\phi_y : \frac{\partial M_{yy}}{\partial y} + \frac{\partial M_{xy}}{\partial x} - Q_{yz} + \frac{K_{yz}}{R_y} = 0 \tag{32}$$

$$\delta\phi_x^* : \frac{\partial S_{xx}}{\partial x} + \frac{\partial S_{xy}}{\partial y} - 3T_{xz} + \frac{U_{xz}}{R_x} = 0 \tag{33}$$

$$\delta\phi_y^* : \frac{\partial S_{yy}}{\partial y} + \frac{\partial S_{xy}}{\partial x} - 3T_{yz} + \frac{U_{yz}}{R_y} = 0 \tag{34}$$

where

$$\left\{\begin{matrix} N_{ij} \\ M_{ij} \\ S_{ij} \end{matrix}\right\} = \int_{-h/2}^{h/2} \sigma_{ij} \left\{\begin{matrix} 1 \\ z \\ z^3 \end{matrix}\right\} dz; \quad \left\{\begin{matrix} Q_{iz} \\ K_{iz} \\ T_{iz} \\ U_{iz} \end{matrix}\right\} = \int_{-h/2}^{h/2} \tau_{iz} \left\{\begin{matrix} 1 \\ z \\ z^2 \\ z^3 \end{matrix}\right\} dz; \quad i,j = x,y \tag{35}$$

$$\left\{\begin{matrix} \{N\} \\ \{M\} \\ \{S\} \end{matrix}\right\} = \begin{bmatrix} [A] & [B] & [E] \\ [B] & [F] & [G] \\ [E] & [G] & [H] \end{bmatrix} \left\{\begin{matrix} \{\varepsilon^{(0)}\} \\ \{\varepsilon^{(1)}\} \\ \{\varepsilon^{(3)}\} \end{matrix}\right\}; \tag{36}$$

$$\left\{\begin{matrix} \{Q\} \\ \{K\} \\ \{T\} \\ \{U\} \end{matrix}\right\} = \begin{bmatrix} [I] & [K] & [L] & [M] \\ [K] & [N] & [O] & [P] \\ [L] & [O] & [Q] & [R] \\ [M] & [P] & [R] & [S] \end{bmatrix} \left\{\begin{matrix} \{\gamma^{(0)}\} \\ \{\gamma^{(1)}\} \\ \{\gamma^{(2)}\} \\ \{\gamma^{(3)}\} \end{matrix}\right\} \tag{37}$$

The stiffness components in Eqs. (36) and (37) are given by:

$$(A_{ij}, B_{ij}, E_{ij}, F_{ij}, G_{ij}, H_{ij}) = \int_{-h/2}^{h/2} Q(i,j)(1, z, z^3, z^2, z^4, z^6) \, dz; \quad i, j = 1, 2, 3$$

(38)

$$(I_{ij}, K_{ij}, L_{ij}, M_{ij}, N_{ij}, O_{ij}, P_{ij}, Q_{ij}, R_{ij}, S_{ij}) =$$
$$\int_{-h/2}^{h/2} Q(i,j)(1, z, z^2, z^3, z^2, z^3, z^4, z^4, z^5, z^6) \, dz; \quad i, j = 4, 5 \quad (39)$$

The boundary conditions for a simply supported shell for a border along the x axis are:

$$u_0 = 0; \, N_{yy} = 0; \, w_0 = 0; \, \phi_x = 0; \, M_{yy} = 0; \, \phi_x^* = 0; \, S_{yy} = 0 \quad (40)$$

5 Numerical Examples

To demonstrate the advantages of the optimization technique, simply supported homogeneous orthotropic spherical shells of several curvature radius $(R1, R2)$ and total thickness h are analyzed. The mechanical properties of the shells are:

$$\frac{E_x}{E_y} = 25; \quad \frac{G_{xz}}{E_y} = \frac{G_{xy}}{E_y} = 0.5; \quad \frac{G_{yz}}{E_y} = 0.2; \quad v_{xy} = 0.25 \quad (41)$$

Tables 2–4 show the central normalized displacement $\overline{w} = wE_y/q$. The results obtained with the present method are compared with an analytical solution of Khare et al. [1].

In Table 2, the shape parameter $c = 2/\sqrt{n}$ is used. Shells with different curvatures (R/a) and thickness (h/a) are tested, for various grids $n \times n$ (9×9 up to 21×21).

The examples in Table 2 are repeated, now using the optimization technique, with an interval of $]0.01 - 2[$. The obtained results for central transverse displacement are listed in Tables 3 and 4. The value of the optimized shape parameter is indicated by $c = c\_opt$.

To evaluate the results, a relative error is calculated by:

$$\text{error}(\%) = \left| \frac{\overline{w} - \text{analytical solution}[1]}{\text{analytical solution}[1]} \right| \times 100 \quad (42)$$

For all the examples, the error produced using the optimized shape parameter is smaller than the one using $c = 2/\sqrt{n}$. The examples $R/a = 1; h/a = 0.01, 0.1, 0.15$ and $R/a = 10^9; h/a = 0.01, 0.1, 0.15$ are plotted in Figs. 5 and 6, respectively, in a logarithmic scale.

Table 2 Central normalized deformation $\bar{w}$, with $c = 2/\sqrt{n}$

R/a	h/a				n				HOST9 [1]
		9	11	13	15	17	19	21	
1	0.01	61.5101	66.5021	69.6681	71.5703	72.6880	73.3456	73.7402	74.4360
	0.1	3.3314	3.5470	3.6646	3.7291	3.7654	3.7863	3.7967	3.8240
	0.15	1.7463	1.8370	1.8850	1.9110	1.9254	1.9337	1.9386	1.9490
2	0.01	224.1714	249.2038	263.5850	271.6381	276.1972	278.8365	280.4092	283.1800
	0.1	5.1293	5.2930	5.3725	5.4133	5.4353	5.4478	5.4551	5.4698
	0.15	2.3099	2.3620	2.3868	2.3993	2.4060	2.4098	2.4120	2.4166
3	0.01	469.9774	522.0512	550.9072	566.7219	575.5780	580.6783	583.7124	589.0600
	0.1	5.7184	5.8266	5.8772	5.9027	5.9163	5.9239	5.9282	5.9372
	0.15	2.4612	2.4937	2.5087	2.5162	2.5201	2.5223	2.5236	2.5262
4	0.01	773.9410	851.1098	892.8914	915.4865	928.0507	935.2387	939.5521	947.0900
	0.1	5.9598	6.0400	6.0768	6.0952	6.1050	6.1104	6.1136	6.1198
	0.15	2.5194	2.5434	2.5543	2.5596	2.5625	2.5641	2.5650	2.5668
5	0.01	1,109.4837	1,203.6016	1,253.7200	1,280.5575	1,295.4039	1,303.9024	1,309.0144	1,317.8000
	0.1	6.0789	6.1441	6.1739	6.1885	6.1963	6.2007	6.2032	6.2080
	0.15	2.5473	2.5670	2.5759	2.5802	2.5825	2.5838	2.5832	2.5860
10	0.01	2,660.4636	2,698.1106	2,722.7400	2,736.5714	2,744.3869	2,748.9032	2,750.6137	2,756.5000
	0.1	6.2457	6.2888	6.3081	6.3175	6.3224	6.3252	6.3268	6.3297
	0.15	2.5856	2.5993	2.6053	2.6082	2.6097	2.6106	2.6111	2.6120
20	0.01	4,105.3391	3,917.1056	3,852.4588	3,824.2059	3,810.1869	3,802.7527	3,777.9245	3,791.3000
	0.1	6.2889	6.3261	6.3426	6.3506	6.3548	6.3571	6.3585	6.3610
	0.15	2.5954	2.6075	2.6127	2.6153	2.6166	2.6174	2.6178	2.6185
10^9	0.01	5,015.4724	4,612.2833	4,471.0546	4,408.3330	4,376.8579	4,359.7965	4,349.3460	4,333.5000
	0.1	6.3035	6.3386	6.3541	6.3617	6.3656	6.3678	6.3691	6.3713
	0.15	2.5987	2.6102	2.6152	2.6177	2.6189	2.6196	2.6200	2.6210

Table 3 Central normalized deformation $\bar{w}$, with optimization technique, $c = c\_opt$

R/a	h/a	n							HOST9 [1]
		9	11	13	15	17	19	21	
1	0.01	70.3967	73.8631	73.9107	74.2941	74.4166	74.3015	73.9683	74.4360
	c_opt	1.4687	1.3734	1.0027	0.9590	0.9953	0.6925	0.4798	
	0.1	3.7603	3.8071	3.8179	3.8178	3.8152	3.8159	3.8144	3.8240
	c_opt	1.4386	1.3185	1.2399	1.0045	0.7701	0.6552	0.5573	
	0.15	1.9182	1.9404	1.9443	1.9461	1.9451	1.9443	1.9469	1.9490
	c_opt	1.3852	1.2732	1.0938	1.0023	0.7701	0.6259	0.7223	
2	0.01	271.1548	281.7754	282.3713	280.8700	282.4364	283.1098	281.8617	283.1800
	c_opt	1.3852	1.3851	1.1237	0.7701	0.7701	0.8843	0.5131	
	0.1	5.4551	5.4383	5.4674	5.4654	5.4645	5.4668	5.4640	5.4698
	c_opt	1.6757	1.0604	1.5302	0.9713	0.7701	0.7701	0.5573	
	0.15	2.4131	2.4147	2.4154	2.4153	2.4131	2.4155	2.4154	2.4166
	c_opt	1.7097	1.5302	1.2399	1.0042	0.6592	0.7701	0.6313	
3	0.01	554.3072	581.7875	583.8229	587.1612	588.7238	588.9516	585.5976	589.0600
	c_opt	1.2399	1.1918	0.9585	0.8931	0.8228	1.1695	0.4798	
	0.1	5.9228	5.9276	5.9341	5.9347	5.9341	5.9355	5.9361	5.9372
	c_opt	1.5302	1.2399	1.1332	0.9670	0.7701	0.7701	0.8366	
	0.15	2.5180	2.5241	2.5199	2.5256	2.5251	2.5258	2.5252	2.5262
	c_opt	1.3850	1.2732	0.7701	1.0451	0.7368	0.9496	0.5573	
4	0.01	896.5431	937.1712	944.8567	940.4913	944.8655	945.1022	942.0424	947.0900
	c_opt	1.2399	1.2399	1.1343	0.7701	0.7701	0.6592	0.4798	
	0.1	6.0990	6.1164	6.1128	6.1145	6.1177	6.1190	6.1168	6.1198
	c_opt	1.3858	1.3540	0.9162	0.7701	0.7701	1.2399	0.5221	
	0.15	2.5653	2.5659	2.5654	2.5656	2.5653	2.5664	2.5665	2.5668
	c_opt	1.7097	1.3836	1.0128	0.8034	0.6592	0.6944	0.7701	

Table 4 Central normalized deformation $\overline{w}$, with optimization technique, $c = c\_opt$

R/a	h/a		9	11	13	15	17	19	21	HOST9 [1]
5	0.01		1,284.1432	1,312.2216	1,308.3873	1,311.5902	1,316.5757	1,315.3185	1,317.5479	1,317.8000
		c_opt	1.4339	1.3860	0.9477	0.8113	0.8572	0.6569	0.7701	
	0.1		6.1915	6.2069	6.2070	6.2040	6.2075	6.2074	6.2046	6.2080
		c_opt	1.3853	1.5302	1.2399	0.7701	0.9982	1.2399	0.4798	
	0.15		2.5803	2.5853	2.5858	2.5855	2.5847	2.5858	2.5857	2.5860
		c_opt	1.3312	1.3938	1.5988	0.9298	0.6259	0.9589	0.6374	
10	0.01		2,728.2198	2,750.0196	2,746.5251	2,740.8945	2,748.4726	2,755.6627	2,756.1974	2,756.5000
		c_opt	1.5636	1.3873	0.9716	0.6592	0.6592	0.7701	0.7701	
	0.1		6.3192	6.3292	6.3284	6.3272	6.3292	6.3296	6.3289	6.3297
		c_opt	1.3842	1.5636	1.0723	0.7701	0.8368	0.8034	0.5739	
	0.15		2.6086	2.6118	2.6117	2.6116	2.6116	2.6119	2.6114	2.6120
		c_opt	1.35176	1.5302	1.1428	0.8931	0.7278	0.9177	1.5302	
20	0.01		3,761.8437	3,786.3119	3,789.9671	3,787.8027	3,789.4350	3,791.1048	3,791.7542	3,791.3000
		c_opt	1.4969	1.7097	1.3084	0.8034	0.7368	0.5907	0.5626	
	0.1		6.3521	6.3585	6.3607	6.3605	6.3601	6.3603	6.3607	6.3610
		c_opt	1.3857	1.2399	1.3318	1.0153	0.7701	0.6925	1.0144	
	0.15		2.6180	2.6182	2.6184	2.6181	2.6184	2.6184	2.6183	2.6185
		c_opt	1.7097	1.4614	1.2399	0.8390	0.8034	0.7701	0.5749	
10^9	0.01		4,312.2478	4,328.6887	4,344.6314	4,332.8799	4,343.7066	4,334.5079	4,335.3186	4,333.5000
		c_opt	1.2399	1.3539	0.7701	0.9220	0.5902	0.6529	0.5573	
	0.1		6.3616	6.3573	6.3663	6.3710	6.3712	6.3711	6.3710	6.3713
		c_opt	1.3428	0.8428	0.8044	1.3664	0.9958	0.7701	0.6319	
	0.15		2.6195	2.6206	2.6200	2.6202	2.6206	2.6206	2.6204	2.6210
		c_opt	1.5302	1.4969	0.9551	0.7701	0.8131	1.2399	0.5204	

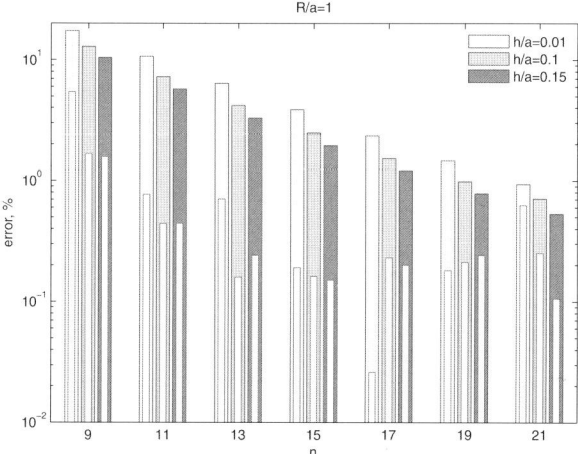

Fig. 5 Relative error for the central deflection, $\overline{w}$, with $R/a = 1$. The large bars correspond to $c = 2/\sqrt{n}$, and the narrow bars to $c = c\_opt$

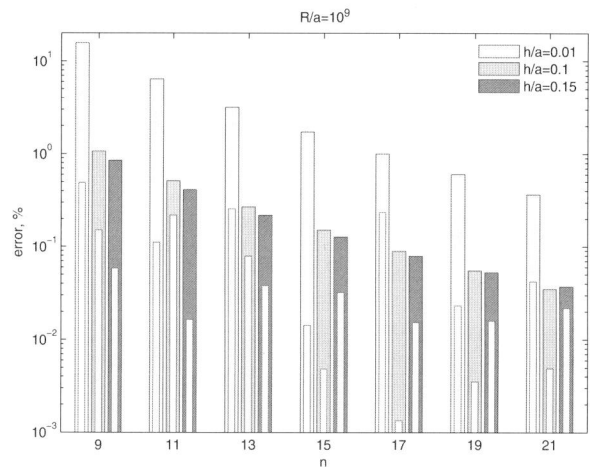

Fig. 6 Relative error for the central deflection, $\overline{w}$, with $R/a = 10^9$. The large bars correspond to $c = 2/\sqrt{n}$, and the narrow bars to $c = c\_opt$

6 Conclusions

In this paper we analyze orthotropic shells by a higher-order theory, using a collocation technique based on an optimized radial basis function method. The optimal shape parameter obtained by the Rippa method always produces better results for static transverse displacements than a user-defined shape parameter. In this paper, we concentrated on statics in an orthotropic shell. In forthcoming papers we will show results for composite shells.

References

1. R. K. Khare, T. Kant, and A. K. Garg. Closed-form thermo-mechanical solutions of higher-order theories of cross-ply laminated shallow shells. Composite Structures, 59(3):313–340, 2003.
2. E. J. Kansa. Multiquadrics: A scattered data approximation scheme with applications to computational fluid-dynamics. I. Surface approximations and partial derivative estimates. Computers & Mathematics with Applications, 19(8–9):127–145, 1990.
3. E. J. Kansa. Multiquadrics: A scattered data approximation scheme with applications to computational fluid-dynamics. II. Solutions to parabolic, hyperbolic and elliptic partial differential equations. Computers & Mathematics with Applications, 19(8–9):147–161, 1990.
4. S. Rippa. An algorithm for selecting a good value for the parameter c in radial basis function interpolation. Advances in Computational Mathematics, 11(2–3):193–210, 1999.
5. A. J. M. Ferreira, C. M. C. Roque, and R. M. N. Jorge. Modelling cross-ply laminated elastic shells by a higher-order theory and multiquadrics. Computers and Structures, 84(19–20):1288–1299, 2006.
6. A. J. M. Ferreira, C. M. C. Roque, and R. M. N. Jorge. Static and free vibration analysis of composite shells by radial basis functions. Engineering Analysis with Boundary Elements, 30(9):719–733, September 2006.
7. B. P. Wang. Parameter optimization in multiquadric response surface approximations. Structural and Multidisciplinary Optimization, 26(3–4):219–223, 2004.
8. A. H. D. Cheng, M. A. Golberg, E. J. Kansa, and Q. Zammito. Exponential convergence and h-c multiquadric collocation method for partial differential equations. Numerical Methods for Partial Differential Equations, 19(5):571–594, 2003.
9. R. L. Hardy. Multiquadric equations of topography and other irregular surfaces. Journal of Geophysical Research, 176:1905–1915, 1971.
10. R. Franke. Scattered data interpolation tests of some methods. Mathematics of Computation, 38:181–200, 1982.
11. G. E. Fasshauer. Newton iteration with multiquadrics for the solution of nonlinear PDEs. Computers and Mathematics with Applications, 43:423–438, 2002.
12. J. N. Reddy. Mechanics of Laminated Composite Plates and Shells. CRC Press, Boca Raton, FL 2004.

An Order-N Complexity Meshless Algorithm Based on Local Hermitian Interpolation

David Stevens, Henry Power(✉), and Herve Morvan

Abstract This work describes a new numerical method utilising RBF interpolants. Based on local Hermitian interpolation of function values and boundary operators, and using an explicit time-advancement formulation, the method is of order N complexity. Computational cost to advance the solution in time is minimal, and is largely dependent on local system support size.

The performance of the method is examined for a variety of linear convection-diffusion-reaction problems, featuring both steady and unsteady solutions. The technique is named the Local Hermitian Interpolation (LHI) method.

Keywords: Local Hermitian Interpolation · Radial Basis Functions · Meshless Methods

1 Rbf Methods

The use of a connected mesh is a basic characteristic of traditional numerical approaches for the solution of partial differential equations (PDEs). It is the case for domain methods such as the finite difference, element and volume methods, and also for boundary methods such as the boundary element method.

The construction of an effective numerical mesh in two or more dimensions is a non-trivial problem. In practice, only low order approximations are typically employed, resulting in a continuous approximation of the function across the mesh, but not its partial derivatives. In addition, models with data-dependent changes in structure (such as a moving surface) may encounter problems as the solution develops, leading to further loss of accuracy or computationally expensive re-meshing.

D. Stevens, H. Power, and H. Morvan

School of Mechanical, Materials and Manufacturing Engineering, Faculty of Engineering Room B111 Coaks University park, University of Nottingham, Nottingham, NG7 2RD, UK, e-mail: henry.power@nottingham.ac.uk

A.J.M. Ferreira et al. (eds.) *Progress on Meshless Methods, Computational Methods in Applied Sciences.*
© Springer Science + Business Media B.V. 2009

Table 1 Commonly used Radial basis functions

Radial basis functions (RBFs)		
Generalized thin plate spline	Generalized multiquadric	Gaussian
$r^{2m-2}\ln(r)$	$\left(r^2+c^2\right)^{m/2}$	$\exp(-r^2/c)$
	where m is an integer and $r = \|x - \xi_j\|$.	

In recent years, considerable effort has been given to the development of numerical methods which do not require a complex, inter-connected mesh in order to operate – relying instead on a set of data collocation points which may be distributed quasi-randomly throughout the domain. Radial Basis Function (RBF) based methods have been successfully used to solve a wide variety of PDEs in this fashion.

A radial basis function $\Psi(\|x - \xi_j\|)$ depends upon the separation distances of a set of functional centres, or *trial points* $\{\xi_j \in \Re^n; \ j = 1, 2, \ldots, N\}$, and exhibits spherical symmetry around these centres. The most commonly used RBFs are given in Table 1.

The Gaussian and the inverse multiquadric ($m < 0$ in the generalized multiquadric function), are positive definite functions. The thin-plate splines and the multiquadric ($m > 0$) are conditionally positive definite functions of order m, which require the addition of a polynomial term of order $m - 1$, together with a homogeneous constraint condition, in order to obtain an invertible interpolation matrix. The 'c' term is known as a 'shape parameter', and describes the relative width of the RBF functions about their centres. In practice, tuning of this parameter can dramatically affect the quality of the solution obtained. Increasing the value of c will lead to a flatter RBF. This will, in general, improve the rate of convergence at the expense of increased numerical ill-conditioning of the resulting linear system [1].

The meshless RBF method for solving PDEs, as described by Kansa [2, 3], constructs the (continuous) solution $u(x)$ of the PDE from a distinct set of N quasi-randomly distributed functional centres ξ_j.

$$u(x) = \sum_{j=1}^{N} \lambda_j \Psi(\|x - \xi_j\|) + \sum_{j=1}^{NP} \lambda_{j+N} P_{m-1}^{j}(x) \qquad x \in \Re^n \qquad (1)$$

Where P_{m-1}^{j} is the jth term of an order $(m-1)$ polynomial, under the constraint:

$$\sum_{j=1}^{N} \lambda_j P_{m-1}^{k}(x_j) = 0 \qquad k = 1, \ldots, NP \qquad (2)$$

With NP being the total number of terms in the polynomial (determined by the polynomial order and the number of spatial dimensions).

Consider a typical boundary value problem:

$$\begin{aligned} L[u] &= f(x) \quad \text{on } \Omega \\ B[u] &= g(x) \quad \text{on } \partial\Omega \end{aligned} \qquad (3)$$

Collocating the system at N distinct locations, known as *test points, x_j,* coinciding with the trial centres ξ_j, leads to a system of equations:

$$\begin{bmatrix} B[\Psi] & B[P_{m-1}] \\ L[\Psi] & L[P_{m-1}] \\ P_{m-1} & 0 \end{bmatrix} \lambda = \begin{bmatrix} g \\ f \\ 0 \end{bmatrix} \tag{4}$$

Which is fully populated and non-symmetric. This approach, known as *Kansa's method* or *the unsymmetric method*, has been applied to a wide range of problems with great success. However, for the standard formulation there is no formal guarantee that the collocation matrix will be non-singular [4]. In the special case of a numerically singular system, a small perturbation of functional centre locations (or of the shape parameter) will in general return a non-singular collocation matrix [5]; however, the perturbed collocation matrix may suffer from numerical ill-conditioning issues.

An alternative approach proposed by Fasshauer [6] uses the PDE operator applied to the RBF as basis functions:

$$u(x) = \sum_{j=1}^{NB} \lambda_j B_\xi \Psi \left(\|x - \xi_j\| \right) + \sum_{j=NB+1}^{N} \lambda_j L_\xi \Psi \left(\|x - \xi_j\| \right) + \sum_{j=1}^{NP} \lambda_{j+N} P_{m-1}^j(x) \tag{5}$$

Note: Operators with ξ subscript are applied at the trial points. Operators with x subscript are applied to test points

Collocating in a similar way leads to the system of equations:

$$\begin{bmatrix} B_x B_\xi[\Psi] & B_x L_\xi[\Psi] & B_x P_{m-1} \\ L_x B_\xi[\Psi] & L_x L_\xi[\Psi] & L_x P_{m-1} \\ B_x P_{m-1}^T & L_x P_{m-1}^T & 0 \end{bmatrix} \lambda = \begin{bmatrix} g \\ f \\ 0 \end{bmatrix} \tag{6}$$

This is known as the *Hermitian*, or *Symmetric* method, producing a collocation matrix which is symmetric (hence attractive for storage benefits), and was shown by Wu [7, 8] to be non-singular provided that no two collocation points sharing the same operator are placed in the same location.

While the Hermitian method has many benefits over the native Kansa's formulation, both methods suffer from the same fundamental problem – described by Robert Schaback [9] as "the uncertainty relation": Better conditioning is associated with worse accuracy, and worse conditioning is associated with improved accuracy. As the system size is increased, this problem becomes more pronounced. In recent years much research has focussed on attempting to circumvent this problem, for example by the formulation of RBF-specific preconditioners [5], and through adaptive selection of functional centres and collocation points [4].

Still, at present, the only reliable method of controlling numerical ill-conditioning and computational cost as problem size increases is through domain decomposition (see, for example, [10–13]). Since the full-domain collocation matrices are fully populated, domain decomposition must be problem-specific rather than automatic, and so generally requires heavy user intervention. However, by taking the domain decomposition principle and applying it to very small and heavily overlapping local systems, a "hybrid" method can be formulated, which retains the flexibility of working on a set of scattered data at the expense of introducing some localised connectivity. However, this connectivity allows much more flexibility than the rigid control volumes described by traditional numerical meshes, and allows an essentially meshless approach to be applied to arbitrarily large problems cheaply and without numerical conditioning issues.

2 Mathematical Formulation

Recent works by Vertnik et al. [14], and also by Wright and Fornberg [15], have shown how RBF interpolation techniques can be applied at a local level to produce solutions over a continuous domain. Such techniques show great promise for the solution of large-scale, practical problems. In this work, a general formulation is described which is capable of operating on scattered data-sets with arbitrary boundary conditions, to maintain the truly meshless nature of full-domain RBF techniques.

2.1 Local System Definition

For a general initial-value PDE:

$$\frac{\partial u}{\partial t} = L_1[u] \quad in \quad \Omega$$
$$B[u(\mathbf{x}, t)] = g(\mathbf{x}, t) \quad on \quad \Gamma$$
$$u(\mathbf{x}, 0) = f(\mathbf{x}) \quad in \quad \Omega$$

The solution space Ω is covered with a series of (quasi-)scattered data centres, with additional centres placed on all boundary surfaces Γ (see Fig. 1). Each centre has associated with it a local system of points. Local system sizes are generally in the region of around 7 to 25 points in 3D.

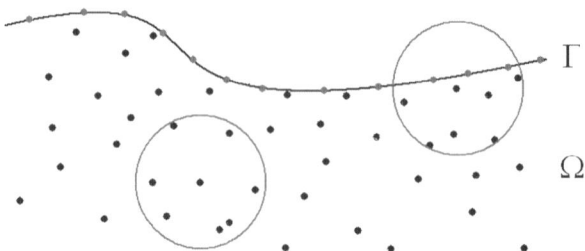

Fig. 1 Schematic representation

The solution is modelled via a series of Hermitian interpolations on each of the local systems, using the functional values for internal points, and the boundary operator values at boundary points (where present in the local system):

$$u^{(k)}(x) = \sum_{j=1}^{NB} \alpha_j^{(k)} B_\xi [\Psi(\|x - \xi_j\|)] + \sum_{j=NB+1}^{NB+NI} \alpha_j^{(k)} \Psi(\|x - \xi_j\|) + \sum_{j=1}^{NP} \alpha_{j+N} P_{m-1}^j(x)$$
(7)

Where: k is the local system index

NB is the number of boundary points

NI is the number of internal points

NP is the number of terms in a 3D polynomial of degree $m - 1$

This description is collocated at each of the points within the local system, applying once again the boundary operator at boundary points, and leading to a series of small symmetric linear systems which can be solved independently to interpolate the (global) initial solution:

$$A^{(k)} \alpha^{(k)} = d^{(k)}$$
(8)

Where:

$$d^{(k)} = \begin{bmatrix} f^{(k)} \\ g^{(k)} \\ 0 \end{bmatrix} \quad A = \begin{bmatrix} \Psi & B_\xi[\Psi] & P_{m-1} \\ B_x[\Psi] & B_x B_\xi[\Psi] & B_x[P_{m-1}] \\ P_{m-1}{}^T & B_\xi[P_{m-1}]^T & 0 \end{bmatrix}$$
(9)

In this way the value of u close to point k can be written as:

$$u^{(k)}(x) = [H]^T \alpha^{(k)} \text{ where}: [H]^T = [\Psi, B_\xi[\Psi] P_{m-1}]$$
(10)

The above systems are solved, with the $[A^{(k)}]^{-1}$ inverse matrices explicitly calculated and stored for later use.

2.2 Time Advancement

The initial data can now be advanced in time according to:

$$\frac{\partial u}{\partial t} = L[u] \quad \text{For a general linear operator } L$$
(11)

Time advancement can take place either explicitly or implicitly. The implicit method generally allows much larger timesteps to be employed, at the expense of having to solve a sparse global matrix system.

2.2.1 Explicit Time Advancement

For a general order s explicit linear-multistep time advancement formula:

$$u^{n+1} = u^n + \Delta t \sum_{j=0}^{s} b_i L^{n-j} \left[u^{n-j} \right] \qquad (b_0 = 1)$$

$$= \tilde{L}[u^n] + \sum_{j=1}^{s} \hat{L}_j \left[u^{n-j} \right] \qquad \text{where} \quad \hat{L}_j = \Delta t \, b_j L^{n-j} \qquad (12)$$

$$\tilde{L} = (1 + \Delta t L^n)$$

Reformulating in terms of the local systems of Eq. (10):

$$u^{n+1} = \left[\tilde{L}H \right]^T \alpha^n + \sum_{j=1}^{s} \left[\hat{L}_j H \right]^T \alpha^{n-j} \qquad (13)$$

It is worth noting that the collocation matrix, A, is independent of the PDE ('L') operator. As such, it will change only if the distribution of points is changed, or a time-dependent boundary condition is present within the local system. If this is the case, the small local linear system is re-inverted to find α^n.

However, in cases where the particular local system does not alter its distribution of points or contain a time-dependent boundary, α^n can be calculated at each stage via forward multiplication of the (stored) A^{-1} inverse collocation matrix with the local solution:

$$\alpha^n = \left[A^{-1} \right] d^n \qquad (14)$$

In cases where the collocation matrix does not change over time, the α^{n-j} vectors can be reused from previous timesteps. Each local system is used to advance its centre-point in time, represented by computing $\left[\tilde{L}H \right]$ and $\left[\hat{L}_j H \right]$ arrays at the centre-point only.

2.2.2 Implicit Time Advancement

The current implementation using implicit timestepping is directly comparable with the explicit procedure described above: The local systems are formed with the solution values and boundary operators, and the PDE operator is applied at the global level. The major difference is that the local right hand side vectors, d, now contain the current (unknown) solution value:

$$d^n = \begin{bmatrix} u^n_i \\ g(x_i) \\ 0 \end{bmatrix} \qquad (15)$$

For steady problems, $L[u] = 0$, the PDE operator can be applied directly to the reconstruction array, H, for each of the global collocation points:

$$L[u]_{centre} = \left(L_x[\Psi], L_x B_\xi[\Psi], L_x[P_{m-1}]\right)\left([A]^{-1}d\right) = 0 \qquad (16)$$

This formulation (Eq. (16)) describes a single row in a sparse, scattered global linear system. This system can then be solved efficiently using any of the popular sparse linear system solution techniques. The GMRES algorithm from the SPARSKIT toolbox [16] is currently used to solve such systems.

For unsteady problems, a modified PDE operator is formed. For example, using the first-order forward Euler technique:

$$\frac{u^{n+1} - u^n}{\Delta t} = L[u^{n+1}] : u^{n+1} - \Delta t L[u^{n+1}] = u^n$$

$$(1 - \Delta t L)u^{n+1} = u^n$$

$$\tilde{L}u^{n+1} = u^n \qquad (17)$$

Once again, the modified PDE operator is applied to the reconstruction array, H, for each of the collocation points, forming a line in a spare global linear system (Eq. (18)):

$$L[u^{n+1}]_{centre} = \left(\tilde{L}_x[\Psi], \tilde{L}_x B_\xi[\Psi], \tilde{L}_x[P_{m-1}]\right)\left([A]^{-1}d^{n+1}\right) = u^n \qquad (18)$$

2.3 Computational Cost

For the case where the A matrix does not change in time, the biggest expense at every timestep is, for the explicit timestepping method, the matrix-vector multiplication in the calculation of α^n, requiring $o(m^2)$ operations for each system (where m is the local system size). The computation of $[\tilde{L}H]^T$ and $[\hat{L}_j H]^T$ at the centre-point and their multiplication with α_n requires only $o(m)$ operations, as these are vector quantities.

The implicit timestepping method has an additional cost to solve the global matrix.

The interpolation of the initial value requires the explicit calculation of the inverse matrices, at a cost of $o(m^3)$ operations per system. This is also the cost at each timestep for systems which have a varying A matrix.

So the overall cost of the method (once the local systems have been established) can be summarised as follows:

$$\text{Cost} = o(Nm^3) + o(TNm^2) + o(TPm^3) + o(TNlm) \qquad (19)$$

Where: N = Total number of local systems
          m = Representative local system size

P = Number of systems with a time-dependent boundary or dynamic point distribution ($\leq$N and often equal to zero)

T = Number of time-steps

I = Number of iterations required to solve global matrix (=0 for explicit time advancement method)

Leading to a truly order-N RBF algorithm, which can scale to the largest problems.

As well as vastly reducing computational cost in comparison to $o(N^2)$ full-domain methods, the localised approach allows certain other flexibilities. For example, the user is able to apply different shape parameters for the multiquadric RBF to each local system. Research into c-parameter optimisation across a given solution domain is ongoing. The localised approach also allows the system sizes to be altered in a data dependent way; for example, using larger local systems in more sensitive areas and vice-versa. The relatively small size of the local systems ensures a reasonably well conditioned collocation matrix.

3 Numerical Results for Linear Problems

A code is in development to implement the method described above. The code is fully 3D, written in Fortran 90, and is designed to handle convection-diffusion type problems on complex geometries. Currently this program uses only a single thread on one CPU, however, a parallel implementation using MPI is also in development.

3.1 Diffusive Shock

Initial validation is performed via a simple mono-dimensional 'diffusive shock' problem, with a steady, uniform velocity field.

The following steady convection-diffusion equation is considered:

$$D\frac{\partial^2 \phi}{\partial x_i^2} - u_i \frac{\partial \phi}{\partial x_i} = 0 \qquad (20)$$

over the domain:

$$0 < x < 1.0$$
$$0 < y < 0.2$$
$$0 < z < 0.2$$

with boundary conditions:

$$\varphi(x = 0) = 1, \quad \varphi(x = 1) = 2$$

Zero-flux is imposed in the y, z directions to retain a 1D solution.

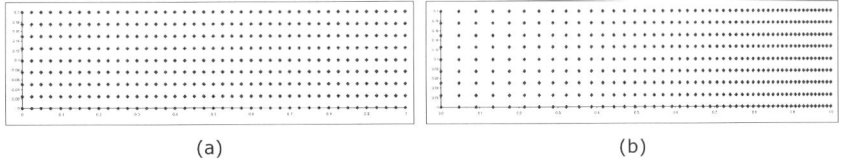

Fig. 2 Mesh representation: (**a**) uniform mesh, (**b**) refined mesh

This formulation has a steady analytical solution for u constant in the x-direction, given by:

$$\phi(x) = 2 - \frac{1 - e^{(x-1)Pe}}{1 - e^{-Pe}} \quad Pe = \frac{u}{D} = \text{Peclet number} \tag{21}$$

It should be noted that using the explicit timestepping formulation the LHI method is not able to solve steady problems directly. Instead, the equivalent time dependent equation is solved from a realistic initial state, $\phi(x) = 1.0$ in this case, and run until a steady state is reached. The explicit time advancement technique is utilised for each of the 1D test cases presented in this work.

The model is run, using the explicit timestepping formulation, on a uniform mesh of $(41 \times 9 \times 9)$ points for Peclet numbers of 50 and 100, and on a refined mesh of $(49 \times 9 \times 9)$ points for Peclet numbers of 200 and 400. All tests are run to convergence (defined here as an L_2 norm of less than 10^{-10} between successive iterations).

An equivalent finite volume mesh is constructed for each test, using the point distributions represented in Fig. 2 as their vertices. A simple finite volume program (also utilising a first-order explicit time advancement method) is run using the first-order upwind and the fifth-order WENO schemes to provide a comparison with established numerical methods. A constant c-parameter value is taken, and results are shown for the parameter which minimises the global L_2 error norm. A compact support radius of 0.026 is used, to recover a seven-point local stencil. For clarity, results are shown only for $x \geq 0.75$.

The LHI method is capable of accurately reproducing the analytical solution in each of the tests performed. A stable solution can be found in each instance, although this requires optimisation of the c-parameter for the tests at higher Peclet numbers.

The pointwise nature of the calculations performed in the LHI method allows, in general, a much more precise capturing of the shock than does the cell-averaging used in the finite volume methods. The highly diffusive first order upwind scheme consistently under-predicts the intensity of the shock. Conversely, the 5th order WENO scheme over-predicts the intensity of the shock in all but the Pe = 400 test. The LHI method reproduces a qualitatively good approximation to the solution in each case (Figs. 3–6), and predicts an L_2 error a factor of ten lower than either of the finite volume schemes in the first three tests, as shown in Table 2.

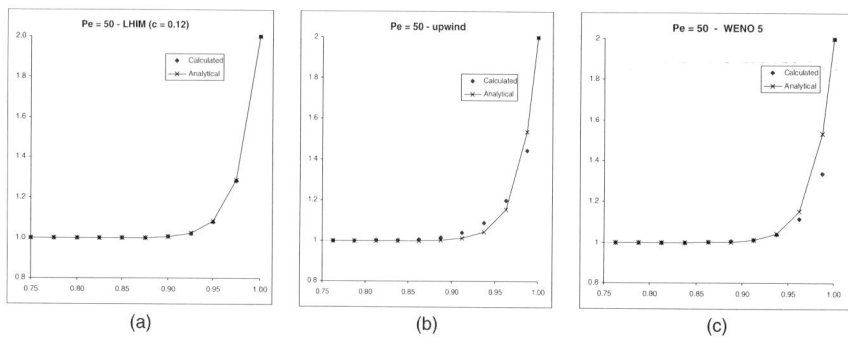

Fig. 3 Shock capturing at Pe = 50 (uniform mesh): (**a**) LHI method, (**b**) finite volume upwind, (**c**) finite volume WENO 5th order

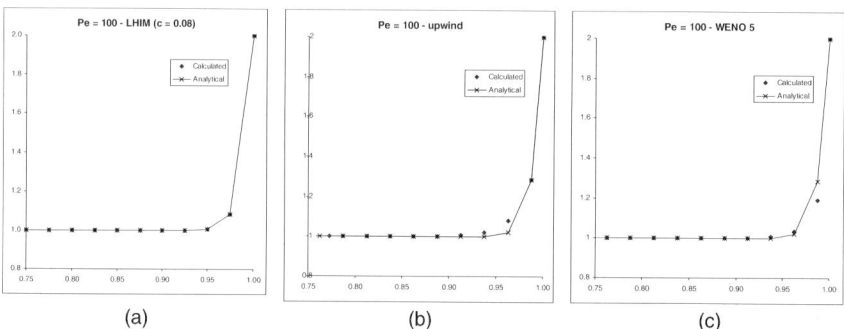

Fig. 4 Shock capturing at Pe = 100 (uniform mesh): (**a**) LHI method, (**b**) finite volume upwind, (**c**) finite volume WENO 5th order

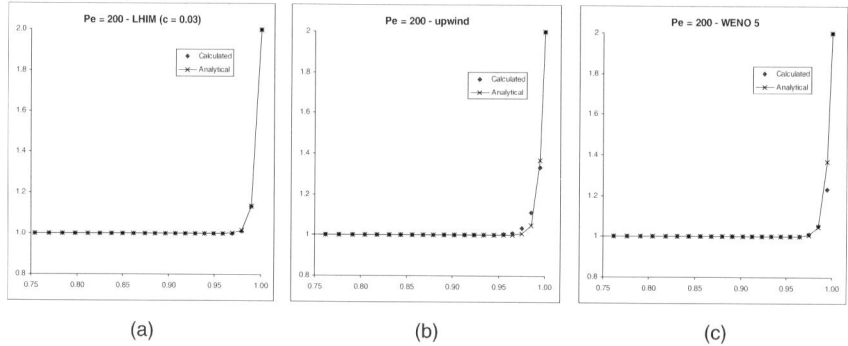

Fig. 5 Shock capturing at Pe = 200 (refined mesh): (**a**) LHI method, (**b**) finite volume upwind, (**c**) finite volume WENO 5th order

A study of the performance of full-domain Kansa's and Hermitian RBF solution techniques has been carried out by Power and Barraco [17]. These calculations were performed using a 2D rather than a 3D distribution of points, at similar point densities. However the LHI method is able to solve the problem with greater accuracy

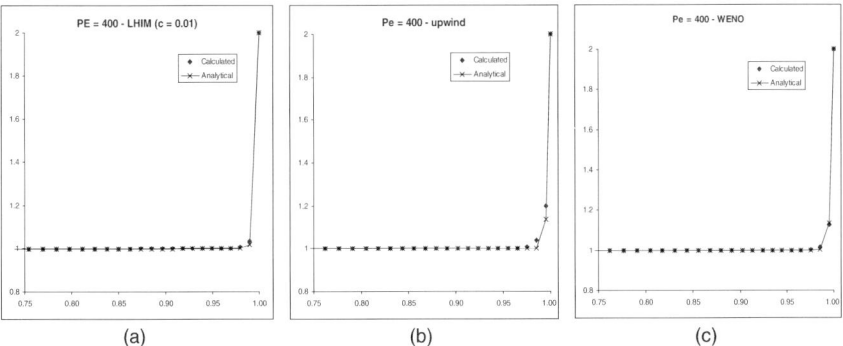

Fig. 6 Shock capturing at Pe $= 400$ (refined mesh): (**a**) LHI method, (**b**) finite volume upwind, (**c**) finite volume WENO 5th order

Table 2 L_2 error calculated at data centres

	LHI	Upwind	WENO 5
Pe $= 50$	0.000151	0.005506	0.005845
Pe $= 100$	0.000020	0.002128	0.002716
Pe $= 200$	0.000229	0.002853	0.002969
Pe $= 400$	0.000530	0.002235	0.000478

and stability, and to higher Peclet numbers than either of the full-domain methods tested in [17].

3.2 Cylindrical Diffusion Problem

The problem of diffusion in a cylindrical annulus is studied:

$$\phi(r_1) = 1.0$$
$$\phi(r_2) = 2.0$$
$$\nabla^2 \Phi = 0 \quad \text{i.e.} \quad \frac{K}{r}\frac{\partial}{\partial r}\left(r\frac{\partial \phi}{\partial r}\right) + \frac{1}{r^2}\frac{\partial^2 \phi}{\partial \theta^2} + \frac{\partial^2 \phi}{\partial z^2} = 0 \quad (22)$$

Given axisymmetry and uniformity in the z-direction, this reduces to:

$$\frac{\partial^2 \phi}{\partial r^2} + \frac{1}{r}\frac{\partial \phi}{\partial r} = 0 \quad \text{for} \quad r_1 \leq r \leq r_2 \quad (23)$$

This can be viewed as a steady, one-dimensional convection-diffusion problem, with an effective velocity of $-1/r$ arising from the coordinate transformation. This leads to a singularity when $r_1 = 0$ (in which case the inner cylinder is a point-sink). As such, we examine the behaviour of the numerical solution in the limit $r_1 \to 0$, to provide a 'stress test' for the method using a variable velocity field.

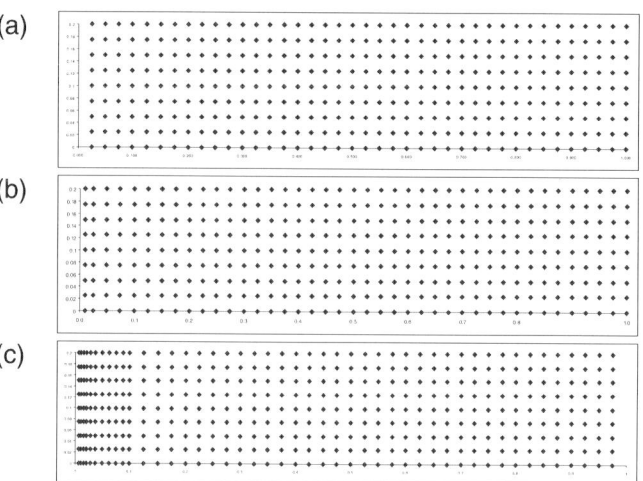

Fig. 7 Mesh representation: (**a**) $r_1 = 0.025$, (**b**) $r_1 = 0.01$, (**c**) $r_1 = 0.003$

For this problem there exists a simple analytical solution. Taking $r_2 = 1.0$:

$$\phi(r) = A + B\ln(r) \quad A = 2.0 \quad B = -\frac{1}{\ln(r_1)} \tag{24}$$

The one-dimensional solution is modelled on a three-dimensional domain of size:

$$r_1 \leq x \leq 1.0, \ 0 \leq y \leq 0.2, \ 0 \leq z \leq 0.2$$

Three different values of r_1 (namely $0.025, 0.01, 0.003$) are studied on three different 'meshes' (Fig. 7). Each mesh has 9 points in the y and z directions with a uniform point separation of 0.025. The standard separation distance in the x-direction is also 0.025, however the mesh used for $r_1 = 0.003$ has extra points in the region $x < 0.1$ to allow the solution to be captured. A compact support radius of 0.026 is used throughout.

Zero flux conditions applied at the y and z boundaries to retain a 1D solution. Each model is run to convergence from an initial state of $\phi(x) = 2.0$. No finite volume comparison is given here since this model uses a compressible flow field, which the finite volume program is not at present capable of handling.

For $r_1 = 0.025$ and 0.01 the method is able to reproduce the analytical solution with a good degree of accuracy (Fig. 8), indicating its ability to handle problems with a strongly varying velocity field. At $r_1 = 0.003$ the method is no longer able to accurately resolve the solution approaching the inner cylinder, even with the refined mesh. Accurately resolving the solution in this case would likely require a variable compact support radius and/or c-parameter.

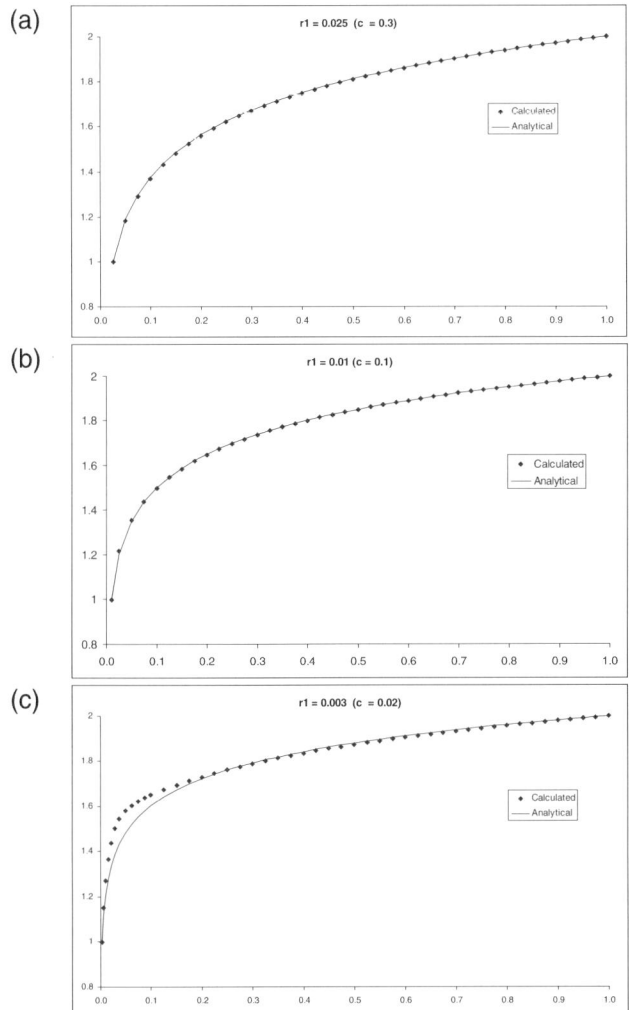

Fig. 8 Calculated solution: (**a**) $r_1 = 0.025$, (**b**) $r_1 = 0.01$, (**c**) $r_1 = 0.003$

3.3 Variable Velocity with a Reaction Term

This problem tests the ability of the code to handle a reaction term. Additionally, the nature of the solution allows an initial examination into the effect of a spatially varying c-parameter. This test case was used by Portapila et al. [18], and proved a difficult solution to accurately represent using full-domain RBF techniques.

The steady convection-diffusion-reaction equation is solved:

$$\frac{\partial^2 \phi}{\partial x^2} - (A + kx)\frac{\partial \phi}{\partial x} - k\phi = 0 \tag{25}$$

on the domain:

$$0 \leq x \leq 1.0$$
$$0 \leq y \leq 0.2$$
$$0 \leq z \leq 0.2$$

with:

$$\phi(0) = 300$$
$$\phi(1) = 100$$
$$A = \ln\left(\frac{\phi_1}{\phi_0}\right) - \frac{k}{2} \tag{26}$$

As previously, zero-flux conditions are applied to the y and z boundaries to retain a 1D solution. The problem has a simple, steady analytical solution given by:

$$\phi(x) = \phi_0 e^{\left(\frac{kx^2}{2} + Ax\right)} \tag{27}$$

For moderately large k ($k > 2 \ln (3)$ in this instance), the velocity field changes sign within the domain. In this case, two diffusive shocks are formed at either end of the domain, with the central region left relatively empty. This effect is magnified as k is increased, and presents a numerical difficulty as both the large values of ϕ around the shocks and the very small values around the centre of the domain must be modelled accurately.

A uniform mesh of $(41 \times 9 \times 9)$ points is used for each of the three k-values tested. The program starts from an initial state of $\phi(x) = 0$, and runs until convergence is reached.

At $k = 40$ the solution is reproduced reasonably well throughout the domain (Fig. 9). At $k = 80$ the shocks are well captured, but the internal region is captured less well (Fig. 10), approaching a relative error of 0.2. The plots in Figs. 9 and 10 were produced with an 'optimal' c parameter (optimal in that it minimises the

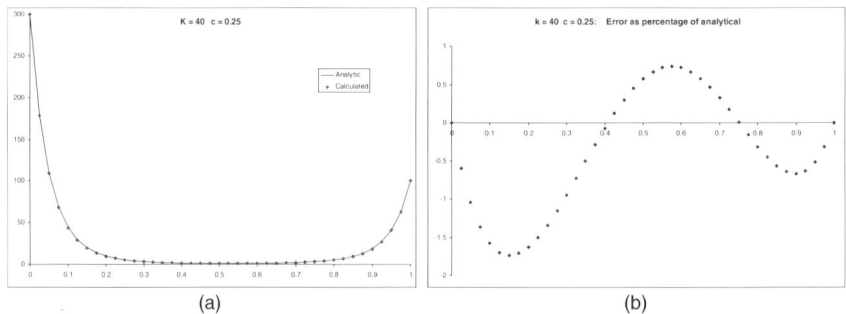

(a)   (b)

Fig. 9 (a) Calculated solution at $k = 40$, (b) relative error plot at $k = 40$

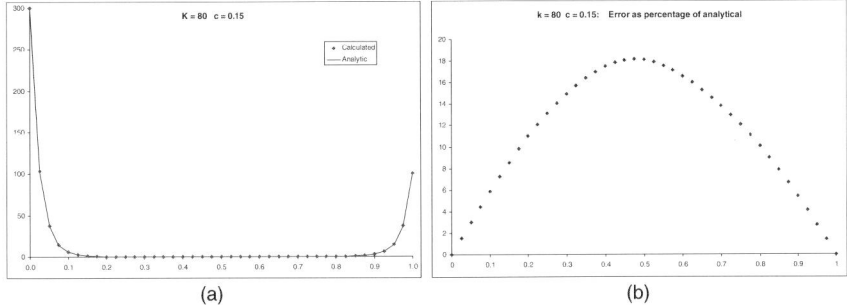

Fig. 10 (**a**) Calculated solution at $k = 80$, (**b**) relative error plot at $k = 80$

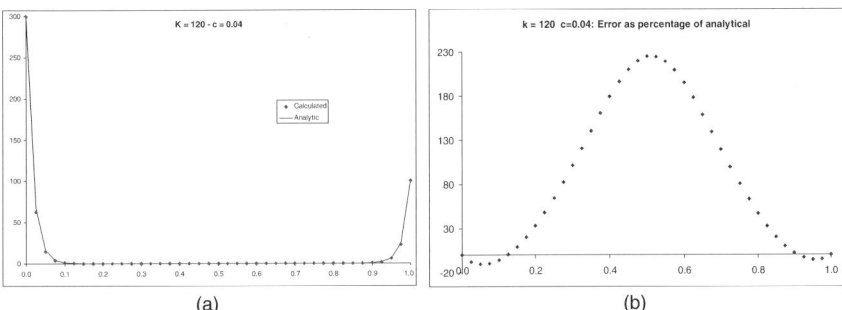

Fig. 11 (**a**) Calculated solution at $k = 120$ and $c = 0.04$, (**b**) relative error plot at $k = 120$ and $c = 0.04$

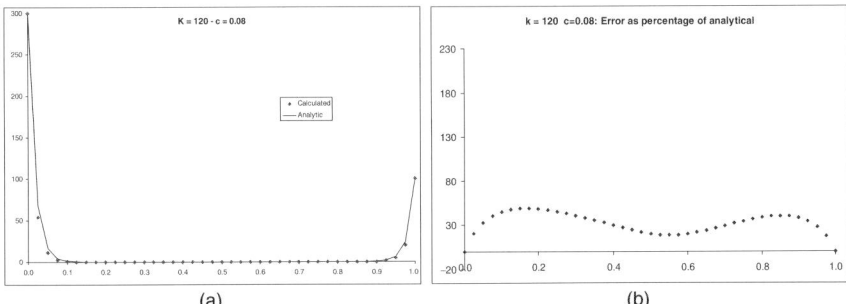

Fig. 12 (**a**) Calculated solution at $k = 120$ and $c = 0.08$, (**b**) relative error plot at $k = 120$ and $c = 0.08$

global L_2 error norm). However, altering the value of c changes the balance between accurately capturing the shocks and the central region. This effect is clearly visible when setting $k = 120$.

Using a smaller value of c allows the shocks at either end of the domain to be reasonably well captured, whereas using a larger value of c allows a much better description of the central region, at the expense of the shocks (Fig. 11 and Fig. 12). This shows that the optimal value of c for a given local system is largely dependent

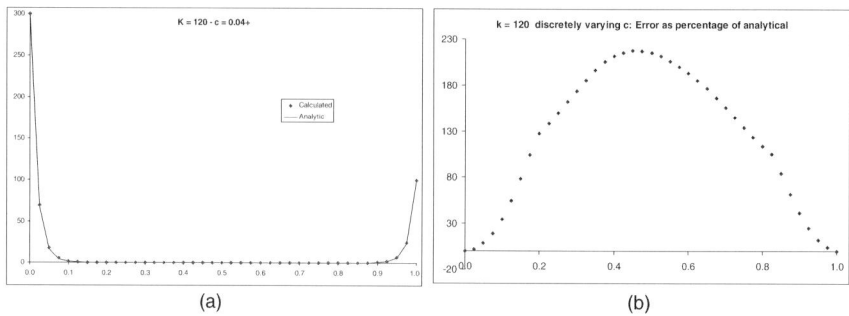

Fig. 13 (**a**) Calculated solution at $k = 120$ and discretely varying c, (**b**) relative error plot at $k = 120$ and discretely varying c

on local flow characteristics – as well as the local point distribution. In this instance the solution can be improved by using a larger value of c towards the centre of the domain, and a smaller value close to the ends. This is demonstrated by considering two different distributions of shape parameter; a discrete and a continuous variation of c:

$$\text{Discretely varying c: } c = 0.12 : 0 < |0.5 - x| < 0.1$$
$$c = 0.08 : 0.1 < |0.5 - x| < 0.3$$
$$c = 0.04 : 0.3 < |0.5 - x| < 0.5$$

$$\text{Continuously varying c: } c = 0.04 + 0.16x \times \leq 0.5$$
$$c = 0.2 - 0.16x \, x > 0.5$$

Using three different values of c in discrete zones (Fig. 13) offers no improvement, and it is clear from the relative error plot that the discrete changes in local basis function shape are causing problems. There are obvious 'kinks' in the solution here, whereas with a constant c value the error is always smoothly varying. Using a continuous function to increase the c parameter towards the centre of the domain (Fig. 14) offers improvements over both of the constant-c results, throughout the domain, and indicates the importance of creating a smooth transition for the solution between adjacent local systems.

Determining an appropriate and efficient technique for shape parameter optimisation of this method is the subject of ongoing research. However, it is clear from the results produced in this work that shape parameter optimisation must take into account the solution data and not just the distribution of collocation points. Analysis of shape parameter optimisation in full-domain Kansa's RBF methods is performed by Fornberg and Zuev [19]. An analogous technique may eventually be applicable to localised RBF methods such as the LHI method described here.

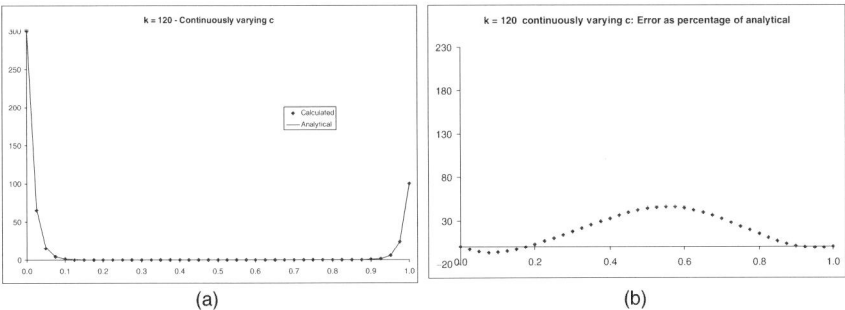

Fig. 14 (**a**) Calculated solution at $k = 120$ and continuously varying c, (**b**) relative error plot at $k = 120$ and continuously varying c

3.4 Transient Inflow Shock

A transient problem will allow validation of the time advancement capability of the code.

The unsteady convection-diffusion equation is solved:

$$\frac{\partial \phi}{\partial t} = D \frac{\partial^2 \phi}{\partial x_i^2} - u_i \frac{\partial \phi}{\partial x_i} \tag{28}$$

Over the domain:

$$0 \le x \le 1.0$$
$$0 \le y \le 0.2$$
$$0 \le z \le 0.2$$

With initial conditions:

$$\phi(0) = 2.0$$
$$\phi(x > 0) = 1.0$$

Zero flux is applied in the y and z directions. In all cases the velocity field is $(1, 0, 0)$, and the simulation is run until $t = 0.5$. The solution is formed over the same uniform point distribution depicted in Fig. 2(a). For this test case, the analytical solution for $D > 0$ is provided by [20]. For $D = 0$ the analytical solution is a simple travelling wave.

The case of pure convection $(D = 0)$ is the examined first. The LHI method is run with two different compact support radii; 0.0251 and 0.0501 (giving local system sizes of 7 and 33 respectively). A range of (constant) c-values are shown. Here, a stable solution is not attainable with either support size (Fig. 15). However, with a larger support size the diffusive component of the error in capturing the shock is greatly reduced. The value of the shape parameter appears to have an effect on the

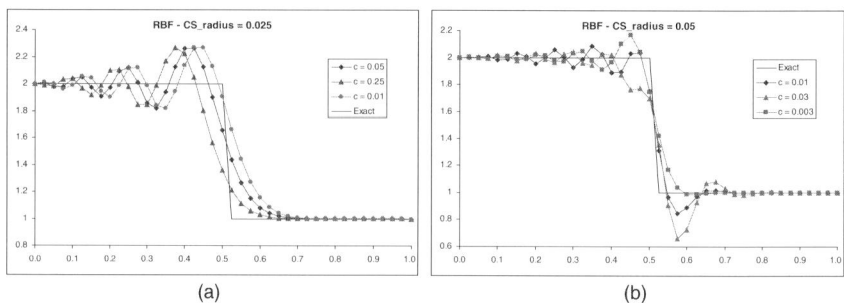

Fig. 15 Travelling shock, pure convection: (**a**) CS radius = 0.025, (**b**) CS radius = 0.05

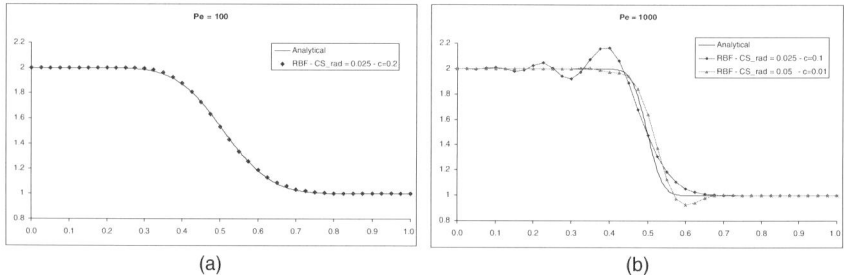

Fig. 16 Travelling shock, finite Peclet number: (**a**) Pe = 100, (**b**) Pe = 1000

phase of the instability in this case. Note, however, that there is no upwinding or artificial viscosity present in these solutions.

At Pe = 100 the LHI method is able to represent the high-resolution FV results with reasonable accuracy, using a variety of compact support radii and c-values (and one such example is shown in Fig. 16(a). At Pe = 1,000 instabilities can be seen developing, and the use of a larger compact support radius again reduces the diffusive error in reproducing the shock front (Fig. 16).

A paper by Cecil et al. [21] investigated the use of the ENO principle to stabilise such shocks at high Peclet numbers, with a localised RBF-interpolation method. A similar method could be applied here, although in practice this would involve defining several possible stencils for each local system, at an increased cost in both computational and memory resources. A more attractive alternative is to investigate bringing the PDE operator itself into the solution construction (Eq. (7)), as with the boundary operator. This would allow information about the local flow field to be directly encoded into the solution construction, providing a kind of 'analytical upwinding' effect, and improving the stability of the resulting solutions.

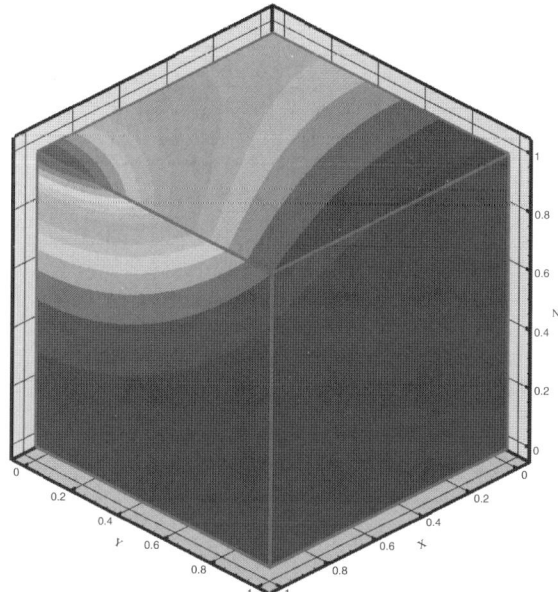

Fig. 17 Analytical solution of 3D linear test-case

3.5 3d Linear Test Case

To demonstrate the capability of the method to work in 3D, the following test case
is formulated:

$$\frac{\partial^2 \phi}{\partial x^2} + \frac{\partial^2 \phi}{\partial y^2} + \frac{\partial^2 \phi}{\partial z^2} = \left(2x\frac{\partial \phi}{\partial x} - 4y\frac{\partial \phi}{\partial y} + 4\frac{\partial \phi}{\partial z}\right) - 2\phi \tag{29}$$

Equation (22) is solved over the unit cube, applying $\phi(\mathbf{x}) = e^{(x^2 - 2y^2 + 4z)}$ at all
boundary faces. The velocity field (varying linearly in the x and y directions, and
constantly in z) pushes the scalar field towards the corner at $(1, 0, 1)$ (see Fig. 17).
    The analytical solution for this case is given by:

$$\phi(\mathbf{x}) = e^{(x^2 - 2y^2 + 4z)} \tag{30}$$

The calculation is performed on a relatively coarse distribution of $(21 \times 21 \times 21)$
uniformly spaced collocation points. A value of 0.1 is used for the c-parameter, and
the simulation is run to convergence from an initial state of $\phi(\mathbf{x}) = 1$ at the internal
points.
    Two slices at constant z are compared against the analytical solution (Figs. 18
and 19). In both cases the calculated solution agrees well with the analytical solution, demonstrating the capability of the code to operate in 3D. This is further
highlighted by Fig. 20, where the solution at the collocation point is plotted against
the analytical solution, along $x = 0.8$, $z = 0.8$.

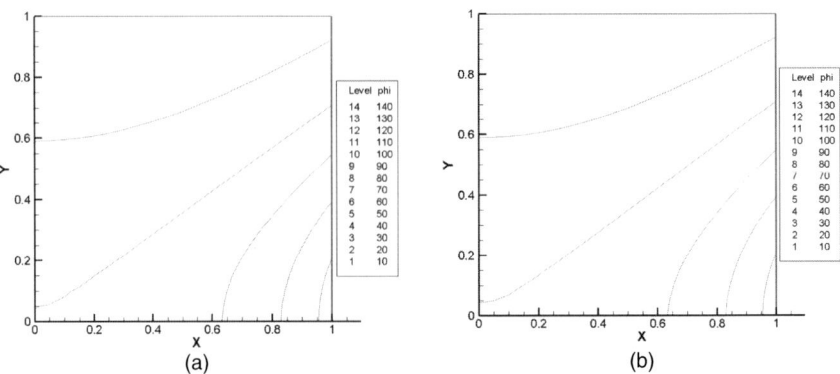

Fig. 18 Solution at z = 0.75: (**a**) analytical solution, (**b**) calculated solution

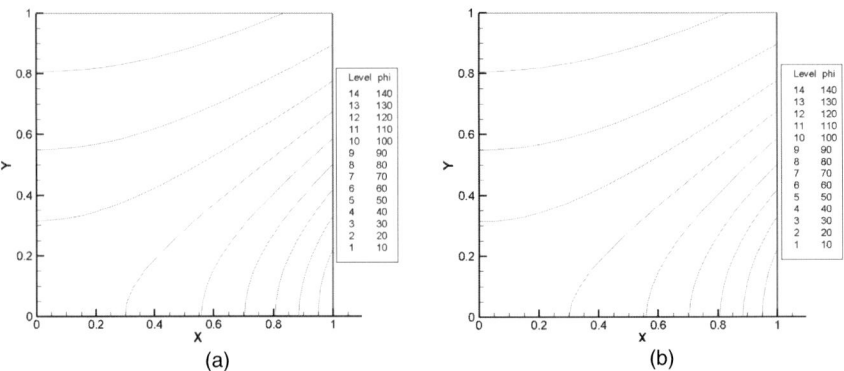

Fig. 19 Solution at z = 0.9: (**a**) analytical solution, (**b**) calculated solution

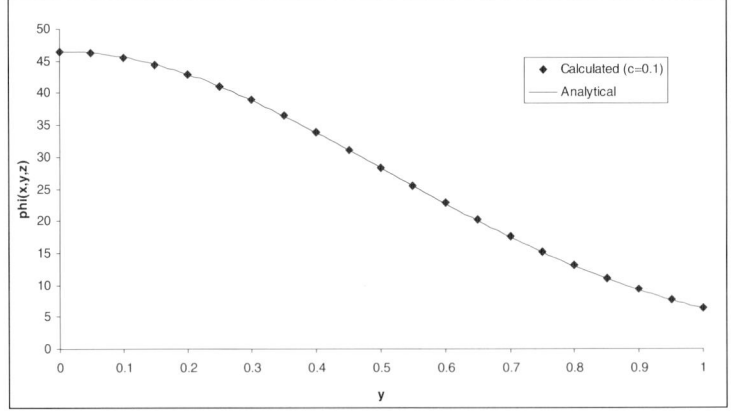

Fig. 20 Profile along x = 0.8, z = 0.8

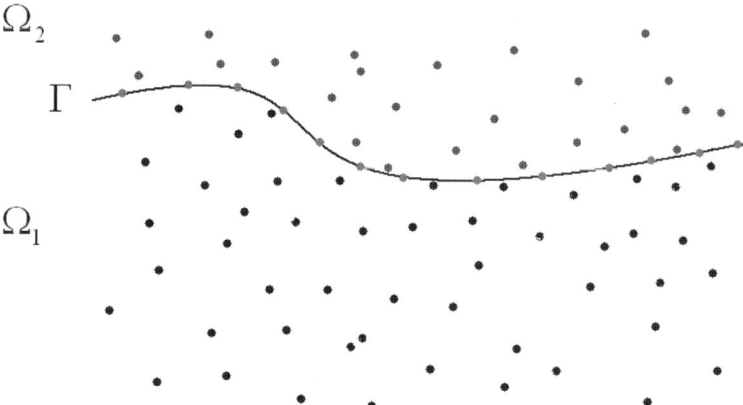

Fig. 21 Initial distribution of collocation points for phreatic surface problems

4 Moving Boundary Surfaces

One of the main practical advantages of the method is its ability to describe moving surfaces without the need for dynamic 'remeshing' of the internal point distribution. Initially, the moving surface boundary condition is modelled in the same way as all other boundaries – via a distribution of surface points and an associated boundary operator. Subsequently, the surface position is updated by moving each surface point based on the flow characteristics at that point (which can be computed from neighbouring local systems). The double-collocation principle [22] of the full-domain Hermitian RBF method can also be applied, ensuring non-singular local systems for coincident surface and internal points. In this way a (non-Dirichlet) moving boundary surface can travel through a fixed distribution of points, without the need to avoid existing internal points.

The initial configuration sees a 2D surface placed inside a scattered distribution of collocation points. A number of boundary points are used to describe the location of the surface.

In the setup stage, the points are divided into two sets, Ω_1 and Ω_2; those inside and outside the domain defined (in part or in whole) by the moving surface (see Fig. 21). Those points which lie outside the domain are deactivated inside the code – they have no local system, and do not appear in any other local systems. Determination of whether a point is inside or outside the physical domain is performed by first calculating the displacement from a given internal point to its closest moving surface point. This displacement is checked against the surface normal at that surface point. If and only if the vectors lie in the same direction (i.e. their dot-product is positive) the point is internal.

As the surface moves through the domain, certain points may either move into or out of the active domain. The reconfiguration of the local systems is handled at a local level, examining the connectivity of neighbouring points to transfer the moving surface points from system to system. In this way, the cost of the local system reconfiguration remains o(N).

The movement of the phreatic surface itself is determined from the local flow properties. The kinematic and dynamic boundary conditions for the moving surface height, h, are:

$$(\overline{\overline{K}}\nabla\phi)\nabla F = n_e\frac{\partial h}{\partial t} \tag{31}$$

$$F = \phi(h) - z$$

Taking a first-order approximation to the time derivative, each surface point is moved via:

$$h^{n+1} = h^n + \frac{\Delta t}{n_e}\left(K_i\frac{\partial\phi}{\partial x_i}\right) \tag{32}$$

For: h^n Surface height at time-level n

n_e Effective porosity

To demonstrate the procedure, a simple problem is formulated, representing pumping within a saturated porous medium:

Darcy's equation in an anisotropic porous medium:

$$K_i\frac{\partial^2\phi}{\partial x_i^2} = 0 \quad K = (1.0 \quad 1.0 \quad 0.1)\ mh^{-1} \tag{33}$$

is solved on a domain:

$$-100\,\text{m} \le x \le 100\,\text{m}$$
$$-100\,\text{m} \le y \le 100\,\text{m}$$
$$0 \le z \le 50\,\text{m}$$

with an impermeable boundary in the y-direction and a constant head boundary in the x-direction:

$$\phi(x = \pm100m) = 0$$

$$n_i\frac{\partial\phi}{\partial x_i}(y = \pm100m) = 0$$

A pump is placed at the centre of the domain, injecting water at 50 m³h⁻¹. In this model the pump is modelled as a simple cuboid, in the region:

$$-2.5\,\text{m} \le x \le 2.5\,\text{m}$$
$$-2.5\,\text{m} \le y \le 2.5\,\text{m}$$
$$20 \le z \le 25\,\text{m}$$

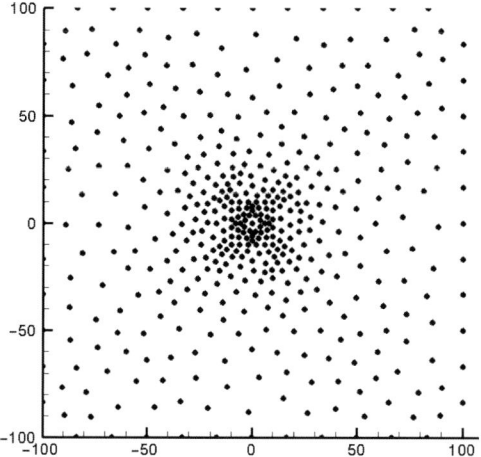

Fig. 22 Representation of horizontal point distribution

The influx of water is modelled via a uniform flux applied over the pumping surface. Boundary conditions on the pumping surface are modelled such that the pumping properties are physical:

$$-\frac{1}{\eta}\left(n_i K_{ij}\right)\frac{\partial\phi}{\partial x_j} = P \tag{34}$$

where

$$P = \frac{Total\,Flux}{Pump\,surface\,area}$$

$$\eta = 0.2\ (\text{porosity})$$

A point distribution of 9,224 points was obtained by creating an unstructured 2D mesh, and extruding to create 21 uniformly spaced layers in the z-direction. Local systems are formed by linking each point in the domain with its immediate neighbours, as connected by the prism-shaped elements. A representation of the horizontal point distribution is given in Fig. 22.

The model is run under the assumption of quasi-static phreatic surface movement. The implicit timestepping scheme is used, requiring relatively few timesteps (in this case, 60 timesteps were required for pressure-field convergence down to 10^{-8}).

The computed phreatic surface height is represented by Fig. 23. The moving surface procedure is yet to undergo formal validation. However, for the present a comparison is provided with results from the CV-RBF method developed by Orsini et al. [23], which has undergone extensive testing with moving surface problems (Table 3). The agreement of phreatic surface height appears reasonable in this case, with the maximum predicted surface height consistent to within 6% between the two methods.

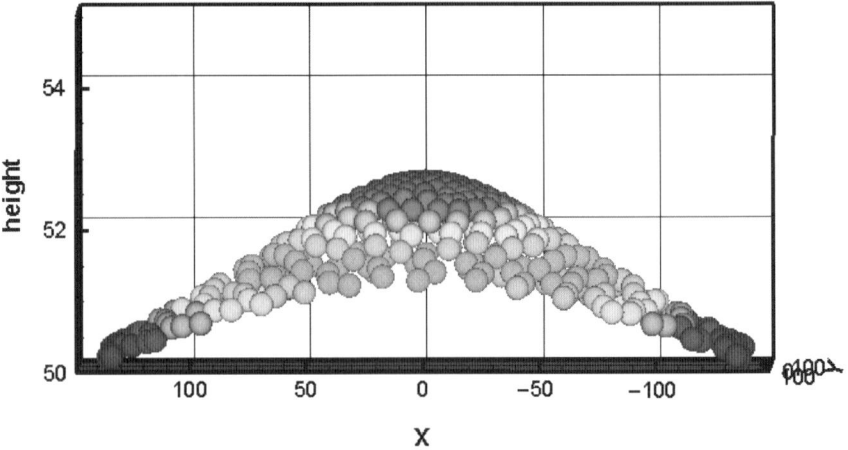

Fig. 23 Phreatic surface height representation

Table 3 Comparison with CV-RBF method

	LHIM	CV-RBF 17,000 cells	CV-RBF 50,000 cells
Maximum phreatic surface height	52.58 m	52.52 m	52.44 m

5 Conclusions

This work has described a fully meshless local decomposition of the Hermitian RBF solution technique, using functional values and boundary operators. A numerical code is in place to apply the method to 3D scattered data sets. The method has order-N complexity, and can scale to the largest problem sizes without numerical conditioning issues.

The technique is validated against a variety of simple steady and transient 1D problems with known analytical solutions. It has been demonstrated that, for this type of localised decomposition problem, spatial variation of the c-parameter will be required for optimal results. This variation must be performed in a data-dependent way.

Current work is largely focusing on finding suitable empirical relationships between shape-parameter and solution data, as well as implementing stabilisation methods for the capturing of discontinuities. In addition, further development is underway on the implicit timestepping technique, providing the option to incorporate the PDE operator into the Hermitian interpolation formulation. This technique has proved successful with other local-RBF methods, for example the CV-RBF method described by Orsini et al. [23], providing a stabilising 'analytical upwinding' effect through incorporating flow-field information into the basis function construction.

References

1. Schaback, R., *Convergence of unsymmetric kernal-based meshless collocation methods.* SIAM Journal of Numerical Analysis, 2007. **45**(1): p. 333–351.
2. Kansa, E.J., *Multiquadrics - A scattered data approximation scheme with applications to computational fluid-dynamics-I: Surface approximations and partial derivatives estimates.* Computers & Mathematics with Applications, 1990. **19**: p. 127–145.
3. Kansa, E.J., *Multiquadrics - A scattered data approximation scheme with applications to computational fluid dynamics-II: Solution to parabolic, hyperbolic and elliptic partial differential equations.* Computers & Mathematics with Applications, 1990. **19**: p. 147–161.
4. Ling, L., R. Opfer, and R. Schaback, *Results on meshless collocation techniques.* Engineering Analysis with Boundary Elements, 2006. **30**(4): p. 247–253.
5. Brown, D., *On approximate cardinal preconditioning methods for solving PDEs with radial basis functions.* Engineering Analysis with Boundary Elements, 2005. **29**(4): p. 343–353.
6. Fasshauer, G.E., *Solving partial differential equations by collocation with radial basis functions.* Surface fitting and Multiresolution Methods, 1997.
7. Wu, Z., *Hermite-Birkhoff interpolation of scattered data by radial basis functions.* Approximation Theory, 1992. **8**(2): p. 1–11.
8. Wu, Z., *Solving PDEs with radial basis functions and the error estimation;*, in *Advances in Computational Mathematics*, Z. Chen, et al., Editors. 1998.
9. Schaback, R., *Multivariate interpolation and approximation by translates of a basis function*, in *Approximation Theory VIII*, C. Chui and L. Schumaker, Editors. 1995.
10. Munoz-Gomez, J., P. Gonzalez-Casanova, and G. Rodriguez-Gomez. *Domain decomposition by radial basis functions for time dependent partial differential equations*, in *Proceedings of the 2nd IASTED international conference on Advances in computer science and technology.* 2006. Puerto Vallarta, Mexico.
11. Ling, L. and E.J. Kansa, *Preconditioning for Radial Basis Functions with Domain Decomposition Methods.* Mathematical and Computer modelling, 2004. **38**(5): p. 320–327.
12. Hernandez Rosales, A. and H. Power, *Non-overlapping domain decomposition algorithm for the Hermite radial basis function meshless collocation approach: applications to convection diffusion problems.* Journal of Algorithms and Technology, 2007 (preprint).
13. Ingber, M., C. Chen, and J. Tanski, *A mesh free approach using radial basis functions and parallel domain decomposition for solving three-dimensional diffusion equations.* International Journal for Numerical Methods in Engineering, 2004. **60**: p. 2183–2201.
14. Vertnik, R., M. Zaloznik, and B. Sarler, *Solution of transient direct-chill aluminium billet casting problem with simultaneous material and interphase moving boundaries by a meshless method.* Engineering Analysis with Boundary Elements, 2006. **30**: p. 847–855.
15. Wright, G. and B. Fornberg, *Scattered node compact finite difference-type formulas generated from radial basis functions.* Journal of Computational Physics, 2006. **212**(1): p. 99–123.
16. Saad, Y. *SPARSKIT: A basic tool-kit for sparse matrix computations.* 1988–2000 [cited; Available from: http://www-users.cs.umn.edu/~saad/software/SPARSKIT/sparskit.html].
17. Power, H. and V. Barraco, *A comparison analysis between unsymmetric and symmetric radial basis function collocation methods for the numerical solution of partial differential equations.* Computers and Mathematics with Application, 2002. **43**: p. 551–583.
18. Portapila, M. and H. Power, *Iterative schemes for the solution of a system of equations arising from the DRM in multi domain approach, and a comparative analysis of the performance of two different radial basis functions used in the interpolation.* Engineering Analysis with Boundary Elements, 2005. **29**: p. 107–125.

19. Fornberg, B. and J. Zuev, *The Runge phenomenon and spatially variable shape parameters in RBF interpolation.* Computers and Mathematics with Application, 2007. **54**(3): p. 379–398.

20. Van Genuchten, M. and W. Alves, *Analytical Solutions of the One-Dimensional Convective-Dispersive Solute Transport Equation.* United States Department of Agriculture, Agricultural research service, 1982. **Technical bulletin number 1661**.

21. Cecil, T., J. Qian, and S. Osher, *Numerical methods for high dimensional Hamilton-Jacobi equations using radial basis functions.* Journal of Computational Physics, 2004. **196**: p. 327–347.

22. LaRocca, A. and H. Power, *Comparison between the single and the double collocation methods*, 2007 (preprint).

23. Orsini, P., H. Power, and H. Morvan, The use of local radial basis function meshless scheme to improve the performance of unstructured volume element method, in ICCES Special Symposium on MESHLESS METHODS 2007: Patras, Greece.

On the Determination of a Robin Boundary Coefficient in an Elastic Cavity Using the MFS

Carlos J.S. Alves and Nuno F.M. Martins(✉)

Abstract In this work, we address a problem of recovering a boundary condition on an elastic cavity from a single boundary measurement on an external part of the boundary. The boundary condition is given by a Robin condition and we aim to identify its Robin coefficient (matrix). We discuss the uniqueness question for this inverse problem and present several numerical simulations, based on two different reconstruction approaches: An approach by solving the Cauchy problem and an iterative Newton type approach (that requires the computation of several direct problems). To solve the mentioned (direct and inverse) problems, we propose the Method of Fundamental Solutions (MFS) whose properties will be discussed.

Keywords: Inverse problems · Robin boundary conditions · Lamé system · method of fundamental solutions

1 Introduction

The problem of boundary condition reconstruction in an inaccessible part of the boundary is a well known nonlinear inverse problem arising in nondestructive evaluation. For the Laplace equation, this problem has been studied as a model problem for thermal imaging (e.g. [5]) and detection of corrosion phenomena ([10,11]). More recently the problem have been addressed in [7], where a method based on integral equations was developed and in [6], where the Kohn and Vogelius cost function was

N.F.M. Martins
CEMAT-IST and Departamento de Matemática, Faculdade de Ciências e Tecnologia, Univ. Nova de Lisboa, Quinta da Torre, 2829-516 Caparica, Portugal, e-mail: nfm@fct.unl.pt

C.J.S. Alves
CEMAT-IST and Departamento de Matemática, Instituto Superior Técnico, TULisbon, Avenida Rovisco Pais, 1096 Lisboa Codex, Portugal, e-mail: calves@math.ist.utl.pt

A.J.M. Ferreira et al. (eds.) *Progress on Meshless Methods, Computational Methods in Applied Sciences.*
© Springer Science + Business Media B.V. 2009

used to retrieve the Robin coefficient in an optimization scheme. When the effect of corrosion is modeled by means of a perturbation of the inaccessible part of the boundary and the inverse problem consists in retrieving the shape arising from this perturbation were refer the recent works [3,7] (for Laplace equation) and [2] (for the Lamé system). This last work (and also [3]) used already, as reconstruction method, an MFS adaptation of the Kirsch-Kress method (introduced in the context of an exterior problem in acoustic scattering).

The MFS was introduced by Kupradze forty years ago ([12]) and it is a meshfree method that, besides sharing all the advantages of the conventional boundary element method (BEM) over domain discretization methods, avoids the computation of the singular integrals on the boundary (arising in the BEM) by introducing an artificial boundary. It provides very accurate solutions with small implementation effort and has been used as a solver for both direct (e.g. [4,9]; see also [1] for an extension to nonhomogeneous problems) and more recently inverse problems ([2,14]). In this paper, we propose two numerical approaches for the aforementioned inverse problem that relies on MFS as direct and Cauchy problem solver. In Section 2 we provide the mathematical setting for the problem and discuss the uniqueness question of the inverse problem. In Sect. 3, we derive the Gâteaux derivative of the map that assigns the Robin coefficient to the generated displacement data on the accessible part of the boundary and use it to define a Newton type method with a penalization term, in order to obtain admissible iterates. Section 4 is dedicated to the MFS as solver for both direct and Cauchy problems and density results are presented. In the last section, we present several numerical simulations for the 2D case and discussion.

2 Direct and Inverse Problem

Consider an elastic, isotropic and homogeneous body $\Omega \subset \mathbb{R}^d$, $d = 2, 3$ and that the stress tensor σ verifies (Hooke's law)

$$\sigma(\mathbf{u}) = (\lambda(\nabla \cdot \mathbf{u})\mathbf{I} + 2\mu\varepsilon)$$

where $\lambda, \mu > 0$ are the *Lamé constants*, $\mathbf{I}$ is the identity tensor, $\nabla \cdot \mathbf{u}$ is the divergence of the displacement vector $\mathbf{u}$ and ε is the stress-strain tensor of $\mathbf{u}$ given by

$$\varepsilon(\mathbf{u})_{ij} = \frac{1}{2}\left(\frac{\partial u_i}{\partial x_j} + \frac{\partial u_j}{\partial x_i}\right) = \frac{1}{2}\left(\nabla \mathbf{u} + \nabla \mathbf{u}^\top\right)_{ij}.$$

When the body is in static equilibrium, and no force term exists, the displacement verifies the homogeneous Lamé system

$$\nabla \cdot \sigma(\mathbf{u}) = \mathbf{0}.$$

We write $\Delta^* \mathbf{u} := \nabla \cdot \sigma(\mathbf{u})$, noticing that

$$\Delta^* \mathbf{u} = \mu \nabla \cdot \nabla \mathbf{u} + (\lambda + \mu)\nabla\nabla \cdot \mathbf{u}.$$

Assuming that Ω is open and simply connected with regular C^1 boundary $\Gamma = \partial\Omega$ let ω be open and simply connected[1] with regular boundary $\gamma = \partial\omega$ such that $\overline{\omega} \subset \Omega$. The domain of elastic propagation is defined by $\Omega_c = \Omega \setminus \overline{\omega}$ and note that $\partial\Omega_c = \Gamma \cup \gamma$.

Direct Problem:

Determine $\mathbf{u}|_\Gamma$, given $\mathbf{g} \in H^{-1/2}(\Gamma)^d$ and $\mathbf{Z} \in \mathcal{M}_d^+(\gamma)$ with

$$\mathcal{M}_d^+(\gamma) := \{\mathbf{Z} \in \mathcal{M}_d(L^\infty(\gamma)) \setminus \{\mathbf{0}\} : \mathbf{Z} \text{ is semi-positive definite}\}$$

such that $\mathbf{u}$ verifies

$$(\mathcal{P}) \begin{cases} \Delta^*\mathbf{u} = 0 & \text{in } \Omega_c \\ \partial_n^*\mathbf{u} = \mathbf{g} & \text{on } \Gamma \\ \partial_n^*\mathbf{u} + \mathbf{Z}\mathbf{u} = 0 & \text{on } \gamma \end{cases} \tag{1}$$

where we used the notation $\partial_n^*\mathbf{u} := \sigma(\mathbf{u})\mathbf{n}$ for the traction vector ($\mathbf{n}$ is the outward normal vector). This problem is well posed with solution in $H^1(\Omega_c)^d$. For a pure traction problem, ie. $\mathbf{Z} = \mathbf{0}$, further assumptions on $\mathbf{g}$ must be considered ([8]).

Inverse Problem:

Assuming that the geometry of the inaccessible part γ is known, we want to recover $\mathbf{Z}$ from the displacement and traction data on an accessible part $\Sigma \subset \Gamma$.

2.1 Uniqueness

We can not retrieve a general Robin matrix coefficient from a single boundary measurement. In fact, consider for instance, $\Omega = B(\mathbf{0}, r) = \{\mathbf{x} \in \mathbb{R}^2 : |\mathbf{x}| < r\}$ with $r > 1$, $\omega = B(\mathbf{0}, 1)$ and Ω_c is the annular domain $\Omega \setminus \overline{\omega}$. Now consider the function (see Fig. 1)

$$\mathbf{u}(\mathbf{x}) = (-x_1(\rho + |\mathbf{x}|^{-2}), x_2(1 - |\mathbf{x}|^{-2})), \quad \rho = \frac{\lambda + 4\mu}{\lambda}. \tag{2}$$

A forward computation shows that $\Delta^*\mathbf{u} = 0$ in $\mathbb{R}^2 \setminus \{\mathbf{0}\}$ and with $\mathbf{Z}_\psi = \text{diag}(2\mu, \psi)$ we have

$$\begin{cases} \Delta^*\mathbf{u} = 0 & \text{in } \Omega_c \\ \partial_n^*\mathbf{u} = \mathbf{g} & \text{in } \Gamma = \partial\Omega \\ \partial_n^*\mathbf{u} + \mathbf{Z}_\psi\mathbf{u} = 0 & \text{in } \gamma = \partial\omega \end{cases}$$

where $\mathbf{g}$ is the non constant function $\partial_n^*\mathbf{u}|_\Gamma$ and $\psi > 0$. This means that at least two matrices of the form $\mathbf{Z}_\psi$ generates the same Cauchy data on Γ. It is then clear that, we must have some a priori knowledge about the impedance $\mathbf{Z}$.

---

[1] The case where ω is an union of simply connected domains with regular boundaries can be treated in the same way.

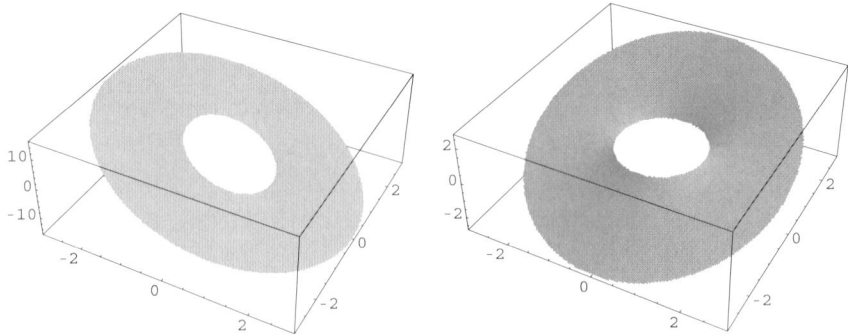

Fig. 1 Example of a function satisfying the Lamé system with vanishing second coordinate on part of the boundary for the displacement and traction vectors. *Left* – plot of the first coordinate, *right* – second coordinate

2.2 Uniqueness Class

We start this section with the Lemma (see [2] for the proof) that extends to the elastic case the Holmgren Lemma.

Lemma 1. *Let $\mathbf{u}$ be the solution of the Lamé system $\Delta^*\mathbf{u} = \mathbf{0}$ in an open domain Ω_c with C^1 boundary $\partial\Omega_c$ such that $\mathbf{u}|_\Sigma = \partial_\mathbf{n}^*\mathbf{u}|_\Sigma = \mathbf{0}$ on some open set Σ in $\partial\Omega_c$. Then, $\mathbf{u} = \mathbf{0}$ on the connected component that contains Σ.*

Given an open subset σ of γ and a matrix $\mathbf{M}$ consider the non linear class

$$\mathscr{A} = \{\mathbf{Z} \in \mathscr{M}_d^+(\gamma) : \mathbf{Z}|_\sigma = z\mathbf{M}, \, z \in C(\sigma) \wedge \mathbf{M} \text{ is invertible}\}.$$

This is thus the class of coefficients with known values in $\gamma \setminus \overline{\sigma}$ and in σ is the product of a (unknown) scalar and continuous function z by an invertible (known) matrix $\mathbf{M}$.

Theorem 2. *Suppose that the prescribed traction data is not null on some open part of Γ, Σ. If the Robin coefficient $\mathbf{Z}$ is known to be in the class $\mathscr{A}$ then is fully identified by the Cauchy data on Σ.*

Proof. Let $\mathbf{u}_1$ and $\mathbf{u}_2$ be two solutions of the direct problem for the Robin coefficient $\mathbf{Z}_1$ and $\mathbf{Z}_2$ respectively and with the same Cauchy data on Σ. Since Ω_c is connected, the previous Lemma gives $\mathbf{u}_1 = \mathbf{u}_2$ in Ω_c and, in particular,

$$\partial_\mathbf{n}^*\mathbf{u}_1|_\gamma = \partial_\mathbf{n}^*\mathbf{u}_2|_\gamma \wedge \mathbf{u}_1|_\gamma = \mathbf{u}_2|_\gamma.$$

Thus, on γ

$$(\mathbf{Z}_1 - \mathbf{Z}_2)\mathbf{u}_1 = \mathbf{0}.$$

If $\mathbf{Z}_1, \mathbf{Z}_2 \in \mathscr{A}$ then, on an open arc σ, we have $\mathbf{Z}_1 = z_1\mathbf{M}$ and $\mathbf{Z}_2 = z_2\mathbf{M}$, for some continuous functions z_1 and z_2. By contradiction, suppose that $z_1 - z_2 \neq 0$, on some point in σ. Continuity implies the existence of an open set σ^* where $z_1 - z_2 \neq 0$ and we conclude that on this set, $\mathbf{Z}_1 - \mathbf{Z}_2 = (z_1 - z_2)\mathbf{M}$ is invertible. This gives $\mathbf{u}_1|_{\sigma^*} = \mathbf{0}$ and the Robin boundary condition $\partial_n^* \mathbf{u}_1|_{\sigma^*} = \mathbf{0}$. By Lemma 1 we conclude that $\mathbf{u}_1 = \mathbf{0}$ in Ω_c which contradicts the hypothesis $\mathbf{g} \neq \mathbf{0}$ and we conclude that $z_1 = z_2$ on σ. The result follows since on $\gamma \setminus \overline{\sigma}$ we are assuming that $\mathbf{Z}_1$ coincides with $\mathbf{Z}_2$.   □

3 Newton Based Method

Consider the inverse problem written in terms of the non linear system of equations

$$\mathbf{F}(\mathbf{Z}) = \mathbf{u}|_\Sigma \tag{3}$$

where $\mathbf{F} : \mathscr{A} \longrightarrow L^2(\Sigma)^d$ is defined by

$$\mathbf{Z} \mapsto \mathbf{u}_{\mathbf{Z}}|_\Sigma$$

and $\mathbf{u}_{\mathbf{Z}}$ is the solution of the direct problem with impedance $\mathbf{Z}$. Note that by theorem 2 the map $\mathbf{F}$ is one to one.

Linearizing Eq. (1) we obtain, in the direction of $\mathbf{W}$,

$$\mathbf{F}'(\mathbf{Z})\mathbf{W} = \mathbf{F}(\mathbf{Z}) - \mathbf{u}|_\Sigma.$$

Theorem 3. *The map $\mathbf{F}$ is Gâteaux differentiable, i.e., for $\varepsilon \in \mathbb{R}^+$ and $\mathbf{Z} \in \mathscr{A}$ such that $\mathbf{Z} + \varepsilon \mathbf{W} \in \mathscr{A}$ we have*

$$\frac{\mathbf{F}(\mathbf{Z} + \varepsilon \mathbf{W}) - \mathbf{F}(\mathbf{Z})}{\varepsilon} \longrightarrow \mathbf{F}'(\mathbf{Z})\mathbf{W} \quad in \;\; L^2(\Sigma)^d$$

where $\mathbf{F}'(\mathbf{Z})\mathbf{W} = \mathbf{u}'_{\mathbf{W}}|_\Sigma$ with

$$\begin{cases} \Delta^* \mathbf{u}'_{\mathbf{W}} = 0 & in \;\; \Omega_c \\ \partial_n^* \mathbf{u}'_{\mathbf{W}} = 0 & on \;\; \Gamma \\ (\partial_n^* + \mathbf{Z})\mathbf{u}'_{\mathbf{W}} = -\mathbf{W}\mathbf{u}_{\mathbf{Z}} & on \;\; \gamma \end{cases}.$$

Proof. The variational formulation of the problem ($\mathscr{P}$) is:
- Find $\mathbf{u}_{\mathbf{Z}} \in H^1(\Omega_c)$ such that

$$\int_{\Omega_c} \sigma(\mathbf{u}_{\mathbf{Z}}) : \varepsilon(\mathbf{v}) = \int_\Gamma \mathbf{g} \cdot \mathbf{v} - \int_\gamma \mathbf{Z}\mathbf{u}_{\mathbf{Z}} \cdot \mathbf{v}, \quad \forall \mathbf{v} \in H^1(\Omega_c)^d,$$

where: stands for the tensor dot product. Consider the linear map $S : H^1(\Omega_c)^d \longrightarrow (H^1(\Omega_c)^d)^*$ defined by $\mathbf{u} \mapsto S_\mathbf{u}$, where $S_\mathbf{u} : H^1(\Omega_c)^d \longrightarrow \mathbb{R}$ is given by $S_\mathbf{u}(\mathbf{v}) = \int_{\Omega_c} \sigma(\mathbf{u}) : \varepsilon(\mathbf{v}) + \int_\gamma \mathbf{Z}\mathbf{u} \cdot \mathbf{v}$ and $(H^1(\Omega_c)^d)^*$ is the dual of $H^1(\Omega_c)^d$. Note that,

$$S_{\mathbf{u}_{\mathbf{Z}}}(\mathbf{v}) = \int_\Gamma \mathbf{g} \cdot \mathbf{v} \quad and \quad S_{\mathbf{u}_{\mathbf{Z}+\varepsilon\mathbf{W}}}(\mathbf{v}) = \int_\Gamma \mathbf{g} \cdot \mathbf{v} - \varepsilon \int_\gamma \mathbf{W}\mathbf{u}_{\mathbf{Z}+\varepsilon\mathbf{W}} \cdot \mathbf{v}$$

hence,

$$S_{\mathbf{u}_{\mathbf{Z}+\varepsilon\mathbf{W}}-\mathbf{u}_{\mathbf{Z}}}(\mathbf{v}) = -\varepsilon \int_\gamma \mathbf{W}\mathbf{u}_{\mathbf{Z}+\varepsilon\mathbf{W}} \cdot \mathbf{v}. \tag{4}$$

Using Riesz representation theorem and the fact that S is an isomorfism we can establish that $\mathbf{u}_{\mathbf{Z}+\varepsilon\mathbf{W}} \longrightarrow \mathbf{u}_{\mathbf{Z}}$ in $H^1(\Omega_c)$. Thus, condition (2) gives

$$\frac{\mathbf{u}_{\mathbf{Z}+\varepsilon\mathbf{W}} - \mathbf{u}_{\mathbf{Z}}}{\varepsilon} \rightharpoonup \mathbf{u}'_{\mathbf{W}} \quad \text{in} \quad H^1(\Omega_c)^d$$

where $\mathbf{u}'_{\mathbf{W}}$ is such that

$$S_{\mathbf{u}'_{\mathbf{W}}}(\mathbf{v}) = -\int_\gamma \mathbf{W}\mathbf{u}_{\mathbf{Z}} \cdot \mathbf{v}, \quad \forall \mathbf{v} \in H^1(\Omega_c)^d$$

and the result follows.   □

Theorem 4.

> If $\mathbf{F}'(\mathbf{Z})\mathbf{W} = \mathbf{0}$ then $\mathbf{W}|_\sigma$ is not invertible.

Proof. Suppose that $\mathbf{F}'(\mathbf{Z})\mathbf{W} = \mathbf{0}$. Then, $\mathbf{u}'_{\mathbf{W}}$ has null Cauchy data on Σ and we conclude by Lemma 1 that $\mathbf{u}'_{\mathbf{W}} = \mathbf{0}$ on Ω_c. This gives,

$$\mathbf{W}\mathbf{u}_{\mathbf{Z}} = \mathbf{0} \quad on \ \gamma.$$

By contradiction, assume that $\mathbf{W}|_\sigma$ is invertible. Then, $\mathbf{u}_{\mathbf{Z}}|_\sigma = \mathbf{0}$ and the fact that, on γ, $\mathbf{u}_{\mathbf{Z}}$ satisfies a homogeneous Robin condition implies $\partial_{\mathbf{n}}^* \mathbf{u}_{\mathbf{Z}}|_\sigma = \mathbf{0}$. It follows that $\mathbf{u}_{\mathbf{Z}}$ has null Cauchy data on part of the boundary and again, by Lemma 1, we must have $\mathbf{u}_{\mathbf{Z}} = \mathbf{0}$ on Ω_c which contradicts the hypothesis that $\partial_{\mathbf{n}}^* \mathbf{u}_{\mathbf{Z}}|_\Sigma$ is not null.   □

The iterative method is now defined by

$$\mathbf{Z}^{(k)} = \mathbf{Z}^{(k-1)} + \mathbf{W}^{(k)}$$

where $\mathbf{F}'(\mathbf{Z}^{(k-1)})\mathbf{W}^{(k)} = \mathbf{F}(\mathbf{Z}^{(k-1)}) - \mathbf{u}$ and $\mathbf{Z}^{(0)} \in \mathscr{A}$ is given. Representing the updating directions $\mathbf{W}^{(k)}$ in some linear space E_n with $\dim E_n = n < \infty$ in terms of an ordered basis $\mathscr{B}_n$ we arrive at the linear system of equations

$$\underbrace{\left[\frac{\partial F_i}{\partial z_j}\right]}_{\mathbf{A}^{(k-1)}}\mathbf{W}^{(k)} = \underbrace{\left[\mathbf{u}_{\mathbf{Z}^{(k-1)}}(\mathbf{x}_i) - \mathbf{u}(\mathbf{x}_i)\right]}_{\mathbf{B}^{(k-1)}}, \quad \mathbf{x}_i \in \Sigma \tag{5}$$

and $\frac{\partial F_i}{\partial z_j} = \mathbf{v}'(\mathbf{x}_i)$ with

$$\begin{cases} \Delta^* \mathbf{v}' = \mathbf{0} & \text{in } \Omega_c \\ \partial_{\mathbf{n}}^* \mathbf{v}' = \mathbf{0} & \text{on } \Gamma \\ \partial_{\mathbf{n}}^* \mathbf{v}' + \mathbf{Z}^{(k-1)}\mathbf{v}' = -e_j \mathbf{u}_{\mathbf{Z}^{(k-1)}} & \text{on } \gamma \end{cases}$$

where e_j is the jth vector of $\mathscr{B}_n$. However, several numerical situations may lead the updates to fall outside the admissible domain $\mathscr{A}$. Thus, we compute instead the updating direction $\mathbf{W}^{(k)}$ by solving the penalized system

$$\left(\mathbf{A}^{(k-1)^\top}\mathbf{A}^{(k-1)} + \mu\mathbf{I}\right)\mathbf{W}^{(k)} = \mathbf{A}^{(k-1)^\top}\mathbf{B}^{(k-1)} \tag{6}$$

in order to assure updates inside the admissible domain and also to reduce the norm of the residual. Note that this iterative procedure requires, for each step, the computation of $n+1$ direct problems (in an n dimensional space for the representation of $\mathbf{W}$). Thus, depending on n, we may have a big computational cost and a fast (and accurate) solver is required.

On the next section, we present the method of fundamental solutions as a solver for this type of problems.

4 The Method of Fundamental Solutions

4.1 The MFS as Solver for the Direct Problem

Recall that $\Omega_c = \Omega \setminus \overline{\omega}$, where Ω, ω are simply connected open sets with regular boundaries Γ and γ respectively, and $\partial\Omega_c = \Gamma \cup \gamma$. The complementary set $\mathbb{R}^d \setminus \overline{\Omega_c}$ has two connected components, one exterior $\Omega^C = \mathbb{R}^d \setminus \overline{\Omega}$ and one interior ω_1. To apply the MFS, we will consider the artificial set $\widehat{\gamma} = \partial\widehat{\omega}$, the internal regular boundary of the simply connected open set $\widehat{\omega}$ such that $\overline{\widehat{\omega}} \subset \omega$. Finally, we define an external boundary $\widehat{\Gamma} = \partial\widehat{\Omega}$ with $\widehat{\Omega}$ an open unbounded set $\overline{\overline{\Omega}} \subset \Omega^C$ with a boundary that encloses the domain, $\Omega \subset \mathbb{R}^d \setminus \widehat{\Omega}$.

Consider the single layer potential representation for the Lamé system,

$$\mathbf{u}(\mathbf{x}) = \int_{\widehat{\Gamma}} \Phi_{\mathbf{x}}(\mathbf{y})\Phi(\mathbf{y})dS_{\mathbf{y}} + \int_{\widehat{\gamma}} \Phi_{\mathbf{x}}(\mathbf{y})\psi(\mathbf{y})dS_{\mathbf{y}}, \quad \mathbf{x} \in \mathbb{R}^d \setminus (\widehat{\Gamma} \cup \widehat{\gamma}) \tag{7}$$

where $\Phi_{\mathbf{x}}$ is the elastic source tensor defined by $\Phi_{\mathbf{x}}(\mathbf{y}) = \Phi(\mathbf{x}-\mathbf{y})$ and Φ is the fundamental solution for the Lamé system,

$$\Phi(\mathbf{x}) = \begin{cases} \dfrac{\lambda+3\mu}{4\pi\mu(\lambda+2\mu)}\left(-\log|\mathbf{x}|\mathbf{I} + \dfrac{\lambda+\mu}{\lambda+3\mu}\dfrac{x\otimes x}{|\mathbf{x}|^2}\right) & \text{2D case} \\[3mm] \dfrac{\lambda+3\mu}{8\pi\mu(\lambda+2\mu)}\left(\dfrac{1}{|\mathbf{x}|}\mathbf{I} + \dfrac{\lambda+\mu}{\lambda+3\mu}\dfrac{x\otimes x}{|\mathbf{x}|^3}\right) & \text{3D case} \end{cases}.$$

We assume further that the Robin boundary condition on γ may be non homogeneous. Discretizing the layer representation we obtain

$$\mathbf{u}(\mathbf{x}) = \sum_{j=1}^{m} \Phi_{\mathbf{x}}(\mathbf{y_j})\alpha_j + \sum_{j=m+1}^{M} \Phi_{\mathbf{x}}(\mathbf{y_j})\beta_{j-m}$$

where $\mathbf{y}_1,\ldots,\mathbf{y}_m \in \widehat{\Gamma}$, $\mathbf{y}_{m+1},\ldots,\mathbf{y}_M \in \widehat{\gamma}$ and α_i, β_j are vectorial coefficients. On the boundary $\partial\Omega_c$, we have (see [2] for proof).

Theorem 5. *The set*

$$\mathscr{S} = span\left\{(\partial_{\mathbf{n}}^*\mathbf{\Phi}_{\mathbf{y}_1}, \partial_{\mathbf{n}}^*\mathbf{\Phi}_{\mathbf{y}_2} + \mathbf{Z}\mathbf{\Phi}_{\mathbf{y}_2}) : \mathbf{y}_1 \in \widehat{\Gamma}, \ \mathbf{y}_2 \in \widehat{\gamma}\right\}$$

is dense in $H^{-1/2}(\Gamma)^2 \times H^{-1/2}(\gamma)^2/\mathbb{R}^p$ *in 2D (or* $H^{-1/2}(\Gamma)^3 \times H^{-1/2}(\gamma)^3$ *in 3D), where*

$$p = \begin{cases} 0 & \text{if } \int_\gamma \mathbf{Z} \text{ is not invertible} \\ 2 & \text{otherwise} \end{cases}$$

and the tensor $\partial_{\mathbf{n}}^*\mathbf{\Phi}$ *is given by* $[\partial_{\mathbf{n}}^*\mathbf{\Phi}]_{i,j} = [\partial_{\mathbf{n}}^*(\mathbf{\Phi}\mathbf{e}_j)_i]_{i,j}$, *with* $\mathbf{e}_j$ *the jth vector of the standard basis of* $\mathbb{R}^d$.

Thus, the vectorial coefficients α, β can be computed by solving, on some collocation points $\mathbf{x}_1,\ldots,\mathbf{x}_n \in \Gamma$ and $\mathbf{x}_{n+1},\ldots,\mathbf{x}_N \in \gamma$, the linear system

$$\begin{bmatrix} \partial_{\mathbf{n}}^*\mathbf{\Phi}_{\mathbf{x}_1}(\mathbf{y}_1) & \cdots & \partial_{\mathbf{n}}^*\mathbf{\Phi}_{\mathbf{x}_1}(\mathbf{y}_M) \\ \vdots & \cdots & \vdots \\ \partial_{\mathbf{n}}^*\mathbf{\Phi}_{\mathbf{x}_n}(\mathbf{y}_1) & \cdots & \partial_{\mathbf{n}}^*\mathbf{\Phi}_{\mathbf{x}_n}(\mathbf{y}_M) \\ \partial_{\mathbf{n}}^*\mathbf{\Phi}_{\mathbf{x}_{n+1}}(\mathbf{y}_1) + \mathbf{Z}(\mathbf{x}_{n+1})\mathbf{\Phi}_{\mathbf{x}_{n+1}}(\mathbf{y}_1) & \cdots & \partial_{\mathbf{n}}^*\mathbf{\Phi}_{\mathbf{x}_{n+1}}(\mathbf{y}_M) + \mathbf{Z}(\mathbf{x}_{n+1})\mathbf{\Phi}_{\mathbf{x}_{n+1}}(\mathbf{y}_M) \\ \vdots & \cdots & \vdots \\ \partial_{\mathbf{n}}^*\mathbf{\Phi}_{\mathbf{x}_N}(\mathbf{y}_1) + \mathbf{Z}(\mathbf{x}_N)\mathbf{\Phi}_{\mathbf{x}_N}(\mathbf{y}_1) & \cdots & \partial_{\mathbf{n}}^*\mathbf{\Phi}_{\mathbf{x}_N}(\mathbf{y}_M) + \mathbf{Z}(\mathbf{x}_N)\mathbf{\Phi}_{\mathbf{x}_N}(\mathbf{y}_M) \end{bmatrix}$$

$$\times \begin{bmatrix} \alpha_1 \\ \vdots \\ \alpha_m \\ \beta_1 \\ \vdots \\ \beta_{M-m} \end{bmatrix} = \begin{bmatrix} \mathbf{g}(\mathbf{x}_1) \\ \vdots \\ \mathbf{g}(\mathbf{x}_n) \\ \mathbf{0} \\ \vdots \\ \mathbf{0} \end{bmatrix}.$$

4.2 The MFS for the Cauchy Problem

We now address the inverse problem, i.e., to retrieve the impedance $\mathbf{Z} \in \mathscr{A}$ from the Cauchy data on Σ. Again, consider the single layer representation (7) for the solution of the inverse problem and it's discretization

$$\mathbf{u}^I(\mathbf{x}) = \sum_{j=1}^m \mathbf{\Phi}_{\mathbf{x}}(\mathbf{y}_\mathbf{j})\alpha_j + \sum_{j=m+1}^M \mathbf{\Phi}_{\mathbf{x}}(\mathbf{y}_\mathbf{j})\beta_{j-m}.$$

Following [2], we have:

Theorem 6. *The set*

$$S_n = span\{(\mathbf{\Phi_y}, \partial_n^* \mathbf{\Phi_y}) : \mathbf{y} \in \widehat{\Gamma} \cup \widehat{\gamma}\}$$

is dense in $H^{1/2}(\Gamma)^2/\mathbb{R}^2 \times H^{-1/2}(\Sigma)^2$ *for the 2D case* $(H^{1/2}(\Gamma)^3 \times H^{-1/2}(\Sigma)^3$ *in 3D)*

and thus, for the inverse problem, the coefficients α, β are given by

$$\begin{bmatrix} \partial_n^* \mathbf{\Phi_{x_1}}(\mathbf{y}_1) & \dots & \partial_n^* \mathbf{\Phi_{x_1}}(\mathbf{y}_M) \\ \vdots & \dots & \vdots \\ \partial_n^* \mathbf{\Phi_{x_n}}(\mathbf{y}_1) & \dots & \partial_n^* \mathbf{\Phi_{x_n}}(\mathbf{y}_M) \\ \mathbf{\Phi_{x_{n+1}}}(\mathbf{y}_1) & \dots & \mathbf{\Phi_{x_{n+1}}}(\mathbf{y}_M) \\ \vdots & \dots & \vdots \\ \mathbf{\Phi_{x_N}}(\mathbf{y}_1) & \dots & \mathbf{\Phi_{x_N}}(\mathbf{y}_M) \end{bmatrix} \begin{bmatrix} \alpha_1 \\ \vdots \\ \alpha_m \\ \beta_1 \\ \vdots \\ \beta_{M-m} \end{bmatrix} = \begin{bmatrix} \mathbf{g}(\mathbf{x}_1) \\ \vdots \\ \mathbf{g}(\mathbf{x}_n) \\ \mathbf{u}(\mathbf{x}_{n+1}) \\ \vdots \\ \mathbf{u}(\mathbf{x}_N) \end{bmatrix}. \qquad (8)$$

Moreover, since the inverse problem is ill–posed, a Tikhonov regularization scheme is used.

The matrix $\mathbf{Z}$ can now be given by

$$\partial_n^* \mathbf{u}^l + \mathbf{Z}\mathbf{u}^l \approx \mathbf{0} \text{ on } \gamma.$$

5 Numerical Simulations

In the following numerical simulations, $\lambda = \mu = 1$ and:

- $\mathbf{M} = \mathbf{I}$ and $\sigma = \gamma$ hence the admissible domain is

$$\mathscr{A} = \{z\mathbf{I} : z \in C(\gamma) \setminus \{0\} \wedge z \geq 0\}.$$

- The domain of propagation is the annulus

$$\Omega_c = \{\mathbf{x} \in \mathbb{R}^2 : 1 < |\mathbf{x}| < 3\}.$$

For the first example we considered the explicit solution

$$\mathbf{u}(\mathbf{x}) = \left(-x_1(5 + |\mathbf{x}|^{-2}), x_2(1 - |\mathbf{x}|^{-2})\right)$$

which is the function (2), for $\lambda = \mu = 1$ and in this case the Robin coefficient is given by

$$\mathbf{Z}_1 \equiv 2\mathbf{I}.$$

For examples two and three, the prescribed data was $\mathbf{g}(\mathbf{x}) = \mathbf{x}$ and $\mathbf{Z}_i = z_i \mathbf{I}$ with

$$
z_2(t) = \begin{cases}
\frac{1}{2} + \frac{t}{\pi} & t \in \left[0, \frac{\pi}{2}\right] \\
1 + \frac{4(t - \pi/2)}{\pi} & t \in \left]\frac{\pi}{2}, \pi\right] \\
3 - \frac{4(t - \pi)}{\pi} & t \in \left]\pi, \frac{3\pi}{2}\right] \\
1 - \frac{t - 3\pi/2}{\pi} & t \in \left]\frac{3\pi}{2}, 2\pi\right[
\end{cases}
$$

and

$$
z_3(t) = \begin{cases}
1.2 + 0.5 \sin(3t) \cos(t/4) & t \in [0, \pi] \\
0.9 + \cos(t/2) \sin^2 t & t \in [\pi, 2\pi[
\end{cases}.
$$

z_2 is non smooth and z_3 is discontinuous (hence is not in the admissible space). We measured the displacement on Γ on 60 (80 for the third example) equally spaced points and for examples 2 and 3 this data was generated synthetically. To simulate the effect of noise on the measured data, we introduced two levels of random noise.

5.1 Reconstruction by Means of the MFS for the Cauchy Problem

Recall that this approach is based on fitting the available Cauchy data, by representing

$$
\mathbf{u}^I(\mathbf{x}) = \sum_{j=1}^{m} \mathbf{\Phi}_{\mathbf{x}}(\mathbf{y_j}) \alpha_j + \sum_{j=m+1}^{M} \mathbf{\Phi}_{\mathbf{x}}(\mathbf{y_j}) \beta_{j-m}
$$

and computing the coefficients α, β by solving the system (8) using some sort of regularization. The considered artificial curves for the MFS were $\widehat{\Gamma} = \partial B(\mathbf{0}, 5)$ and $\widehat{\gamma} = \partial B(\mathbf{0}, 0.5)$ and we took 120 source and 60 collocation points (160 source and 80 collocation points for the third simulation). For the regularization, we used the well known Tikhonov scheme with the L–curve criterion for the choice of parameter (see [14] for a discussion on the accuracy of the solution in function of the number of the points, distance between curves, etc.). For an example of the data fitting see Fig. 2.

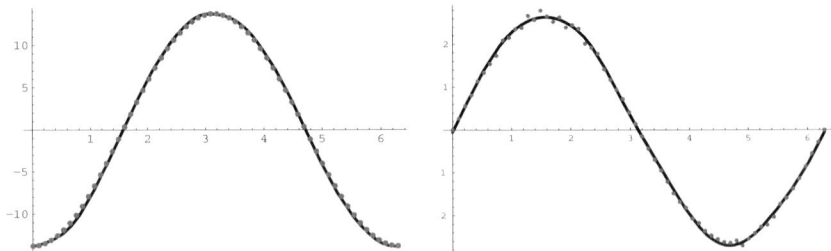

Fig. 2 Fitting of the Cauchy data (on Γ) for example 1 (noise level: 5%). *Left* – comparison between the first coordinate of the traction data (*red dots*) and computed with the MFS (*black line*); *right* – same legend but for the second coordinate of the measured displacement data and the computed with the MFS

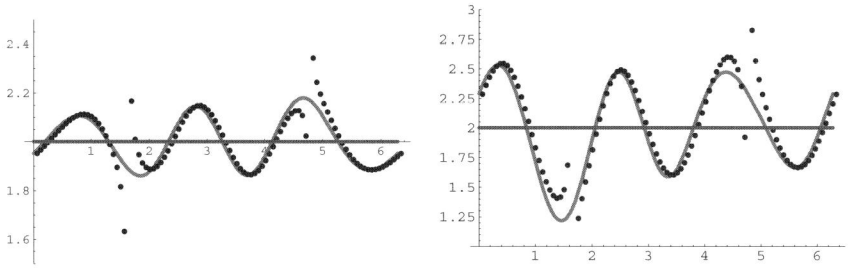

Fig. 3 Comparison between the explicit reconstruction and the implicit for simulation 1. *Blue line* – goal; *black dots* – first coordinate of the explicit reconstruction and *red line* – minimization based recontruction. *Left plot* – data with 2% of noise and *right plot* – 5% of noise

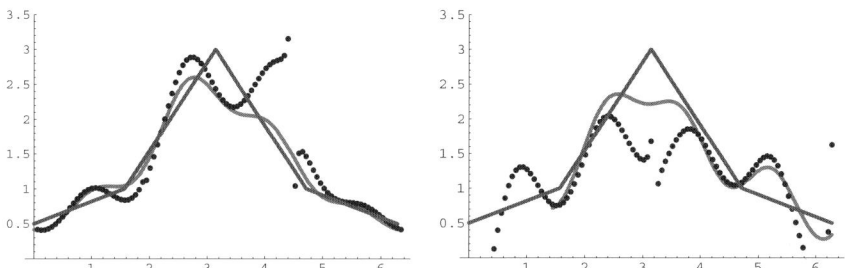

Fig. 4 The same as the previous figure but for the second simulation

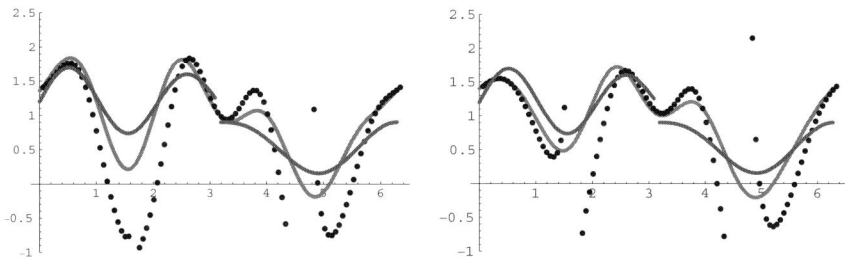

Fig. 5 The same as previous, but for the third simulation

Having computed the coefficients, the Robin function can be given explicitly by

$$-\frac{(\partial_n^* \mathbf{u}^I|_\gamma)_1}{(\mathbf{u}^I|_\gamma)_1} \quad \text{or} \quad -\frac{(\partial_n^* \mathbf{u}^I|_\gamma)_2}{(\mathbf{u}^I|_\gamma)_2}$$

which may have singularities or implicity by

$$\partial_n^* \mathbf{u}^I + \mathbf{Z}\mathbf{u}^I \approx \mathbf{0} \quad \text{on } \gamma$$

which requires a minimization procedure and the representation of $\mathbf{Z}$ in some space. In Figs. 3–5 we present a comparison between the explicit reconstruction and the implicit, using for representation the space of trigonometric polynomials $E_{11}\mathbf{I}$, where

$$E_{2n+1} = \text{span } \mathscr{B}_{2n+1}, \quad \mathscr{B}_{2n+1} = \{1, \cos(t), \sin(t), \ldots, \cos(nt), \sin(nt)\},$$

for data with up to 2% and 5% of noise.

Note the blowing up occurring in the explicit reconstruction (due to the zeros of $\mathbf{u}^I$ on γ) leading, in the third simulation, to an approximation outside the admissible domain (Fig. 5). For the first case we can observe some numerical instabilities arising from the fact that the representation space is E_{11}, whereas we are recovering a constant. Considering lower dimensional spaces an extra regularization effect is achieved and better results can be obtained. In fact, having the a priori information that we are retrieving a constant then the reconstruction in E_1 yields an excellent result (in this case we obtained the approximation $\widetilde{z}_1 \approx 2.029$).

Nonetheless, we obtained accurate reconstructions with this fast numerical approach. We now compare this results with the results produced by the iterative method.

5.2 Reconstruction with the Newton–MFS Method

Recall that (see Sect. 3), for a given $z^{(0)}\mathbf{I} \in \mathscr{A}$ we aim to construct a sequence defined by

$$z^{(n)} = z^{(n-1)} + w^{(n)}$$

and considering the updates $w^{(n)}$ in $E_p \cong \mathbb{R}^p$ we arrive at the penalized system

$$\left(\mathbf{A}^{(n-1)\top}\mathbf{A}^{(n-1)} + \mu\mathbf{I}\right) w^{(n)} = \mathbf{A}^{(n-1)\top}\mathbf{B}^{(n-1)}.$$

We follow [13] to automatic update the damping term μ. Note that in the space of admissible functions $\mathscr{A}$, theorem 4 is the local Lipschitz stability that ensures $w^{(n)} \approx 0$ whenever the residual vector $\mathbf{B}^{(n-1)}$ is approx. zero.

We consider that the update space E_p may vary with p. This phased approximation scheme is achieved by computing several iterations in some E_p and then use this information as initialization for further computations in E_q, $p < q$. As stopping criterion we use

$$\frac{\left(\mathbf{B}^{(n-1)} - \mathbf{B}^{(n)}\right) \cdot \left(\mathbf{B}^{(n-1)} - \mathbf{B}^{(n)}\right)}{\mathbf{B}^{(n-1)} \cdot \mathbf{B}^{(n-1)}} < 0.01.$$

The presented simulations regards the same computed displacement data with 5% of noise. For the arising direct problems, we used 120 points on $\partial\Omega_c$ and the same amount on $\widehat{\Gamma} \cup \widehat{\gamma}$ with $\widehat{\Gamma} = \partial B(\mathbf{0}, 5)$ and $\widehat{\gamma} = \partial B(\mathbf{0}, 0.9)$.

For the first simulation, the starting function was $z^{(0)}(t) = 10$ and we took the updates in E_1. Here, the algorithm stopped at $\widetilde{z}(t) \approx 2.02$. Next step is to start from this value and proceed with the updates in E_{11}. We can see in Fig. 6 (right) that in this case the iterative method performs better. On the left of the same figure, we

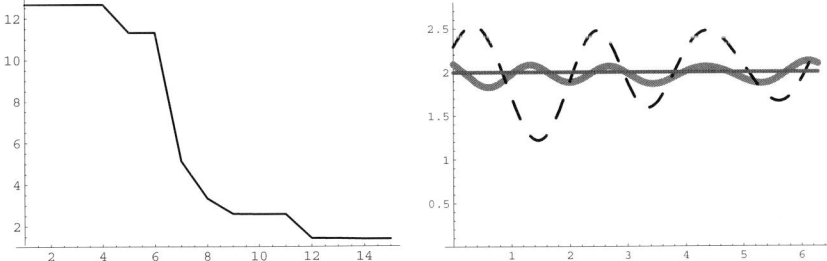

Fig. 6 *Left* – Evolution in the ℓ^2–norm of the residual with the iterations. *Right* – Comparison between the MFS–Newton (*red bold line*) and the implicit reconstruction (*black dashed line*); the blue line is the goal and the level of noise is 5%

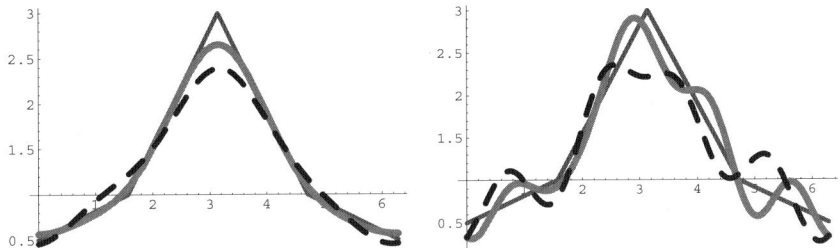

Fig. 7 Reconstructions in the update spaces E_7 (*left*) and E_{11} (*right*)

have the evolution of the residual's norm. Note that no evolution is achieved in the first four iterations. This is due to the adjustment of the damping parameter in order to avoid non descent directions and it can be justified by the "bad" starting guess we took. Nonetheless, we were able to achieve convergence.

For the second simulation we started from the initial guess $z^{(0)} = 5$ and took the first updates in E_1. Next, we updated in E_7 and E_{11} and Fig. 7 shows the results. Here, we can see the effect of the arising numerical instabilities with the increasing dimensions of the update space. Still, the iterative reconstruction is better than the approximation computed by solving the Cauchy problem.

For the third simulation, we present the iterations computed in E_1 and E_{11} (Fig. 8). Starting from the initial function $z^{(0)} = 5$ the algorithm stopped at the 4th iteration in E_1 and in E_{11} six iterations were computed. Again, the iterative method performed better than the first approach (see Fig. 9).

6 Conclusions

In this work, we discussed the identifiability question of a Robin boundary coefficient in an inaccessible part of an elastic medium. We presented two methods for this inverse problem:

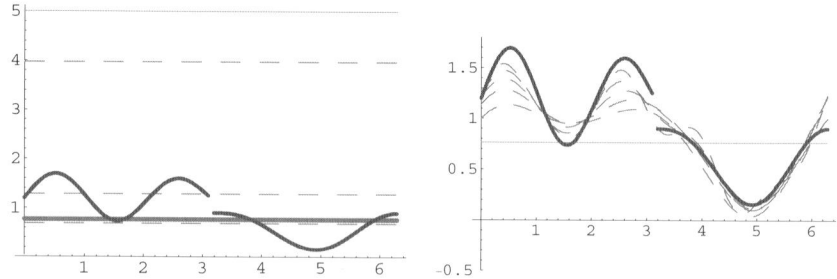

Fig. 8 Evolution of the iterations in E_1 (*left*) and E_{11} (*right*). *Red line* – starting function; *dashed lines* – intermediate iterations; *bold red line* – last iteration and *blue line* – goal

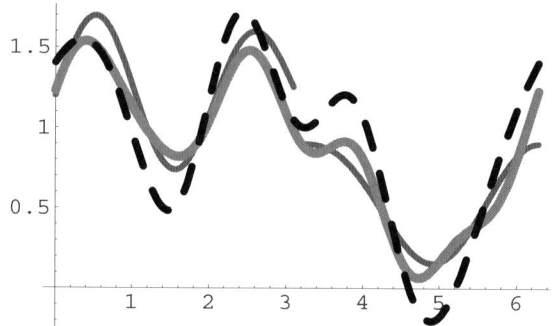

Fig. 9 Comparison between the two methods for the 3rd simulation

- The first method was based on solving the Cauchy problem on part of the boundary. This revealed to be fast an produced accurate results. Moreover, the ill–posed nature of this problem requires a regularization technique which usually depends on the choice of some parameter.
- The second method proposed was based on a Newton type of method and the simulations showed that it performed better. However, the computational effort is much bigger and a combination of both approaches should be considered in order to reduce the computational time.

  Overall, the good features of the MFS as a solver for this system of PDES (such as low implementation cost, good accuracy and speed) makes it an excellent numerical approach for the mentioned problem.

Acknowledgments Authors acknowledge the support of Fundação para a Ciência e Tecnologia (FCT), through projects POCI MAT/61792/2004, MAT/60863/2004 and ECM/58940/2004. N. Martins is also partially supported by FCT through the scholarship SFRH/BD/27914/2006.

References

1. C. J. S. Alves and C. S. Chen, *A new method of fundamental solutions applied to nonhomogeneous elliptic problems*, Adv. Comp. Math., **23**, 125–142, 2005.
2. C. J. S. Alves and N. F. M. Martins, *The Direct Method of Fundamental Solutions and the Inverse Kirsch-Kress Method for the Reconstruction of Elastic Inclusions or Cavities*, Preprint (22), Department of Mathematics, FCT/UNL, Caparica, Portugal, 2007.
3. C. J. S. Alves and N. F. M. Martins, *Reconstruction of inclusions or cavities in potential problems using the MFS, Submitted.*
4. A. Bogomolny, *Fundamental solutions method for elliptic boundary value problems*, SIAM J. Numer. Anal., **22**, 644–669, 1985.
5. K. Bryan and L. Caudill, *An inverse problem in thermal imaging*, SIAM J. Appl. Math., **56**, 715–735, 1996.
6. S. Chaabane, C. Elhechmi and M. Jaoua, *A stable recovery method for the Robin inverse problem*, Math. Comput. Simulat., **66**, 367–383, 2004.
7. F. Cakoni and R. Kress, *Integral equations for inverse problems in corrosion detection from partial Cauchy data*, Inverse Problems and Imaging, **1**(2), 229–245, 2007.
8. G. Chen and J. Zou, *Boundary Element Methods*, Academic, London, 1992.
9. G. Fairweather and A. Karageorghis, *The method of fundamental solutions for elliptic boundary value problems*, Adv. Comput. Math., **9**, 69–95, 1998.
10. G. Inglese, *An inverse problem in corrosion detection*, Inverse Probl., **13**, 977–994, 1997.
11. P. Kaup and F. Santosa, *Nondestructive evaluation of corrosion damage using electrostatic measurements*, J. Nondestruct. Eval., **14**, 127–136, 1995.
12. V. D. Kupradze and M. A. Aleksidze, *The method of functional equations for the aproximate solution of certain boundary value problems*, Zh. vych. mat., **4**, 683–715, 1964.
13. K. Madsen, H. B. Nielsen and O. Tingleff, *Methods for non-linear least squares problems*, IMM, 60 pages, Denmark, 2004.
14. L. Marin and D. Lesnic, *The method of fundamental solutions for the Cauchy problem in two-dimensional linear elasticity*, Int. J. Solids Struct., **41**(13), 3425–3438, 2004.

Several Meshless Solution Techniques for the Stokes Flow Equations

Csába Gáspár

Abstract The permanent 2D Stokes flow is considered. The applied solution technique is the classical pressure correction method, which converts the original problem to a sequence of Poisson equations. These Poisson equations are discretized and solved in a meshless way, using local interpolation based on radial basis functions. Further improvement can be achieved by using a direct multi-elliptic approach instead of local interpolation, which results in re-globalized, quadtree-based schemes. The number of unknowns can be reduced by applying the method of fundamental solutions. A special regularization technique is introduced which uses higher order fundamental solutions without singularities. This regularization is combined with the direct multi-elliptic interpolation, which significantly reduces the computational cost and makes it possible to avoid the use of dense and ill-conditioned matrices.

Keywords: Stokes Flow · Radial Basis Functions · Regularization Technique

1 The Stokes Flow Equations

The motion of slow and viscous incompressible fluid can be approximately described by the Stokes equations [14]:

$$\text{div } \mathbf{u} = 0 \tag{1}$$

$$\frac{\partial \mathbf{u}}{\partial t} - \nu \Delta \mathbf{u} + \frac{1}{\rho} \text{grad} p = f,$$

where $\mathbf{u} = (u, v, w)$ is the velocity, p is the pressure, ν denotes the kinematic viscosity and ρ is the density. ν and ρ are assumed to be constant both in space and time. The function f describes the acceleration due to external forces.

C. Gáspár
Széchenyi István University, P.O. Box 701, H-9007 Győr, Hungary, e-mail: gasparcs@sze.hu

A.J.M. Ferreira et al. (eds.) *Progress on Meshless Methods, Computational Methods in Applied Sciences.*

In contrast to the full Navier-Stokes system, Eq. (1) are *linear*, which makes it possible to construct simpler solution techniques. Moreover, several algorithms for the Navier-Stokes equations require an efficient Stokes solver. In this paper, we restrict ourselves to the permanent 2D equations without external forces. The governing equations are as follows:

$$\frac{\partial u}{\partial x} + \frac{\partial v}{\partial y} = 0$$

$$\Delta u = \frac{1}{\mu}\frac{\partial p}{\partial x} \tag{2}$$

$$\Delta v = \frac{1}{\mu}\frac{\partial p}{\partial y}$$

where $\mu := \nu\rho$.

For simplicity, we assume that Eq. (2) are supplied with Dirichlet boundary condition, i.e. the velocities are prescribed along the boundary Γ of the bounded flow domain Ω. This means that the velocity components are supposed to be known both at the inflow and at the outflow boundaries. Along solid walls, no-slip boundary condition is imposed i.e. both velocity components are equal to zero. For the pressure p, no boundary condition is prescribed: instead, the global condition $\int_\Omega p\, d\Omega = 0$ is required. As it is well known, this problem has a unique solution in the space $H^1(\Omega) \times H^1(\Omega) \times L_2(\Omega)$ (provided that the velocity components prescribed along the boundary Γ belong to the space $H^{1/2}(\Gamma)$, and satisfy the continuity condition $\int_\Gamma \mathbf{u}\cdot\mathbf{n}\, d\Gamma = 0$). The solution minimizes the quadratic functional

$$\int_\Omega (||\mathrm{grad}\, u||^2 + ||\mathrm{grad}\, v||^2)\, d\Omega$$

with respect to the boundary condition and the condition div $\mathbf{u} = 0$. In this context, the pressure p can be considered the Lagrange multiplier. This is the basis of the classical finite element solution techniques based on variational principles. In the following, however, no finite element solution is used.

2 Iterative Solution Techniques Based on Uzawa's Method

For the Stokes equation (2), a number of numerical techniques have been developed. Here we use the classical Uzawa algorithm [3, 4] which produces the approximate solution of (2) by the following iteration procedure:

$$\Delta u_{k+1} := \frac{1}{\mu}\cdot\frac{\partial p_k}{\partial x},$$

$$\Delta v_{k+1} := \frac{1}{\mu}\cdot\frac{\partial p_k}{\partial y}, \tag{3}$$

$$p_{k+1} := p_k - \omega\mu \cdot \left(\frac{\partial u_{k+1}}{\partial x} + \frac{\partial v_{k+1}}{\partial y} \right),$$

where $\omega > 0$ is an iteration parameter.

The first two Poisson equations are supplied with Dirichlet boundary conditions.

It is well known [3] that the sequence defined by the above iteration converges to the (unique) solution of the Stokes equations (2) provided that the iteration parameter ω is sufficiently small.

The essential idea of the iteration (3) is to convert the solution of the Stokes equations to the solution of a sequence of Poisson equations.

In practice, there is no need to exactly solve the Poisson equations in (3). After a suitable discretization, it is sufficient to apply a few (under)relaxation steps. Thus we obtain the simple *pressure correction method*:

- Apply several (under)relaxation steps to the (discretized) moment equations:

$$\Delta u = \frac{1}{\mu} \cdot \frac{\partial p}{\partial x},$$

$$\Delta v = \frac{1}{\mu} \cdot \frac{\partial p}{\partial y}.$$

- Improve the pressure by the computed divergence:

$$p := p - \omega\mu \cdot \left(\frac{\partial u}{\partial x} + \frac{\partial v}{\partial y} \right),$$

and repeat the procedure from the previous step.

As a simple consequence of Eq. (2), the pressure itself satisfies the Laplace equation: $\Delta p = 0$ (however, no physically correct boundary conditions can be prescribed). This idea can be easily incorporated in the algorithm, which results in the smoothed pressure correction method. Here the previous algorithm is completed by an additional step:

- Apply several (under)relaxations to the equation $\Delta p = 0$, and modify the pressure by

$$p := p - \frac{1}{|\Omega|} \int_{\Omega} p \, d\Omega$$

(where $|\Omega|$ denotes the area of the domain Ω), in order that the pressure satisfies the global condition $\int_{\Omega} p \, d\Omega = 0$.

These methods are well known. However, they are usually applied in a structured discretization i.e. finite difference (Cartesian or curvilinear) or finite element context. To avoid the often sophisticated grid (mesh) generation, it is necessary that the relaxations are performed in a meshless way. For this reason, the Laplace operator has to be discretized over an unstructured set of points. Here we apply the idea of the local interpolation technique and the re-globalized schemes based on the biharmonic interpolation [9, 10]. Later, we show a boundary version of the technique, which requires boundary interpolation points only.

3 Meshless Poisson Solvers Based on Local and Re-globalized Interpolation

Two usual techniques to solve Poisson equations in a meshless way are the direct substitution method (Kansa's method [13]) and the method of particular solutions. In the first approach, the solution itself is to be approximated by a scattered data interpolation, usually by the method of radial basis functions [11]:

$$u(x) \sim \sum_{j=1}^{N+M} \alpha_j \Phi_j(x - x_j), \tag{4}$$

where $x_1, x_2, ..., x_N \in \Omega$ and $x_{N+1}, x_{N+2}, ..., x_{N+M} \in \Gamma$ are interpolation points scattered in Ω and along Γ, respectively, and $\Phi_1, \Phi_2, ..., \Phi_{N+M}$ are predefined circularly symmetric functions (radial basis functions, RBFs). Substituting the expression (4) into the given Dirichlet problem of the Poisson equation

$$\Delta u = f, \tag{5}$$

$$u|_\Gamma = u_0,$$

the coefficients $\alpha_1, ..., \alpha_{N+M}$ can be determined by solving the following linear system of equations:

$$\sum_{j=1}^{N} \alpha_j \Delta \Phi_j(x_k - x_j) = f_k \quad (k = 1, 2, ...N) \tag{6}$$

$$\sum_{j=1}^{N} \alpha_j \Phi_j(x_k - x_j) = u_0(x_k) \quad (k = N+1, N+2, ..., N+M).$$

In the method of particular solutions, the function f is approximated by interpolation:

$$f(x) \sim \sum_{j=1}^{N} \alpha_j \Phi_j(x - x_j),$$

and a particular solution of the Poisson equation (5) is expressed by:

$$u(x) := \sum_{j=1}^{N} \alpha_j \Psi_j(x - x_j), \tag{7}$$

where $\Psi_1, ..., \Psi_N$ are radial basis functions defined by $\Delta \Psi_j = \Phi_j$ (they have analytical forms in general). Thus, the problem is converted to a *homogeneous* one, and

the solution u of (5) can be decomposed in the form $u = v + w$, where v is a particular solution and w is the solution of the homogeneous problem supplied with the
modified boundary condition $w|_\Gamma = u_0 - v|_\Gamma$.

The homogeneous problem can be efficiently solved by e.g. a boundary technique
like Boundary Element Method. However, this method still requires a boundary
discretization. This can be completely avoided by applying a meshless boundary
technique e.g. the boundary knot method [5] or the method of fundamental solutions
(MFS, see [1, 12]).

Both approaches lead to a linear system with a full, often nonsymmetric and ill-
conditioned matrix, which causes computational problems. This disadvantage can
be reduced by localizing the interpolation. For each interpolation point x_m ($m =
1, 2, ..., N$), a local interpolation function is constructed based on the interpolation
points located in a well-defined neighborhood of the point x_m only. We have used
radial basis functions with polynomial augmentation:

$$\tilde{u}^{(m)}(x) := \sum_{j=1}^{N_m} \alpha_j^{(m)} \Phi(x - x_j^{(m)}) + \sum_{j=1}^{M} a_j^{(m)} p_j(x), \tag{8}$$

where $x_1^{(m)}, x_2^{(m)}, ..., x_{N_m}^{(m)}$ denote the neighboring interpolation points. The coefficients $\alpha_1^{(m)}, ..., \alpha_{N_m}^{(m)}, a_1^{(m)}, ..., a_M^{(m)}$ can be computed by solving the *local interpolation equations*:

$$\sum_{j=1}^{N_m} \alpha_j^{(m)} \Phi(x_k^{(m)} - x_j^{(m)}) + \sum_{j=1}^{M} a_j^{(m)} p_j(x_k^{(m)}) = u_k^{(m)} \quad (k = 1, 2, ...N_m), \tag{9}$$

completed with the orthogonality conditions:

$$\sum_{j=1}^{N_m} \alpha_j^{(m)} p_k(x_j^{(m)}) = 0 \qquad (k = 1, 2, ...M) \tag{10}$$

Equations (9–10) can be written in a more compact form:

$$\begin{pmatrix} A^{(m)} & B^{(m)} \\ B^{(m)*} & 0 \end{pmatrix} \begin{pmatrix} \alpha^{(m)} \\ a^{(m)} \end{pmatrix} = \begin{pmatrix} \mathbf{u}^{(m)} \\ 0 \end{pmatrix}, \tag{11}$$

where $A^{(m)}$ is a symmetric matrix. The elements of the matrices $A^{(m)}$ and $B^{(m)}$ are
as follows:
$$A_{kj}^{(m)} = \Phi(x_k^{(m)} - x_j^{(m)}) \qquad (k, j = 1, ..., N_m),$$

$$B_{kj}^{(m)} = p_j(x_k^{(m)}) \qquad (k = 1, ..., N_m, \quad j = 1, ..., M),$$

$\alpha^{(m)}$ and $a^{(m)}$ are unknown vectors with the components $\alpha_1^{(m)}, \alpha_2^{(m)}, ..., \alpha_{N_m}^{(m)}$ and
$a_1^{(m)}, a_2^{(m)}, ..., a_M^{(m)}$, respectively. The vector $\mathbf{u} = u_1^{(m)}, u_2^{(m)}, ..., u_{N_m}^{(m)}$ contains the values
which are attached to the neighboring points $x_1^{(m)}, x_2^{(m)}, ..., x_{N_m}^{(m)}$.

It should be pointed out that the above local system cannot be solved in every case. Solvability depends on the structure of the interpolation points taken into account, which is difficult to check. However, our experience is that if the radius of influence is large enough (i.e. the number N_m is large enough) and the interpolation points are defined in a quasi-random way, then the local interpolation equations are singular in very exceptional cases only. In the following, we assume that the local systems have always unique solutions.

Having solved the local interpolation systems, the Laplace operator can be discretized in the following way. Define four fictitious points around x_m in the main coordinate directions at the distance h_m (where h_m is the square mean value of the distances $||x_j^{(m)} - x_m||$). Denoting these fictitious points by $x_m^N, x_m^W, x_m^S, x_m^E$, define:

$$(\Delta u)_m \sim \frac{\tilde{u}^{(m)}(x_m^N) + \tilde{u}^{(m)}(x_m^W) + \tilde{u}^{(m)}(x_m^S) + \tilde{u}^{(m)}(x_m^E) - 4u_m}{h_m^2}. \tag{12}$$

Based on the above formula, the Poisson equation (5) can be discretized as follows (written in Seidel iteration form):

$$u_m := \frac{\tilde{u}^{(m)}(x_m^N) + \tilde{u}^{(m)}(x_m^W) + \tilde{u}^{(m)}(x_m^S) + \tilde{u}^{(m)}(x_m^E)}{4} - \frac{h_m^2 \cdot f(x_m)}{4} \tag{13}$$

Now we show that the first term in the right-hand side can be expressed by the values $u_j^{(m)}$:

$$\sum_{j=1}^{N_m} \alpha_j^{(m)} \frac{\Phi(x_m^N - x_j^{(m)}) + \Phi(x_m^W - x_j^{(m)}) + \Phi(x_m^S - x_j^{(m)}) + \Phi(x_m^E - x_j^{(m)})}{4}$$

$$+ \sum_{j=1}^{M} a_j^{(m)} \frac{p_j(x_m^N) + p_j(x_m^W) + p_j(x_m^S) + p_j(x_m^E)}{4}$$

$$=: \sum_{j=1}^{N_m} \alpha_j^{(m)} \beta_j^{(m)} + \sum_{j=1}^{M} a_j^{(m)} b_j^{(m)} =: \left\langle \begin{pmatrix} \alpha^{(m)} \\ a^{(m)} \end{pmatrix}, \begin{pmatrix} \beta^{(m)} \\ b^{(m)} \end{pmatrix} \right\rangle,$$

where

$$\beta_j^{(m)} := \frac{\Phi(x_m^N - x_j^{(m)}) + \Phi(x_m^W - x_j^{(m)}) + \Phi(x_m^S - x_j^{(m)}) + \Phi(x_m^E - x_j^{(m)})}{4}$$

($j = 1, 2, ..., N_m$), and

$$b_j^{(m)} := \frac{p_j(x_m^N) + p_j(x_m^W) + p_j(x_m^S) + p_j(x_m^E)}{4}$$

$(j = 1, 2, ..., M)$. Since obviously:

$$\begin{pmatrix} \alpha^{(m)} \\ \mathbf{a}^{(m)} \end{pmatrix} = \begin{pmatrix} A^{(m)} & B^{(m)} \\ B^{(m)*} & 0 \end{pmatrix}^{-1} \begin{pmatrix} \mathbf{u}^{(m)} \\ 0 \end{pmatrix},$$

therefore

$$\left\langle \begin{pmatrix} \alpha^{(m)} \\ \mathbf{a}^{(m)} \end{pmatrix}, \begin{pmatrix} \beta^{(m)} \\ \mathbf{b}^{(m)} \end{pmatrix} \right\rangle = \left\langle \begin{pmatrix} \mathbf{u}^{(m)} \\ 0 \end{pmatrix}, \begin{pmatrix} A^{(m)} & B^{(m)} \\ B^{(m)*} & 0 \end{pmatrix}^{-1} \begin{pmatrix} \beta^{(m)} \\ \mathbf{b}^{(m)} \end{pmatrix} \right\rangle$$

$$=: \left\langle \begin{pmatrix} \mathbf{u}^{(m)} \\ 0 \end{pmatrix}, \begin{pmatrix} \mathbf{w}^{(m)} \\ \mathbf{v}^{(m)} \end{pmatrix} \right\rangle,$$

where the vector pair $\mathbf{w}^{(m)}$, $\mathbf{v}^{(m)}$ solves the following local system:

$$\begin{pmatrix} A^{(m)} & B^{(m)} \\ B^{(m)*} & 0 \end{pmatrix} \begin{pmatrix} \mathbf{w}^{(m)} \\ \mathbf{b}^{(m)} \end{pmatrix} = \begin{pmatrix} \beta^{(m)} \\ \mathbf{b}^{(m)} \end{pmatrix} \tag{14}$$

Summarizing the above results, the localized Seidel scheme is constructed by the following algorithm. For each interpolation point x_m:

- Determine the neighboring points $x_1^{(m)}, x_2^{(m)}, ..., x_{N_m}^{(m)}$
- Compute the corresponding vectors $\beta^{(m)}$ and $\mathbf{b}^{(m)}$
- Generate and solve the local system (14)
- Using the vector $\mathbf{w}^{(m)}$ obtained in the previous step, define:

$$u_m := \sum_{j=1}^{N_m} w_j^{(m)} u_j^{(m)} - \frac{h_m^2 \cdot f(x_m)}{4} \tag{15}$$

The coefficients $w_j^{(m)}$ have to be computed only once, at the beginning of the computation.

At the boundary interpolation points, the Dirichlet boundary conditions are taken into account in a natural way. The above defined *localized scheme* leads to a sparse system. The resulting Seidel iteration can be applied as an iterative solver, or as the smoothing procedure of a quite natural multi-level technique (details are omitted here). Our experience is that the Seidel iteration itself converges rapidly in spite of the fact that the above localized scheme does not remain symmetric in general.

From the computational point of view, the method can be improved by using a direct multi-elliptic (e.g. biharmonic) interpolation [7] instead of the local RBF-based interpolation. Thus we obtain the *re-globalized schemes* [9, 10]. Here the interpolation function $\tilde{u}$ is determined by solving the biharmonic equation

$$\Delta\Delta\tilde{u} = 0 \quad \text{in } \Omega_0 \setminus \{x_1, x_2, ..., x_{N+M}\} \tag{16}$$

(where $\Omega_0 \supset \Omega$ is a larger but bounded domain) supplied with the interpolation conditions as special pointwise boundary conditions:

$$\tilde{u}(x_k) = u_k \quad (k = 1, 2, ..., N+M) \tag{17}$$

Along the boundary of Ω_0, any regular boundary condition (e.g. Dirichlet or Neumann) can be prescribed for the biharmonic equation. It was shown [7] that this results in a well-posed problem in a proper subspace of the Sobolev space $H^2(\Omega)$.

To solve the biharmonic problem (16–17), quadtree-based multigrid techniques are proposed, which significantly reduce the computational cost. The number of arithmetic operations is proportional to the *first* power of the number of interpolation points. Moreover, the problems of dense and ill-conditioned matrices are completely avoided.

In the re-globalized schemes, it is sufficient to generate a quadtree cell system controlled by the interpolation points. No fictitious neighboring points have to be generated. Instead, scheme (12) is based on the *neighboring quadtree cells*, and we obtain the following algorithm:

- Perform some relaxation steps to the biharmonic equation (16) in each cell, with the exception of the cells, the center of which are (boundary or inner) interpolation points.
- Update the values belonging to the interpolation points by performing a Seidel iteration for the Poisson equation (based on their neighboring cells) and continue the iteration from the previous step.

That is, the biharmonic re-globalized algorithm consists of relaxations for the Poisson equation (at the cells whose centers are inner interpolation points) and for the biharmonic equation (at the other non-boundary cells).

The resulting scheme can be easily incorporated in a quadtree-based multigrid context. It should be pointed out that the inner interpolation points can be defined automatically by the quadtree cell system as well. In this sense, the re-globalized method is of boundary type.

To illustrate the above approaches, consider the following model problem for the Laplace equation. The exact solution is:

$$u(x,y) := \cosh\left(\pi\left(x - \frac{1}{2}\right)\right) \cdot \sin \pi y \qquad (18)$$

defined on the unit square. The Dirichlet boundary condition is defined in a consistent way. Figure 1 shows the approximate solutions defined by the local scheme

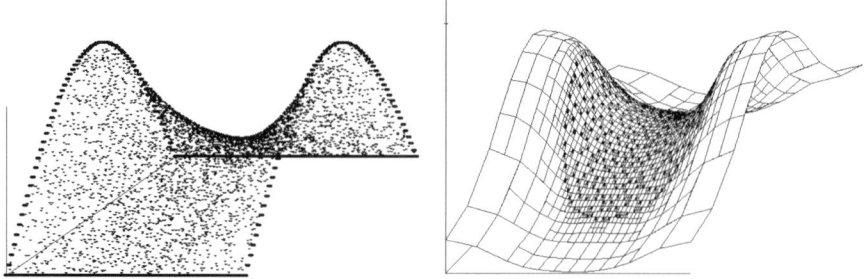

Fig. 1 Solution of the harmonic test problem (18) by local and re-globalized schemes

(12) and the re-globalized scheme. In the first case, 4,096 inner and 256 boundary points were used. The inner points were scattered in the domain in a quasi-random way. The minimal and maximal number of neighboring points taken into account in the construction of the local schemes were 8 and 34, respectively. The applied RBF in the local interpolation was the thin plate spline (TPS) with polynomial augmentation of the first degree. Though the local scheme leads to nonsymmetric equations, the Seidel iteration proved convergent. After 560 iteration steps, the relative L_2-error decreased under 1%; after reaching convergence, the relative L_2-error of the approximate solution was 0.14%. In the second case, only 350 inner and 80 boundary points were used (here the domain of the model problem was a circle with radius $\frac{1}{4}$ in the middle of the unit square). The relative L_2-error was 0.84%.

4 A Numerical Example

The above described pressure correction method combined with a meshless Poisson solver was applied to the Stokes equations (2). The flow was assumed to be horizontal and to have a parabolic velocity profile, i.e. the exact solution is as follows:

$$u(x,y) = c \cdot y(L-y), \quad v(x,y) = 0, \quad p(x,y) = c \cdot \mu \cdot (L-2x). \qquad (19)$$

The flow domain was assumed to be to unit square ($L = 1$). The scaling constant c as well as μ were set to 1. Local schemes (13) for relaxing the moment equations were applied. In the pressure correction step, the divergence was computed also in a meshless way, similarly to the discretization of the Laplace operator:

$$(\operatorname{div} \mathbf{u})(x_m) \sim \frac{\tilde{u}^{(m)}\left(x_m^E\right) - \tilde{u}^{(m)}\left(x_m^W\right)}{2h_m} + \frac{\tilde{v}^{(m)}\left(x_m^N\right) - \tilde{v}^{(m)}\left(x_m^S\right)}{2h_m},$$

where $\tilde{u}^{(m)}$ and $\tilde{v}^{(m)}$ denote the local interpolation functions of the (approximate) velocity components around the interpolation point x_m. In the test solution, 1,024 inner and 128 boundary points were used. The approximate solutions u and p are shown in Fig. 2. The relative L_2-error of the computed velocities was 0.754%.

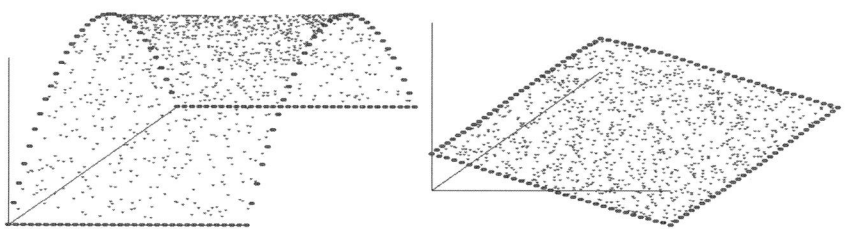

Fig. 2 Computed horizontal velocity components and pressures in the test problem (19)

5 Solution Techniques Based on the Method of Fundamental Solutions

The above Uzawa-type methods require a set on interpolation points *inside* the domain. The computational efficiency can significantly be improved by using a (meshless) boundary technique, which requires *boundary* interpolation points only. The Method of Fundamental Solutions (MFS, see e.g. [1, 12]) is a typical technique with the above property.

A fundamental solution of the 2D Stokes equation (2) is a pair of a matrix function $G := \begin{pmatrix} u_1 & u_2 \\ v_1 & v_2 \end{pmatrix}$ and a vector function $p := \begin{pmatrix} p_1 \\ p_2 \end{pmatrix}$, which satisfies the following pair of Stokes equations:

$$-\mu \Delta u_1 + \frac{\partial p_1}{\partial x} = \delta, \quad -\mu \Delta v_1 + \frac{\partial p_1}{\partial y} = 0, \quad \frac{\partial u_1}{\partial x} + \frac{\partial v_1}{\partial y} = 0,$$

$$-\mu \Delta u_2 + \frac{\partial p_2}{\partial x} = 0, \quad -\mu \Delta v_2 + \frac{\partial p_2}{\partial y} = \delta, \quad \frac{\partial u_2}{\partial x} + \frac{\partial v_2}{\partial y} = 0,$$

(20)

where δ denotes the Dirac distribution concentrated to the origin. Such a fundamental solution can be expressed by the following formulas (see [2, 15]):

$$u_1(x,y) = -\frac{1}{8\pi\mu} \left(\log(x^2 + y^2) + 1 + \frac{2y^2}{x^2 + y^2} \right)$$

$$v_1(x,y) = \frac{1}{8\pi\mu} \cdot \frac{2xy}{x^2 + y^2}$$

$$p_1(x,y) = \frac{1}{4\pi} \cdot \frac{2x}{x^2 + y^2}$$

(21)

$$u_2(x,y) = \frac{1}{8\pi\mu} \cdot \frac{2xy}{x^2 + y^2}$$

$$v_2(x,y) = -\frac{1}{8\pi\mu} \left(\log(x^2 + y^2) + 1 + \frac{2x^2}{x^2 + y^2} \right)$$

$$p_2(x,y) = \frac{1}{4\pi} \cdot \frac{2y}{x^2 + y^2}$$

Simple calculations show that the above functions can be rewritten as:

$$G = \frac{1}{\mu} \begin{pmatrix} -\frac{\partial^2 E_2}{\partial y^2} & \frac{\partial^2 E_2}{\partial x \partial y} \\ \frac{\partial^2 E_2}{\partial x \partial y} & -\frac{\partial^2 E_2}{\partial x^2} \end{pmatrix}, \quad p = \begin{pmatrix} \frac{\partial E_1}{\partial x} \\ \frac{\partial E_1}{\partial y} \end{pmatrix},$$

where the functions E_1 and E_2 are the harmonic and the biharmonic fundamental solution, respectively:

$$E_1(x,y) = \frac{1}{4\pi} \log(x^2 + y^2), \quad E_2(x,y) = \frac{1}{16\pi}(x^2 + y^2)\log(x^2 + y^2).$$

The Method of Fundamental Solutions produces the approximate solution of the Stokes equation (2) in the following form:

$$u(x,y) \sim \sum_{j=1}^{N} a_j^{(1)} u_1(x - \tilde{x}_j, y - \tilde{y}_j) + \sum_{j=1}^{N} a_j^{(2)} u_2(x - \tilde{x}_j, y - \tilde{y}_j)$$

$$v(x,y) \sim \sum_{j=1}^{N} a_j^{(1)} v_1(x - \tilde{x}_j, y - \tilde{y}_j) + \sum_{j=1}^{N} a_j^{(2)} v_2(x - \tilde{x}_j, y - \tilde{y}_j) \qquad (22)$$

$$p(x,y) \sim \sum_{j=1}^{N} a_j^{(1)} p_1(x - \tilde{x}_j, y - \tilde{y}_j) + \sum_{j=1}^{N} a_j^{(2)} p_2(x - \tilde{x}_j, y - \tilde{y}_j),$$

where the points $(\tilde{x}_j, \tilde{y}_j)$ $(j = 1, 2, ..., N)$ are located outside of the domain Ω (*source points*). The a priori unknown coefficients $a_j^{(1)}$, $a_j^{(2)}$ $(j = 1, 2, ..., N)$ can be determined by solving the following system of linear equations derived from the boundary conditions:

$$\sum_{j=1}^{N} a_j^{(1)} u_1(x_k - \tilde{x}_j, y_k - \tilde{y}_j) + \sum_{j=1}^{N} a_j^{(2)} u_2(x_k - \tilde{x}_j, y_k - \tilde{y}_j) = u_k$$

$$\qquad (23)$$

$$\sum_{j=1}^{N} a_j^{(1)} v_1(x_k - \tilde{x}_j, y_k - \tilde{y}_j) + \sum_{j=1}^{N} a_j^{(2)} v_2(x_k - \tilde{x}_j, y_k - \tilde{y}_j) = v_k$$

where the points (x_k, y_k) $(k = 1, 2, ..., N)$ are located on the boundary of Ω (*collocation points*). Thus, compared with the methods of the previous sections, much less unknowns are introduced. However, from the computational point of view, the system (23) is much less comfortable, its matrix is fully populated and often severely ill-conditioned, especially in the case when the source points are located far from the boundary (though in this case, the accuracy increases in general).

The source and collocation points are not allowed to coincide because of the singularity of the fundamental solution at the origin.

As an example, consider again the test problem (19) of the previous section. Table 1 summarizes the relative L_2-errors of the computed velocities for different number of source points and different distances between the source points and the boundary.

It is clearly seen that if the sources are located farther from the boundary, then the error of the approximate solution decreases. It should be pointed out, however, that the system (23) becomes extremely ill-conditioned in this case, and the approximate solution exhibits strong irregularities in the vicinity of the sources, which quickly destroys the approximation inside the domain as well, especially in such a case when

Table 1 Relative L_2-errors of the computed velocities of the test problem (19). N is the number of sources, d is the distance of the sources and the boundary

$N\backslash d$	0.01	0.02	0.05	0.10	0.20
64	4.0575	2.7775	1.1840	0.0801	0.0050
128	1.4028	0.4290	0.0649	0.0012	0.0000
256	0.6300	0.0580	0.0022	0.0000	0.0000

Fig. 3 Computed horizontal velocity components in the test problem (19). Method of Fundamental Solutions, $d = 0.1$. Irregularities in the vicinity of the sources

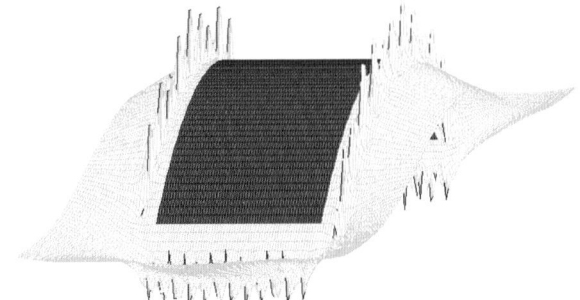

the solution is not so smooth as in this simple example. As an illustration, see Fig. 3, where $d = 0.1$ and $N = 128$.

To avoid the problem of severe ill-conditioning as well as the use of external sources, we have applied a regularization technique proposed in [8]. The essential idea is to approximate the Laplace operator Δ by the fourth-order Laplace-(modified)Helmholtz operator $\Delta(I - \frac{1}{c^2}\Delta)$, where I denotes the identity operator, and $c > 0$ is a scaling constant. Combined this idea with the Method of Fundamental Solutions, we apply an MFS-like approximation (22) but using the fundamental solution of the modified system:

$$-\mu\Delta(I - \frac{1}{c^2}\Delta)u + \frac{\partial p}{\partial x} = 0$$

$$-\mu\Delta(I - \frac{1}{c^2}\Delta)v + \frac{\partial p}{\partial y} = 0 \qquad (24)$$

$$\frac{\partial u}{\partial x} + \frac{\partial v}{\partial y} = 0$$

Theorem 1. *A fundamental solution of the fourth-order system (24) is expressed in the following form:*

$$G = \begin{pmatrix} u_1 & u_2 \\ v_1 & v_2 \end{pmatrix} = \frac{c^2}{\mu} \begin{pmatrix} \frac{\partial^2 E}{\partial y^2} & -\frac{\partial^2 E}{\partial x \partial y} \\ -\frac{\partial^2 E}{\partial x \partial y} & \frac{\partial^2 E}{\partial x^2} \end{pmatrix}, \quad p = \begin{pmatrix} p_1 \\ p_2 \end{pmatrix} = \begin{pmatrix} \frac{\partial E_1}{\partial x} \\ \frac{\partial E_1}{\partial y} \end{pmatrix}, \qquad (25)$$

where

$$E(x,y) = -\frac{1}{2\pi c^4}\left(K_0(cr) + \log(cr) + \frac{(cr)^2}{4}\log(cr) - \frac{(cr)^2}{4}\right)$$

is the fundamental solution of the sixth-order operator $\Delta^2(\Delta - c^2I)$ *(with* $r = \sqrt{x^2 + y^2}$*), and* E_1 *is the fundamental solution of the Laplace operator:*

$$E_1(x,y) = \frac{1}{4\pi}\log(x^2 + y^2).$$

Proof. Using the well-known fact that the fundamental solution of the modified Helmholtz operator $\Delta - c^2I$ is $K_0(cr)$, standard calculations show that the fundamental soution of the operator $\Delta(\Delta - c^2I)$ and $\Delta^2(\Delta - c^2I)$ are

$$-\frac{1}{2\pi c^2}\left(K_0(cr) + \log(cr)\right)$$

and

$$-\frac{1}{2\pi c^4}\left(K_0(cr) + \log(cr) + \frac{(cr)^2}{4}\log(cr) - \frac{(cr)^2}{4}\right),$$

respectively. The last fundamental solution solves the equation

$$\Delta^2(\Delta - c^2I)E = \delta,$$

where δ denotes the Dirac distribution concentrated to the origin. Taking the 2D Fourier transform:

$$(\xi^2 + \eta^2)^2(-\xi^2 - \eta^2 - c^2)\hat{E} = 1,$$

whence

$$\hat{E}(\xi,\eta) = -\frac{1}{(\xi^2 + \eta^2)^2(\xi^2 + \eta^2 + c^2)}.$$

Now consider the following pair of partial differential systems, which defines the functions $u_1, v_1, p_1, u_2, v_2, p_2$ in the fundamental solution of (24):

$$-\mu\Delta(I - \frac{1}{c^2}\Delta)u_1 + \frac{\partial p_1}{\partial x} = \delta$$

$$-\mu\Delta(I - \frac{1}{c^2}\Delta)v_1 + \frac{\partial p_1}{\partial y} = 0 \qquad (26)$$

$$\frac{\partial u_1}{\partial x} + \frac{\partial v_1}{\partial y} = 0$$

$$-\mu\Delta(I - \frac{1}{c^2}\Delta)u_2 + \frac{\partial p_2}{\partial x} = 0$$

$$-\mu\Delta(I - \frac{1}{c^2}\Delta)v_2 + \frac{\partial p_2}{\partial y} = \delta \qquad (27)$$

$$\frac{\partial u_2}{\partial x} + \frac{\partial v_2}{\partial y} = 0$$

Taking the 2D Fourier transform of Eq. (26), we obtain:

$$\mu(\xi^2 + \eta^2)\left(1 + \frac{\xi^2 + \eta^2}{c^2}\right)\hat{u}_1 - i\xi\hat{p}_1 = 1$$

$$\mu(\xi^2 + \eta^2)\left(1 + \frac{\xi^2 + \eta^2}{c^2}\right)\hat{v}_1 - i\eta\hat{p}_1 = 0$$

$$-i\xi\hat{u}_1 - i\eta\hat{v}_1 = 0$$

This system can be solved for $\hat{u}_1$, $\hat{v}_1$, $\hat{p}_1$ without difficulty, yielding:

$$\hat{u}_1 = \frac{c^2}{\mu} \cdot \frac{\eta^2}{(\xi^2 + \eta^2)^2(c^2 + \xi^2 + \eta^2)}$$

$$\hat{v}_1 = -\frac{c^2}{\mu} \cdot \frac{\xi\eta}{(\xi^2 + \eta^2)^2(c^2 + \xi^2 + \eta^2)}$$

$$\hat{p}_1 = \frac{i\xi}{\xi^2 + \eta^2}$$

Utilizing the form of the Fourier transform of the function E and E_1, we obtain:

$$\hat{u}_1 = -\frac{c^2}{\mu^2} \cdot \eta^2\hat{E} = \frac{c^2}{\mu^2} \cdot \frac{\partial^2 \hat{E}}{\partial y^2}$$

$$\hat{v}_1 = \frac{c^2}{\mu^2} \cdot \xi\eta\hat{E} = -\frac{c^2}{\mu^2} \cdot \frac{\partial^2 \hat{E}}{\partial x \partial y}$$

$$\hat{p}_1 = -i\xi\hat{E}_1 = \frac{\partial \hat{E}_1}{\partial x}$$

which implies the statement of the theorem for the function triplet (u_1, v_1, p_1). The system of Equations (27) can be treated in a completely similar way. □

Observe that the matrix function G is *continuous* at the origin, and

$$G(0,0) = -\frac{\log 2 - \gamma}{4\mu\pi}\begin{pmatrix} 1 & 0 \\ 0 & 1 \end{pmatrix},$$

where γ denotes the Euler constant: $\gamma = 0.5772...$ This follows from a standard series expansion of the function E and its derivatives, which can be performed by e.g. the software MAPLE without difficulty.

Because of the continuity of G, now the source and the collocation points may coincide, i.e. there is no need to introduce external fictitious points $(\tilde{x}_j, \tilde{y}_j)$. This makes the system (23) much better conditioned, however, the error of approximation increases.

As an example, consider again the test function of the previous section. Table 2 shows the relative L_2-errors of the approximate solutions obtained by using the

Table 2 Relative L_2-errors of the computed velocities of the test problem (19) for different numbers of collocation points

N	64	128	256
Relative L_2-error (%)	9.7687	3.5155	1.0050

Table 3 Relative L_2-errors of the computed velocities of the test problem (19) for different values of the scaling parameter. N is the number of collocation points

$N\backslash c$	500	400	300	200	100	50	20
64	9.7687	8.8470	7.6473	5.9590	3.3494	2.1172	3.2844
128	3.5155	3.0132	2.3895	1.6182	1.0530	1.7164	3.3757
256	1.0050	0.7980	0.6020	0.5405	0.9626	1.8066	3.4600

fundamental solution (25) for different numbers of boundary points. The applied scaling factor was $c = 500$ in each case. The source and collocation points were the same. It can be seen that the errors are now higher than in the case of using the Stokes fundamental solution (20) (see Table 1).

However, the approximation can be improved by a careful choice of the scaling parameter as can be seen in Table 3.

As a rule of thumb, the scaling parameter c should be inversely proportional to the characteristic distance of the neighboring boundary points (cf. [8]).

Improvement by direct multi-elliptic interpolation.

When the source and collocation points coincide, the unknown coefficients $a_j^{(1)}$, $a_j^{(2)}$ in the formula (22) are determined by the linear system

$$\sum_{j=1}^{N} a_j^{(1)} u_1(x_k - x_j, y_k - y_j) + \sum_{j=1}^{N} a_j^{(2)} u_2(x_k - x_j, y_k - y_j) = u_k$$

$$\sum_{j=1}^{N} a_j^{(1)} v_1(x_k - x_j, y_k - y_j) + \sum_{j=1}^{N} a_j^{(2)} v_2(x_k - x_j, y_k - y_j) = v_k$$

(28)

$(k = 1, 2, ..., N)$, where the matrix function $\begin{pmatrix} u_1 & u_2 \\ v_1 & v_2 \end{pmatrix}$ is defined by Theorem 1.

Though this system is much more moderately ill-conditioned than the system in the traditional MFS (see Eq. (23)), the computational properties are not attractive and its matrix is still dense. Moreover, the computation of the matrix entries requires a lot of special function calls. These inconveniences can be avoided by applying the idea of the direct multi-elliptic interpolation proposed in [6] (see also [7, 8]). Here we utilize the fact that the approximate solution defined by (22) satisfies Eq. (24). Consequently, it can be determined by directly solving the fourth-order system (24) in $\Omega_0 \setminus \{(x_1, y_1), ..., (x_N, y_N)\}$ (where $\Omega_0 \supset \Omega$ is a larger but bounded domain, $(x_1, y_1), ..., (x_N, y_N)$ are the boundary collocation points) supplied with the

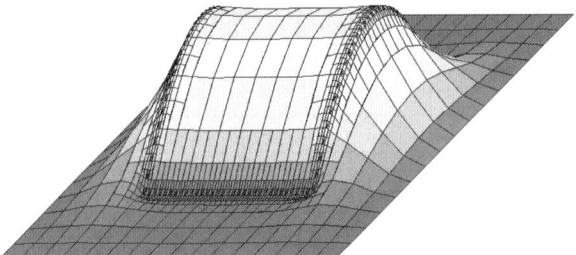

Fig. 4 Computed horizontal velocity components of the test problem (19) over the quadtree cell system obtained by the direct multi-elliptic interpolation combined with the pressure correction method

interpolation conditions:

$$u(x_k, y_k) = u_k, \quad v(x_k, y_k) = v_k \quad (k = 1, 2, \ldots, N)$$

Along the boundary of Ω_0, an arbitrary usual boundary condition (e.g. homogeneous Dirichlet condition) can be prescribed.

To directly solve the system (24), again, quadtree-based multigrid methods can be used. The applied quadtree cell system is generated by the boundary interpolation points $(x_1, y_1), \ldots, (x_N, y_N)$. As to the smoothing procedure of the multigrid algorithm, the pressure correction method is proposed (see Sect. 2) based on Uzawa's algorithm. Since the quadtree cell system as well as the applied schemes are generated purely by the boundary interpolation points, the method can be considered a meshless boundary-only technique, and it exhibits very advantageous computational properties (the necessary number of arithmetic operations is proportional to the *first* power of the number of boundary points). It should also be pointed out that the use of large, dense and ill-conditioned linear systems is completely avoided, and there is no need to evaluate special functions.

To illustrate the method, consider again the test problem of the previous section (19). Now the domain of the problem is embedded into a larger square, along the boundary of which a homogeneous Dirichlet boundary condition is prescribed. The maximal level of subdivision was 8. The number of boundary interpolation points was 128. The scaling parameter was set to the value 100. Figure 4 shows the computed velocity component u over the quadtree cell system. The discrete L_2-error of the computed velocity was 0.991%.

6 Summary and Conclusions

A meshless Stokes solver has been presented. The method is based on the classical Uzawa algorithm and the smoothed pressure correction method. The appearing Poisson equations are solved in a meshless way by either local or re-globalized schemes.

In the latter case, a quadtree based multigrid method is used to define biharmonic interpolation functions. To eliminate the inner interpolation points, the Method of Fundamental Solutions was also used. Unfortunately, the source and the collocation points should not coincide because of the singularity of the Stokes fundamental solution. On the other hand, if the source points are located far from the boundary, the resulting linear system becomes extremely ill-conditioned. To achieve a compromise, a regularized version the Method of Fundamental Solutions was introduced which approximates the Stokes operator by a higher order operator containing a scaling parameter to fine tune the approximation. Thus, the source and collocation points are now allowed to coincide, which improves the condition number. Finally, the regularized MFS-solution was approximated by a direct multi-elliptic interpolation technique using quadtrees and multi-level tools. This makes it possible to completely avoid the use of large, dense and ill-conditioned systems, and significantly reduces the computational cost.

References

1. Alves, C.J.S., Chen, C.S., Sarler, B. "The Method of Fundamental Solutions for Solving Poisson Problems", *International Series on Advances in Boundary Elements*, Vol. **13**. Proceedings of the 24th International Conference on the Boundary Element Method incorporating Meshless Solution Seminar (ed. by C.A. Brebbia A. Tadeu, V. Popov), pp. 67–76, *WitPress*, Southampton, Boston, MA, 2002.
2. Alves, C.J.S., Silvestre, A.L. "Density results using Stokeslets and a method of fundamental solutions for the Stokes equations", *Engineering Analysis with Boundary Elements*, Vol. **28**, pp. 1245–1252, 2004.
3. Benzi, M., Golub, G.H., Liesen, J. "Numerical solution of saddle point problems", *Acta Numerica*, Vol. **14**, pp. 1–137, 2005.
4. Bramble, J.H., Pasciak, J.E., Vassilev, A.T. "Analysis of the inexact Uzawa algorithm for saddle point problems", *SIAM Journal on Numerical Analysis*, Vol. **34**, pp. 1072–1092, 1997.
5. Chen, W., Shen, L.J., Shen, Z.J., Yuan, G.W. "Boundary knot method for Poisson equations", *Engineering Analysis with Boundary Elements*, Vol. **29**, No. 8, pp. 756–760, 2005.
6. Gáspár, C. "Multi-level biharmonic and bi-Helmholz interpolation with application to the boundary element method", *Engineering Analysis with Boundary Elements*, Vol. **24**, No. 7–8, pp. 559–573, 2000.
7. Gáspár, C. "Fast Multi-Level Meshless Methods Based on the Implicit Use of Radial Basis Functions", *Lecture Notes in Computational Science and Engineering*, Vol. **26**, pp. 143–160, *Springer*, Berlin/Heidelberg/New York, 2002.
8. Gáspár, C: "A meshless polyharmonic-type boundary interpolation method for solving boundary integral equations", *Engineering Analysis with Boundary Elements*, Vol. 28, No. 10, pp. 1207–1216, 2004.
9. Gáspár, C. "Global and Local Multi-Level Meshless Schemes Based on Multi-Elliptic Interpolation", *Proceedings of ECCOMAS Thematic Conference on Meshless Methods, held in Lisbon, Portugal, July 11–14* (ed. by V.M.A. Leitao, C.J.S. Alves, C.A. Duarte), pp. B12.1–B12.6, 2005.
10. Gáspár, C. "Meshless Boundary Interpolation: Local and Global Multi-Level Techniques", *Advances in Boundary Element Techniques VII. Proceedings of the International Conference on Boundary Element Techniques VII, held in Paris, France, September 4–6* (ed. by B. Gatmiri, A. Sellers, M.H. Aliabadi), pp. 73–78, 2006.

11. Golberg, M.A., Chen, C.S. "A bibliography on radial basis function approximation", *Boundary Element Communications*, Vol. **7**, No. 4, pp. 155–163, 1996.
12. Golberg, M.A. "The method of fundamental solutions for Poisson's equation", *Engineering Analysis with Boundary Elements*, Vol. **16**, pp. 205–213, 1995.
13. Kansa, E.J. "Multiquadrics - a scattered data approximation scheme with applications to computational fluid dynamics - I.–II.", *Computers Mathematics with Applications* Vol. **19**, No. 8/9, pp. 127–161, 1990.
14. Temam, R. "Navier-Stokes Equations", *North-Holland*, New York, 1977.
15. Young, D.L., Jane, S.J., Fan, C.M., Murugesan, K., Tsai, C.C. "The method of fundamental solutions for 2D and 3D Stokes problems", *Journal of Computational Physics*, Vol. **211**, No. 1, pp. 1–8, 2006.

Orbital HP-Clouds for Quantum Systems

Jiun-Shyan (JS) Chen(✉) and Wei Hu

Abstract Solving Schrödinger equation in quantum mechanics presents a challenging task in numerical methods due to the high order behavior and high dimension characteristics in the wave functions. This work introduces orbital and polynomial enrichment functions to the partition of unity for solution of Schrödinger equation under the framework of HP-Clouds. An intrinsic enrichment of orbital function and extrinsic enrichment of monomial functions are proposed. Due to the employment of higher order basis functions, a higher order stabilized conforming nodal integration is developed. The proposed methods are implemented using the density functional theory for solution of Schrödinger equation. Analysis of hydrogen systems and a spherical quantum dot demonstrates the effectiveness of the proposed method.

Keywords: HP-Clouds · Quantum Systems · Schrodinger equation

1 Introduction

During the last few decades, first principle calculation for electronic structures is primarily performed using basis set calculation under ab initio framework. A number of basis functions have been constructed and used for different quantum systems, such as Gaussian type orbital functions (GTOs) and Slater type orbital functions (STOs) in traditional ab initio calculation [11], and plane-wave orbital functions (PW) for density functional theory [14, 20]. GTOs and STOs are commonly used for the small- and medium-sized atoms and molecules [26]. The advantage of using these orbital functions is that all of the integrals for computing the Hamiltonian matrix elements can be done analytically. However, these orbital basis functions are global functions, which yield a full sized Hamiltonian matrix [24]. PW functions are

J.-S. Chen and W. Hu
Department of Civil and Environmental Engineering, University of Califormia, Los Angeles, CA 90095-1593, USA, e-mail: jschen@seas.ucla.edu

A.J.M. Ferreira et al. (eds.) *Progress on Meshless Methods, Computational Methods in Applied Sciences.*

efficient for density functional theory, especially for crystal structures [20]. Nevertheless, the resolution of PW can not be controlled locally, thus a very large number of PW functions are needed in order to obtain reliable solution near the atomic core. Further, PW approach can only deal with periodic structures, which is not desirable for molecular systems. Finite difference method (FDM) [1–3] and finite element method (FEM) [12, 13, 19, 20, 22–28] have also been introduced in real-space ab initio calculation, but they require large degrees of freedom for reasonable accuracy. Pickett [20] introduced FEM for computing all-electron and pseudo-potential formulations under density functional theory. Tsuchida et al. [23] proposed an efficient scheme by introducing adaptive curvilinear coordinates into FEM to vary the grid logarithmically near the nuclei for accurate representation of wave functions. Recently, Yamakawa et al. [27, 28] introduced Gaussian-FEM to enrich standard FEM with special GTOs orbital basis functions for increased accuracy in describing the core electron, which, however, makes the approximation global.

In this work we introduce orbital intrinsic enrichment function and monomial extrinsic enrichment functions of partition of unity under HP-Clouds framework [7–9], [15–17] for solving Schrödinger equation in quantum mechanics. In the proposed OHPC method, orbital basis functions are reproduced everywhere in the domain of quantum system, while monomial basis functions are introduced as an additional enhancement of orbital basis functions through extrinsic enrichment to allow varying order of p-refinement. Domain integration is an important task in meshfree method constructed based on Galerkin weak form. Chen et al. [5, 6] proposed a stabilized conforming nodal integration (SCNI) as a stabilization of rank instability in nodal integration, and as a mechanism to pass linear patch test. In this work we introduce a higher order extension of SCNI for the proposed orbital HP-Clouds. Application to quantum dot semiconductor is also discussed. An interface enrichment function is introduced for enhanced accuracy in approximating the Ben-Daniel interface conditions.

This paper is arranged as follows. The basic equations in quantum mechanics are outlined in Section 2. The proposed orbital HP-Clouds (OHPC) method and the corresponding stabilized quadrature rule are presented in Section 3. The interface enrichment function for the OHPC method is proposed in Section 4. Applications to hydrogen systems and quantum dot semiconductor are introduced in Sections 5. Concluding remarks are given in Section 6.

2 Basic Equations in Quantum Mechanics

The most fundamental equation in quantum mechanics is the Schrödinger equation that governs the wave function based on energy conservation:

$$\hat{H}(\mathbf{r}_1, \mathbf{r}_2, \ldots, \mathbf{r}_N, t)\bar{\Theta}(\mathbf{r}_1, \mathbf{r}_2, \ldots, \mathbf{r}_N, t) = i\hbar \frac{\partial}{\partial t}\bar{\Theta}(\mathbf{r}_1, \mathbf{r}_2, \ldots, \mathbf{r}_N, t), \tag{1}$$

where $\bar{\Theta}$ is the wave function, $\hat{H}$ is the Hamiltonian operator, $\mathbf{r}_k$ is the position vector of the k-th particle, N is the total number of particles, $i\hbar\frac{\partial}{\partial t}$ is total energy operator, i is the imaginary unit, and $\hbar$ is the Plank constant. In the case where the potential field is time-independent, the static Schrödinger equation for atomic/molecular systems can be derived from Eq. (1) by taking the form of total wave function $\bar{\Theta}(\mathbf{r}_1,\mathbf{r}_2,\ldots,\mathbf{r}_N,t) = \Theta(\mathbf{r}_1,\mathbf{r}_2,\ldots,\mathbf{r}_N)\Pi(t)$ to yield

$$H(\mathbf{r}_1,\mathbf{r}_2,\ldots,\mathbf{r}_N)\Theta(\mathbf{r}_1,\mathbf{r}_2,\ldots,\mathbf{r}_N) = E\Theta(\mathbf{r}_1,\mathbf{r}_2,\ldots,\mathbf{r}_N), \tag{2}$$

where E is the total energy. The Born-Oppenheimer approximation [21] can be used to separate the electron wave function Θ_e from the total wave function Θ by an assumption that the nuclei are nearly stationary in space with respect to the motion of electrons. Therefore, Eq. (2) can be simplified as

$$H_e\Theta_e(\mathbf{r},\mathbf{R}) = E_e\Theta_e(\mathbf{r},\mathbf{R}), \tag{3}$$

where H_e is the electron Hamiltonian operator, and E_e is the total energy of electrons. In the traditional *ab initio* calculation using Hartree-Fock molecular orbital approximation [11], the total electron wave function Θ_e is approximated by a set of single electron wave functions, and Eq. (3) can be simplified by a set of uncoupled single electron Schrödinger equation by introducing an average electron density function to represent the Coulomb interaction between any of two electrons:

$$H_i\Theta_i = E_i\Theta_i, \tag{4}$$

where each i represents a single electron. Alternatively, the density functional theory (DFT) [18] that describes the electron motion by electron density instead of wave function results in a system with greatly reduced degrees of freedom. In the Kohn-Sham approximation of DFT, the governing equation for Kohn-Sham wave function $\Theta_i^{KS}(\rho)$ can be expressed in atomic units as:

$$\left\{-\frac{1}{2}\nabla^2 + \hat{V}_{eff}(\rho(\mathbf{r}))\right\}\Theta_i^{KS}(\rho(\mathbf{r})) = \mathscr{E}_i\Theta_i^{KS}(\rho(\mathbf{r})), \tag{5}$$

where $\mathscr{E}_i$ is the i-th Kohn-Sham energy corresponding to Kohn-Sham wave function Θ_i^{KS}, and $\hat{V}_{eff}(\rho(\mathbf{r}))$ is effective potential expressed by electron density. Both classical Schrödinger equation in (4) and DFT based Schrödinger equation in (5) exhibit highly nonlinear behavior in the region close to the nuclei.

The commonly used basis set calculation approximates the wave function by the linear combination of orbital functions $g_k(\mathbf{r})$

$$\Theta_i(\mathbf{r}) = \sum_k c_{ki}\, g_k(\mathbf{r}), \tag{6}$$

In atomic/molecular systems, Gaussian type orbital functions (GTOs) and Slater type orbital functions (STOs) have shown to provide an accurate description for electron motion. For example, three-dimensional STOs have the following expressions:

$$g_{1s} = \left(\frac{\xi_1^3}{\pi}\right)^{1/2} \exp(-\xi_1 r), \quad g_{2s} = \left(\frac{\xi_2^5}{96\pi}\right)^{1/2} r\exp(\frac{-\xi_2 r}{2}),$$

$$g_{2p_x} = \left(\frac{\xi_2^5}{32\pi}\right)^{1/2} x\exp\left(\frac{-\xi_2 r}{2}\right), \tag{7}$$

where $1s, 2s, 2p_x$ are the standard symbols in quantum physics denoting the first principle orbital (spherically symmetric), the first sub-level (spherically symmetric) of the second principle orbital and the second sub-level (x-symmetric) of the second principle orbital, etc., r is the distance measured from the nucleus to the electron, and ξ_i is the scaling parameter representing the size of the atom/molecule. However, these orbital functions are global functions, leading to a full matrix in basis set calculation.

3 Orbital HP-Clouds and Quadrature Rule

To embed fundamental characteristics of the quantum system into the numerical approximation function, an intrinsic enrichment with orbital functions has been introduced [4]. For demonstration purpose, the first STO function $e^{-\xi r}$ in (7) is used herein. Other orbital functions can also be considered. We start with the following approximation of wave function:

$$\Theta^h(\mathbf{r}) = \sum_I \Psi_I \alpha_I$$

$$\Psi_I(\mathbf{r}) = \Phi_I(\mathbf{r})\left\{b_0(\mathbf{r}) + e^{-\xi r_I}b_1(\mathbf{r})\right\}, \quad r_I = \|\mathbf{r}_I\|, \tag{8}$$

where $\Phi_I(\mathbf{r})$ is a kernel function with finite cover (support). The coefficients $b_0(\mathbf{r})$ and $b_1(\mathbf{r})$ are to be obtained by enforcing reproducibility of 1 and $e^{-\xi r}$:

$$\sum_{I=1}^N \Psi_I(\mathbf{r}) = 1, \quad \sum_{I=1}^N \Psi_I(\mathbf{r})e^{-\xi r_I} = e^{-\xi r}, \tag{9}$$

Thus we have the following two equations:

$$\left\{\sum_{I=1}^N \Phi_I(\mathbf{r})\begin{bmatrix}1\\e^{-\xi r_I}\end{bmatrix}\begin{bmatrix}1 & e^{-\xi r_I}\end{bmatrix}\right\}\begin{bmatrix}b_0(\mathbf{r})\\b_1(\mathbf{r})\end{bmatrix} = \begin{bmatrix}1\\e^{-\xi r}\end{bmatrix}. \tag{10}$$

The resulting intrinsically enriched function is

$$\Psi_I(\mathbf{r}) = \mathbf{q}^T(\mathbf{r}_I)\mathbf{A}^{-1}(\mathbf{r})\mathbf{q}(\mathbf{r})\Phi_I(\mathbf{r}), \tag{11}$$

$$\mathbf{A}(\mathbf{r}) = \sum_{I=1}^N \Phi_I(\mathbf{r})\mathbf{q}(\mathbf{r}_I)\mathbf{q}^T(\mathbf{r}_I), \quad \mathbf{q}^T(\mathbf{r}) = \begin{bmatrix}1, e^{-\xi r}\end{bmatrix}, \mathbf{b}^T(\mathbf{r}) = [b_0(\mathbf{r}), b_1(\mathbf{r})].$$

$$\tag{12}$$

The global approximation of the wave function Θ, denoted as Θ^h, is constructed by using the set of partition of unity functions $\{\Psi_I(\mathbf{r})\}_{I=1}^N$ and monomial extrinsic enrichment functions as

$$\Theta^h(\mathbf{r}) = \sum_{I=1}^N \Psi_I(\mathbf{r}) \left\{ \sum_{i=1}^n P_i(\mathbf{r} - \mathbf{r}_I) \, \alpha_I^i \right\}, \tag{13}$$

where $\{P_i(\mathbf{r} - \mathbf{r}_I)\}_{i=1}^n$ is a set of shifted monomial functions. For easy discussion in the numerical example, we rewrite the proposed orbital HP-Cloud (OHPC) approximation of wave function as follows:

$$\Theta^h(\mathbf{r}) = \sum_{I=1}^N \Psi_I(\mathbf{r}) \, \mathbf{h}_I^{ex^T} \alpha_I, \quad \Psi_I(\mathbf{r}) = \Phi_I(\mathbf{r}) \, \mathbf{h}_I^{in^T} \mathbf{b}, \tag{14}$$

where $\mathbf{h}_I^{in}$ and $\mathbf{h}_I^{ex}$ are the vectors of intrinsic and extrinsic basis functions, respectively, and $\mathbf{b}$ and α_I are the corresponding coefficient vectors.

The Galerkin approximation of this problem is to find $\Theta^h \in H_0^1$, $\forall \Lambda^h \in H_0^1$, such that

$$\int_\Omega \left[\frac{\nabla \Lambda^h(\mathbf{r}) \cdot \nabla \Theta^h(\mathbf{r})}{2} + \Lambda^h(\mathbf{r}) \Theta^h(\mathbf{r}) \hat{V}(\mathbf{r}) \right] d\Omega$$
$$= \varepsilon^h \int_\Omega \left[\Lambda^h(\mathbf{r}) \Theta^h(\mathbf{r}) \right] d\Omega \tag{15}$$

where Θ and Λ are test and trial functions kinematically admissible to the homogeneous boundary conditions on $\partial\Omega$. Using a layer of finite element mesh on the boundary is a straightforward way to impose homogeneous boundary conditions. To achieve a higher order accuracy in domain integration of weak form of Schrödinger equation with orbital HP-Cloud approximation, we introduce a correction to stabilized confirming nodal integration (SCNI) [5]. First, consider the following identity associated with Laplace operator:

$$\int_\Omega \nabla \Lambda^h(\mathbf{r}) \cdot \nabla \Theta^h(\mathbf{r}) d\Omega$$
$$= \underbrace{\sum_{L=1}^N \bar{\nabla} \Lambda^h(\mathbf{r}_L) \cdot \bar{\nabla} \Theta^h(\mathbf{r}_L) w_L}_{\text{SCNI}} \tag{16}$$
$$+ \underbrace{\sum_{L=1}^N \int_{\Omega_L} \left(\nabla \Lambda^h(\mathbf{r}) - \bar{\nabla} \Lambda^h(\mathbf{r}_L) \right) \cdot \left(\nabla \Theta^h(\mathbf{r}) - \bar{\nabla} \Theta^h(\mathbf{r}_L) \right) d\Omega}_{\text{correction term}}$$

Now one could use n-th order Gauss integration on left or right side of (16) and expect to get the same results, provided that $\bar{\nabla} \Theta^h(\mathbf{r}_L)$ (and $\bar{\nabla} \Lambda^h(\mathbf{r}_L)$) is integrated

Fig. 1 Nodal representative
domain

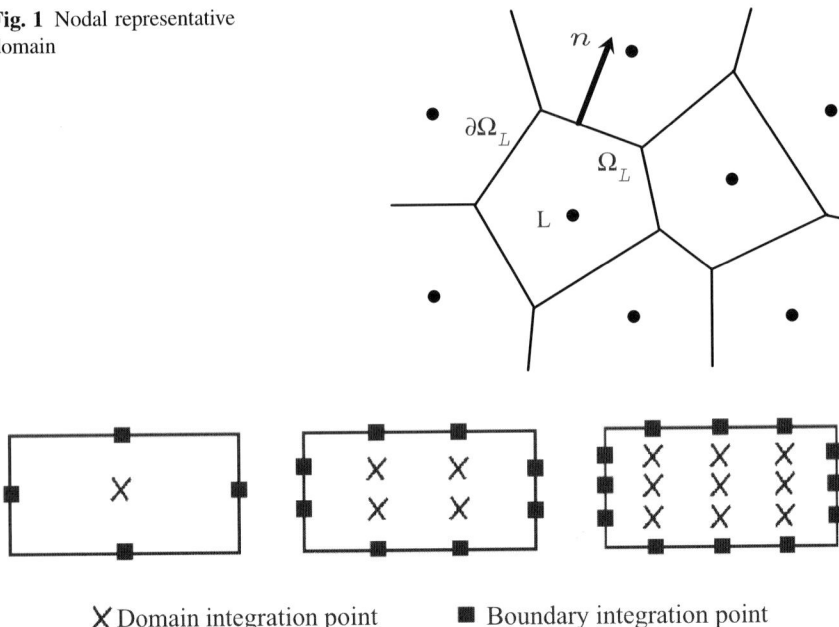

X Domain integration point   ■ Boundary integration point

Fig. 2 Consistent domain and boundary integration points in HSCNI

using the same Gauss rule. If Θ is approximated by polynomial functions, it is well known that the first order patch test can not be recovered using standard Gauss integration in (16). However $\bar{\nabla}\Theta^h(\mathbf{r}_L)$ (and $\bar{\nabla}\Lambda^h(\mathbf{r}_L)$) is computed numerically by its divergence counterpart:

$$\bar{\nabla}\Theta^h(\mathbf{r}_L) = \frac{1}{w_L}\int_{\Omega_L}\nabla\Theta^h d\Omega = \frac{1}{w_L}\int_{\partial\Omega_L}\Theta^h\mathbf{n}d\Gamma, \quad w_L = \int_{\Omega_L}d\Omega, \qquad (17)$$

linear patch test is met. Here Ω_L is the nodal representative domain, which can be obtained from triangulation or Voronoi cell of a set of discrete points as shown in Fig. 1. Since in linear patch test Θ is linear, $\bar{\nabla}\Theta(\mathbf{r}) - \bar{\nabla}\Theta^h(\mathbf{r}_L) = 0$, and (16) reduces to standard SCNI and thus passes linear patch test, regardless of the order of quadrature rule used in the boundary integral of $\bar{\nabla}\Theta^h(\mathbf{r}_L)$ according to Chen et al. [5].

To achieve higher order accuracy, the second term on the right hand side of (16) is integrated using n-th order Gauss integration and $\bar{\nabla}\Theta^h(\mathbf{r}_L)$ and $\bar{\nabla}\Lambda^h(\mathbf{r}_L)$ are computed by the divergence counterpart in Eq. (17). To pass linear patch test, the boundary integral of (17) should be consistent with Gauss quadrature rule used in domain integration for the second term of right hand side of (16) as shown in Fig. 2. For irregular shape nodal domain, triangular quadrature rules can be used.

The Schrödinger equation in (15) is now integrated according to HSCNI in (16) to yield

$$\frac{1}{2}\sum_{L=1}^{N}\bar{\nabla}\Lambda^{h}\left(\mathbf{r}_{L}\right)\cdot\bar{\nabla}\Theta^{h}\left(\mathbf{r}_{L}\right)w_{L}$$

$$+\frac{1}{2}\sum_{L=1}^{N}\int_{\Omega_{L}}\left(\nabla\Lambda^{h}\left(\mathbf{r}\right)-\bar{\nabla}\Lambda^{h}\left(\mathbf{r}_{L}\right)\right)\cdot\left(\nabla\Theta^{h}\left(\mathbf{r}\right)-\bar{\nabla}\Theta^{h}\left(\mathbf{r}_{L}\right)\right)d\Omega \qquad (18)$$

$$+\sum_{L=1}^{N}\int_{\Omega_{L}}\Lambda^{h}\left(\mathbf{r}\right)\hat{V}\Theta^{h}\left(\mathbf{r}\right)d\Omega=\varepsilon^{h}\sum_{L=1}^{N}\int_{\Omega_{L}}\Lambda^{h}\left(\mathbf{r}\right)\Theta^{h}\left(\mathbf{r}\right)d\Omega$$

HSCNI is able to achieve higher order accuracy and optimal convergence rate by increasing the order of Gaussian integration for the correction term in (16) and force vector, and at the same time, pass linear patch test. To demonstrate that, consider a 2-dimensional Poisson equation:

$$\begin{cases} \nabla^2 u\left(x,y\right)=-2\pi^2\sin\left(\pi x\right)\sin\left(\pi y\right) & \left(x,y\right)\in\left(0,1\right)\times\left(0,1\right)\equiv\Omega \\ u=0 & \partial\Omega \end{cases} \qquad (19)$$

The analytical solution of this problem is $\sin\left(\pi x\right)\sin\left(\pi y\right)$. We employed HP-Cloud approximation with complete linear and quadratic monomial bases in the intrinsic enrichment. Third order Gauss quadrature rule is employed in HSCNI. The convergence rate of the L2 error norm in Fig. 3 shows that a better accuracy and convergence rate of HSCNI than those of SCNI for both analyses using linear and quadratic polynomial bases, especially when higher order basis functions are employed.

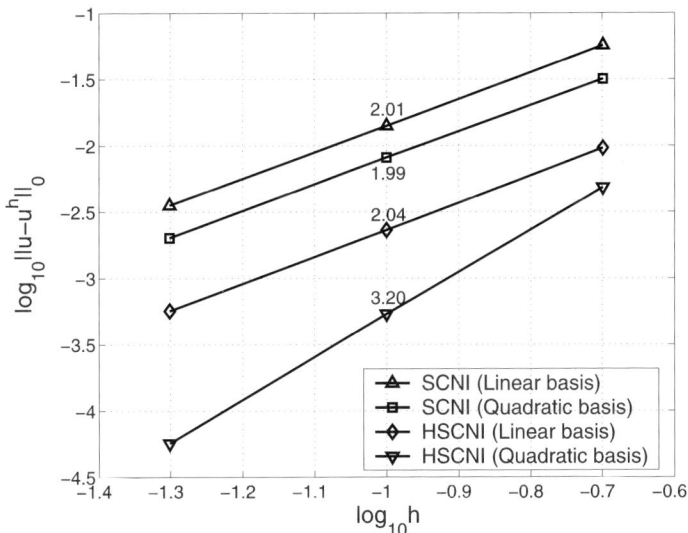

Fig. 3 Convergence rate of L2 error norm in 2D Poisson equation

4 Application to Quantum Dot Semiconductor

In quantum mechanics, the wave functions Θ are continuous everywhere. In quantum dots, due to the difference of the effective mass across the interface of the heterojunction (Fig. 4), the Ben-Daniel interface condition is introduced:

$$\frac{1}{m^{*+}}\frac{\partial\Theta}{\partial\mathbf{n}^{+}}\big|_{\Gamma^I} + \frac{1}{m^{*-}}\frac{\partial\Theta}{\partial\mathbf{n}^{-}}\big|_{\Gamma^I} = 0, \tag{20}$$

where Γ^I is the interface, m^{*+} and m^{*-} are the effective mass of two different materials, and $\mathbf{n}^{+}$ and $\mathbf{n}^{-}$ are normal directions on the interface corresponding to different material domain. Here, a jump in $\frac{\partial\Theta}{\partial\mathbf{n}}$ is necessary to ensure continuous current flow across the interface so as to preserve the Hermitian symmetry of the Hamiltonian operator.

We introduce a modified construction of the OHPC approximation of the wave function to yield a discontinuous derivative in the wave function across the interface in the quantum dot. Consider the set of nodal points S is separated into two sets S_1 and S_2 by the interface, where the nodes in S_1 are inside the material domain and those in S_2 are right on the interface. The OHPC approximation of wave function in Eq. (14) is modified as follows:

$$\Theta^h(\mathbf{r}) = \sum_{I:\mathbf{r}_I\in S_1} \Psi_I(\mathbf{r})\mathbf{h}_I^{ex^T}(\mathbf{r})\alpha_I + \sum_{I:\mathbf{r}_I\in S_2} \hat{\Psi}_I(\mathbf{r})\mathbf{h}_I^{ex^T}(\mathbf{r})\alpha_I \tag{21}$$

where $\hat{\Psi}_I(\mathbf{r}(\tilde{r},\tilde{s})) = \varphi(\tilde{r})\phi_I(\tilde{s})$ is the interface enrichment function [25] defined at the discrete points on the interface, and $\tilde{r}$ and $\tilde{s}$ are the coordinates normal to and along the interface, respectively, as shown in Fig. 5(c). The function $\varphi'(\tilde{r}) \in C^{-1}$ possesses derivative discontinuities across the interface, and $\phi_I(\tilde{s})$ is a smooth kernel function along the interface as demonstrated in Fig. 5(a, b). $\Psi_I(\mathbf{r})$ is the intrinsically enriched function introduced in Eq. (11). The reproducing condition in Eq. (10) is modified as follows:

$$\sum_{I:\mathbf{r}_I\in S_1} \Psi_I(\mathbf{r})\mathbf{q}(\mathbf{r}_I) + \sum_{I:\mathbf{r}_I\in S_2} \hat{\Psi}_I(\mathbf{r})\mathbf{q}(\mathbf{r}_I) = \mathbf{q}(\mathbf{r}) \tag{22}$$

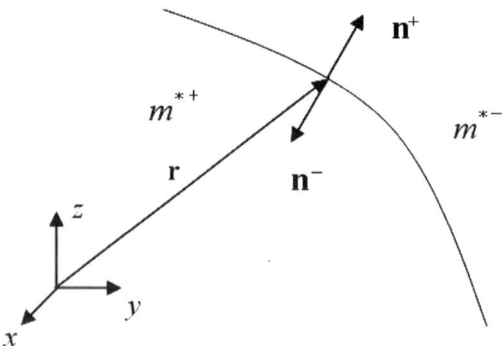

Fig. 4 Interface of heterojunction

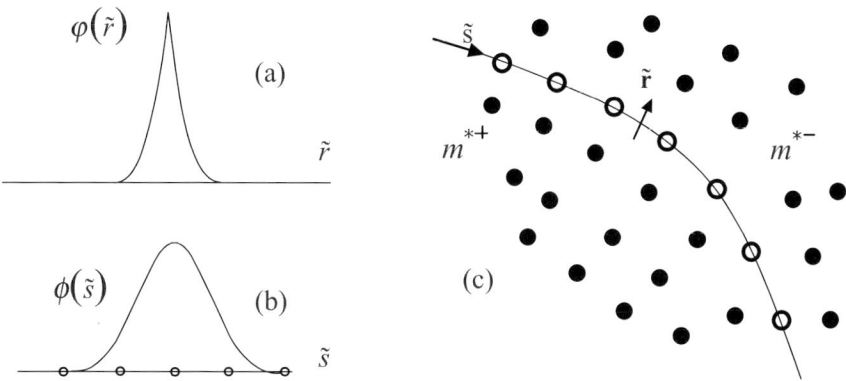

Fig. 5 Discretization of interface and different kernel functions for two perpendicular directions with different orders of smoothness

where $\mathbf{q}(\mathbf{r})$ is basis function vector with orbital function enrichment defined in Eq. (12). Thus, the coefficient vector $\mathbf{b}(\mathbf{r})$ in Eq. (12) is obtained as:

$$\mathbf{b}(\mathbf{r}) = \mathbf{A}^{-1}(\mathbf{r})\left[\mathbf{q}(\mathbf{r}) - \hat{\mathbf{F}}(\mathbf{r})\right] \tag{23}$$

where the moment matrix $\mathbf{A}(\mathbf{r})$ is defined in Eq. (12), and

$$\hat{\mathbf{F}}(\mathbf{r}) = \sum_{I:\mathbf{r}_I \in S_2} \hat{\Psi}_I(\mathbf{r})\mathbf{q}(\mathbf{r}_I) \tag{24}$$

Finally the intrinsically enriched function is

$$\Psi_I(\mathbf{r}) = \mathbf{q}^T(\mathbf{r}_I)\mathbf{A}^{-1}(\mathbf{r})\left[\mathbf{q}(\mathbf{r}) - \hat{\mathbf{F}}(\mathbf{r})\right]\Phi_I(\mathbf{r}) \tag{25}$$

Since $\hat{\Psi}_I(\mathbf{r})$ is introduced locally around the interface, for a point of evaluation $\mathbf{r}$ that is not covered by the influence zone of the interface enrichment function, $\hat{\Psi}_I(\mathbf{r})$ vanishes, and Eq. (25) reduces to the standard intrinsically enriched function in Eq. (11).

5 Numerical Examples

5.1 Hydrogen Systems

Three hydrogen systems as shown in Fig. 6 are analyzed. The solutions obtained from the proposed OHPC are compared with those obtained from third order p-adaptive FEM [26] and FEM approximation enriched with Gaussian function (Gaussian-FEM) [28] in the following study. Note that Gaussian-FEM is a global approximation and thus yields a full matrix in its discrete equation, while the proposed OHPC yields a banded matrix.

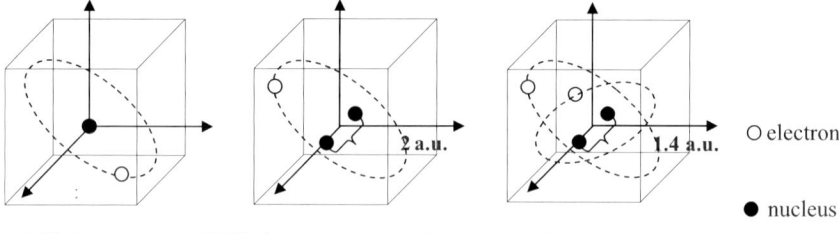

(a) Hydrogen atom   (b) Hydrogen molecular ion   (c) Hydrogen molecule

Fig. 6 Three quantum systems and computational domains

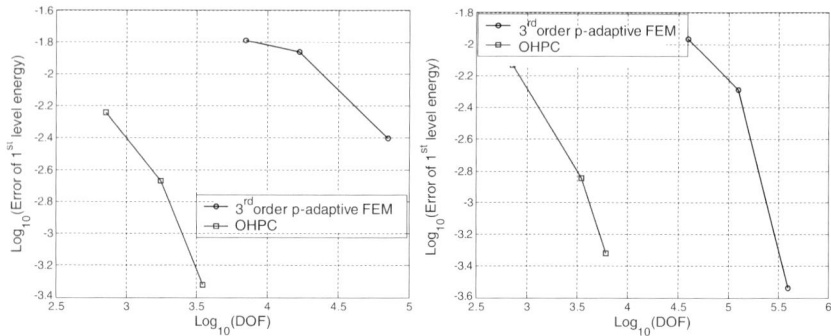

Fig. 7 Numerical error of the first level energy of (**a**) the hydrogen atom (**b**) the hydrogen molecular ion

5.1.1 Hydrogen Atom

The cut-off boundary $|x_i| \leq 10$ a.u. (atomic unit) is used as the numerical domain. The Schrödinger equation (4) is solved using OHPC with the following enrichments:

$$\mathbf{h}_I^{in} = [1,\ e^{-\xi r}], \mathbf{h}_I^{ex} = [1], \tag{26}$$

where $\xi = 1$ for Hydrogen atom. In Fig. 7(a), the orbital HP-Cloud solution is compared to the solution obtained from the third order p-adaptive finite element method [26]. The numerical results clearly show the superior performance of the proposed OHPC method compared to the third order p-adaptive finite element method. Note that since $e^{-\xi r}$ is the 3-dimensional characteristic function of the Hydrogen atom, the numerical error of this problem using HP-Cloud approximation is due to domain integration.

5.1.2 Hydrogen Molecular Ion

Since this quantum system contains two nuclei, we consider the following OHPC approximation:

Table 1 Numerical error of the first level energy for Hydrogen molecule

Numerical methods	No. of nodes	DOFs	Error of first level energy (%)
Gaussian-FEM [28]	12,167	12,167	0.355
OHPC $\mathbf{h}_I^{in} = [1, \frac{x_1-x_{1I}}{c_1}, \frac{x_2-x_{2I}}{c_2}, \frac{x_3-x_{3I}}{c_3}]$	729	979	1.03
$\mathbf{h}_I^{ex} = \begin{cases} [1, e^{-r_1}, e^{-r_2}], & \|x_i\| \leq 3 \text{ a.u.} \\ [1], & \|x_i\| > \text{a.u.} \end{cases}$	1,331	2,017	0.27
OHPC $\mathbf{h}_I^{in} = [1, e^{-r_{1I}}, e^{-r_{2I}}]$	729	1,104	0.29
$\mathbf{h}_I^{ex} = \begin{cases} [1, \frac{x_1-x_{1I}}{c_1}, \frac{x_2-x_{2I}}{c_2}, \frac{x_3-x_{3I}}{c_3}], & \|x_i\| \leq 3 \text{ a.u.} \\ [1], & \|x_i\| > \text{a.u.} \end{cases}$	1,331	2,360	0.16

$$\mathbf{h}_I^{in} = [1, e^{-\xi r_1}, e^{-\xi r_2}], \quad \mathbf{h}_I^{ex} = \left[1, \frac{x_I - x_{1I}}{c_1}, \frac{x_I - x_{2I}}{c_2}, \frac{x_I - x_{3I}}{c_3}\right], \quad (27)$$

where c_i is the dimension of the cover of kernel function in the i-th direction, which is chosen in this example as $1.5a$ where a is nodal distance; $r_1 = \|\mathbf{r} - \mathbf{R}_1\|$ and $r_2 = \|\mathbf{r} - \mathbf{R}_2\|$ are distance measured from the I-th nodal point to the locations of the two nuclei, $\mathbf{R}_1$ and $\mathbf{R}_2$. In this problem, the orbital bases for Hydrogen atom are only approximations, and linear monomial functions are used as an extrinsic enrichment of OHPC. The results of the first level energy obtained from OHPC and the third order p-adaptive FEM compared in Fig. 7(b) show a better solution accuracy by using the proposed OHPC method.

5.1.3 Hydrogen Molecule

This quantum system contains two electrons, and thus Kohn-Sham Schrödinger equation based on the density functional theory with Newton iteration is employed. The proposed OHPC performs superior to Gaussian-FEM that combines Gaussian orbital function with FEM approximation [28] in Table 1. The results also show that using orbital functions intrinsically and polynomial functions extrinsically is more effective than using polynomial functions intrinsically and orbital functions extrinsically. For OHPC methods, the extrinsic enrichment is employed only near the nuclei $\|x_i\| \leq 3$ a.u. Figure 8 shows the electron density distribution $\rho = \Theta^2$ obtained by the proposed OHPC method.

5.2 Spherical Quantum Dot Semiconductor

Consider a spherically symmetric semiconductor composed of two materials *GaAs* and *InAs* as shown in Fig. 9, where R_1 is the radius of dot and R_2 is the radius of boundary where the density of electron is assumed to be zero. In this example, quantum dots with radius $R_1 = 300$, and $R_2 = R_1 + 100$ [10] are considered. Schrödinger

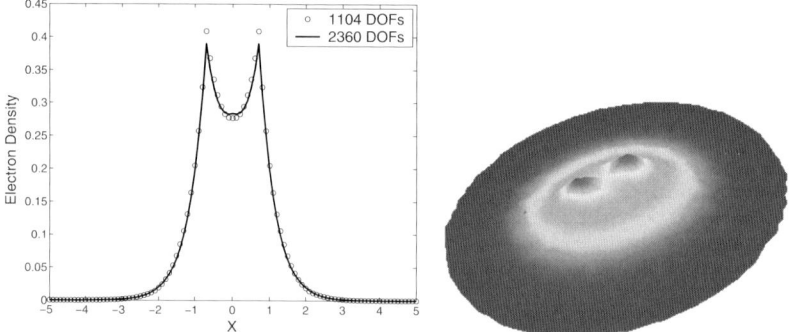

Fig. 8 Electron density distribution of the hydrogen molecule: (**a**) Electron density along the x axis $(y = 0, z = 0)$, (**b**) electron density contour on x-y plane $(z = 0)$

Fig. 9 Spherical quantum dot

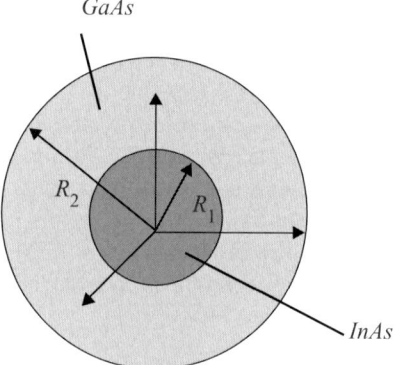

equation for a constant effective mass can be written as

$$-\frac{\hbar^2}{2m^*(r)}\nabla^2\Theta(r) + V(r)\Theta(r) = E\Theta(r), \quad r \in [0, R_1) \cup (R1, R_2),$$

$$\frac{1}{m^*_{InAs}}\frac{\partial\Theta}{\partial\mathbf{n}_{InAs}}\mid_{\Gamma^I} + \frac{1}{m^*_{GaAs}}\frac{\partial\Theta}{\partial\mathbf{n}_{GaAs}}\mid_{\Gamma^I} = 0, \quad r = R_1, \qquad (28)$$

$$\Theta(r) = 0, \quad r = R_2,$$

where Θ is wave function, E is energy, and Γ^I is the interface. The effective mass m^* and the confinement potential V for two materials are given as:

$$m^*(r) = \begin{cases} m^*_{InAs} = 0.023\, m_0, & r < R_1 \\ m^*_{GaAs} = 0.067\, m_0, & R_1 < r < R_2 \end{cases}, \quad V(r) = \begin{cases} 0, & r < R_1 \\ 0.77eV, & R_1 < r < R_2 \end{cases}, \quad (29)$$

where m_0 is the mass of electron in the vacuum. Equation (28) can be simplified to one dimensional equation along radial direction:

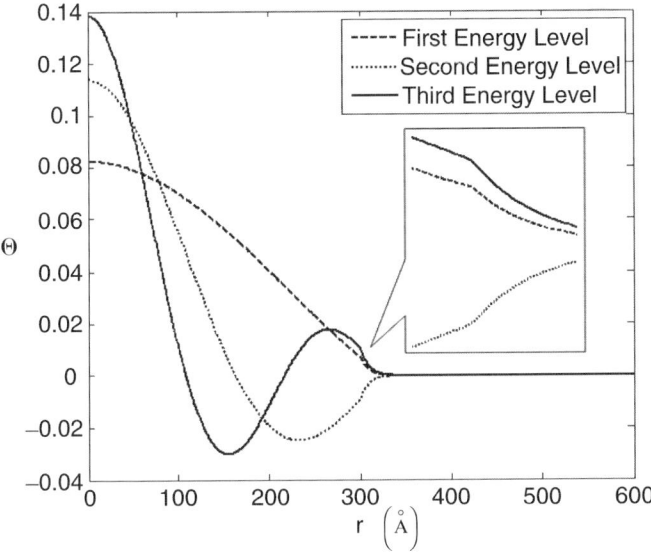

Fig. 10 Wave function of the first three energy levels of spherical QDs solved by OHPC with interface enrichment

$$-\frac{\hbar^2}{2m^*}\left(\frac{2}{r}\frac{\partial}{\partial r}+\frac{\partial^2}{\partial r^2}\right)\Theta(r)+V(r)\Theta(r)=E\Theta(r). \tag{30}$$

Figure 10 shows the solution of the wave function corresponding to the first three energy levels by using OHPC approximation with the interface enrichment function, which well captured the discontinuity of the derivative of wave function across the interface ($r = 300\text{Å}$). Figure 11(a–c) compares the results of OHPC with linear and quadratic intrinsic basis (without extrinsic enrichment) to the results of linear and quadratic FEM, where OHPC solutions are much more accurate than FEM solutions. Figure 11(d) shows the first three energy levels associated with the radius of dot, which varies from 25 to 300, by using OHPC with quadratic intrinsic basis.

6 Conclusion Remarks

We present an orbital HP-Cloud (OHPC) approximation for quantum systems, in which orbital basis functions are reproduced everywhere in the domain of quantum system, while monomial basis functions are introduced as an additional enhancement of orbital basis functions through extrinsic enrichment to allow varying order of p-refinement. Note the the proposed OHPC approximation has finite support and it yields a banded discrete Laplace operator in the Schrödinger equation. It

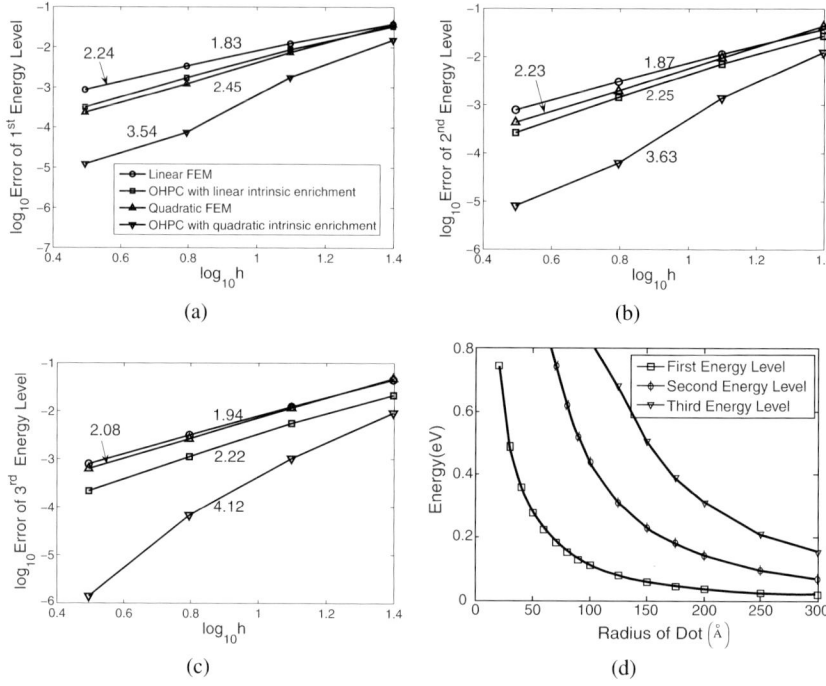

Fig. 11 (**a–c**) Convergence of FEM and OPHC solutions of first, second and third energy levels respectively, (**d**) the first three energy levels associated with the radius of dot predicted by OHPC with quadratic intrinsic basis

is shown that when reproduction of orbital basis functions is enforced in the HP-Cloud approximation as an intrinsic enrichment, higher order extrinsic monomial enrichment is only needed in the vicinity of the nuclei. A construction of OHPC approximation to yield discontinuous derivative across the material interface in the application to quantum dots is introduced. Several quantum systems have been analyzed and the results obtained using the proposed method compare favorably to those obtained using p-adaptive or Gaussian enriched finite element methods.

Acknowledgements The support of this work by Lawrence Livermore National Laboratory (LLNL) under is greatly acknowledged.

References

1. Alemany MMG, Jain M, Kronik L, Chelikowsky JR, *Real-space pseudopotential method for computing the electronic properties of periodic systems*, Physical Review B, 69, 075101–075106, 2004

2. Chelikowsky JR, Troullier N, Saad Y, *Finite-difference-pseudopotential method: Electronic structure calculations without a basis*, Physical Review Letter, 72, 1240–1243, 1994
3. Chelikowsky JR, Troullier N, Wu K, Saad Y, *Higher-order finite-difference pseudopotential method: An application to diatomic molecules*, Physical Review B, 50, 11355–11364, 1994
4. Chen JS, Hu W, Puso MA, *Orbital HP-Cloud for Schrödinger equation in quantum mechanics*, Computer Methods in Applied Mechanics and Engineering, 196, 3693–3705, 2007
5. Chen JS, Wu CT, Yoon S, You Y, *A stabilized conforming nodal integration for galerkin meshfree methods*, International Journal Numerical Methods Engineering, 50, 435–466, 2001
6. Chen JS, Wu CT, Yoon S, You Y, *Nonlinear version of stabilized conforming nodal integration for galerkin meshfree methods*, International Journal for Numerical Methods in Engineering, 53, 2587–2615, 2002
7. Duarte CAM, Oden JT, *An h-p adaptive method using clouds*, Computer Methods in Applied Mechanics and Engineering, 139, 237–262, 1996
8. Duarte CAM, Oden JT, *Hp Clouds – an hp meshless method*, Numerical Methods for Partial Differential Equations, 12, 673–705, 1996
9. Garcia O, Fancello EA, Barcellos CS, Duarte CAM, *Hp-Clouds in Mindlin's Thick Plate Model*, International Journal for Numerical Methods in Engineering, 47, 1381–1400, 2000
10. Harrison P, **Quantum Wells, Wires and Dots**, Wiley, New York, 1999
11. Hehre WJ, Radom L, Schleyer P, Pople JA, **Ab Initio Molecular Orbital Theory**, Willey, New York, 1986
12. Heinemann D, Fricke B, Kolb D, *Solution of the Hartree-Fock-Slater equations for diatomic molecules by the finite-element method*, Physical Review A, 38, 4994–5001, 1988
13. Hermansson B, Yevick D, *Finite-element approach to band-structure analysis*, Physical Review B, 33, 7241–7242, 1986
14. Ihm J, Zunger A, Cohen ML, *Momentum-space formalism for the total energy of solids*, Journal of Physics C: Solid State Physics, 12, 4409–4422, 1979
15. Liszka TJ, Duarte CAM, Tworzydlo WW, *Hp-Meshless cloud method*, Computer Methods in Applied Mechanics and Engineering, 139, 263–288, 1996
16. Mendonca PT, Barcellos CS, Duarte CAM, *Investigations on the Hp-Cloud method by solving timoshenko beam problems*, Computational Mechanics, 25, 286–295, 2000
17. Oden JT, Duarte CAM, Zienkiewicz OC, *A new cloud-based hp finite element method*, Computer Methods in Applied Mechanics and Engineering, 153, 117–126, 1998
18. Parr RG, Yang W, **Density Functional Theory of Atoms and Molecules**, Oxford University Press, New York, 1989
19. Pask JE, Sterne PA, *Real-space formulation of the electrostatic potential and total energy of solids*, Physical Review B, 71, 113101–113104, 2005
20. Pickett WE, *Pseudopotential methods in condensed matter applications*, Computer Physics Reports, 9, 115–197, 1989
21. Szabo A, Ostlund NS, **Modern Quantum Chemistry: Introduction to Advanced Electronic Structure Theory**, McGraw-Hill, New York, 1989
22. Tsuchida E, Tsukada M, *Electronic-structure calculations based on the finite-element method*, Physical Review B, 52, 5573–5578, 1995
23. Tsuchida E, Tsukada M, *Adaptive finite-element method for electronic-structure calculations*, Physical Review B, 54, 7602–7605, 1996
24. Tsuchida E, Tsukada M, *Large-scale electronic-structure calculations based on the adaptive finite-element method*, Journal of the Physical Society of Japan, 67, 3844–3858, 1998
25. Wang D, Chen JS, Sun L, *Homogenization of magnetostrictive particle-filled elastomers using an interface-enriched reproducing kernel particle method*, Finite Elements in Analysis and Design, 39, 765–782, 2003
26. White SR, Wilkins JW, Teter MP, *Finite-element method for electronic structure*, Physical Review B, 39, 5819–5833, 1989
27. Yamakawa S, Hyodo S, *Electronic state calculation of hydrogen in metal clusters based on Gaussian-FEM mixed basis function*, Journal of Alloys and Compounds, 231, 356–357, 2003
28. Yamakawa S, Hyodo S, *Gaussian finite-element mixed-basis method for electronic structure calculations*, Physical Review B, 71, 035113–035121, 2005

The Radial Natural Neighbours Interpolators Extended to Elastoplasticity

Lúcia Maria de Jesus Simas Dinis(✉), Renato Manuel Natal Jorge, and Jorge Belinha

Abstract Considering a small strain formulation a Radial Natural Neighbour Interpolator method is extended to the elastoplastic analysis. Resorting to the Voronoï tessellation the nodal connectivity is obtained. The Delaunay triangulation supplies the integration background mesh. The improved interpolation functions based on the Radial Point Interpolators are provided with the delta Kronecker property, easing the imposition of the essential and natural boundary conditions. The Newton-Raphson method is used for the solution of the nonlinear system of equations and an Hill yield surface is considered. Benchmark examples prove the high accuracy and convergence rate of the proposed method.

Keywords: Radial point interpolation method · Radial basis function · Natural neighbours · Meshfree method · Plasticity

1 Introduction

In this last fifteen years numerous meshless methods [1, 2] have been developed and applied in various engineering fields and applied sciences. In the meshless methods the nodes can be arbitrary distributed, once the field functions are approximated within an influence domain rather than an element, as in the well-known

L.M.J.S. Dinis
Associate Professor, Faculty of Engineering of the University of Porto – FEUP Rua Dr. Roberto Frias, 4200-465 Porto, Portugal, e-mail: ldinis@fe.up.pt

R.M.N. Jorge
Departamento de Engenharia Mecânica e Gestão Industrial, Faculdade de Engenharia da Universidade do Porto, Rua Dr. Roberto Frias, 4200-465 Porto, Portugal, e-mail: rnatal@fe.up.pt

J. Belinha
Researcher, Institute of Mechanical Engineering – IDMEC Rua Dr. Roberto Frias, 4200-465 Porto, Portugal, e-mail: jorge.belinha@fe.up.pt

A.J.M. Ferreira et al. (eds.) *Progress on Meshless Methods, Computational Methods in Applied Sciences.*
© Springer Science + Business Media B.V. 2009

Finite Element Method (FEM) [3, 4]. In meshless methods the influence domains may and must overlap each other, in opposition to the no-overlap rule between elements in the FEM. In the meshless methods the nodes can be arbitrary distributed, once the field functions are approximated within an influence domain rather than an element.

The first meshless method that uses the moving least square approximants (MLS) [5] in the construction of the approximation function was the Diffuse Element Method (DEM) [6]. Belytschko evolved the DEM and developed one of the most popular meshless method, the Element Free Galerkin Method (EFGM) [7–9]. The Smooth Particle Hydrodynamics Method (SPH) [10], which is one of the oldest, it is in the origin of the Reproducing Kernel Particle Method (RKPM) [11]. Other meshless methods such as the meshless local Petrov-Galerkin method (MLPG) [12], the Finite Point Method (FPM) [13] and the Method of Finite Spheres (FSM) [14] were developed as well. These referred meshless methods all use approximation functions instead of interpolation functions, one immediate consequence is the difficulty on imposing the essential and natural boundary conditions, due to the lack of the delta Kronecker property.

To address the above problem several new meshless methods were developed in the last few years, the Point Interpolation Method (PIM) [15, 16], the Point Assembly Method [17], the Radial Point Interpolation Method [18, 19], the Natural Neighbour Finite Elements Method (NNFEM) [20, 21] or Natural Element Method (NEM) [22–25] and the Meshless Finite Element Method (MFEM) [26].

In this paper the analysis of 2D and 3D elastoplastic problems considering a small strain formulation is performed resorting to the Natural Neighbour Radial Point Interpolation Method – NNRPIM [27]. Resorting to Voronoï cells [28], a set of influence cells are created departing from an unstructured set of nodes. The Delaunay triangles [29], which are the dual of the Voronoï cells, are used to create a node-depending background mesh used in the numerical integration of the NNRPIM interpolation functions. Unlike the FEM, where geometrical restrictions on elements are imposed for the convergence of the method, in the NNRPIM there are no such restrictions, which permits a total random node distribution for the discretized problem. The NNRPIM interpolation functions, used in the Galerkin weak form, are constructed in a similar process to the RPIM, with some differences that modify the method performance.

Once the scope of this work is to extend and validate the NNRPIM in the elastoplastic analysis, the used non-linear solution algorithm is the Newton-Rapson initial stiffness method [30] and the efficient "forward-Euler" procedure [31] is used in order to return the stress to the yield surface.

The outline of the paper is as follows: In Sect. 2 are presented the construction of the NNRPIM influence-cells, the node dependent integration schemes and the construction of the NNRPIM interpolation functions. In Sect. 3 the small strain elastoplastic formulation is presented [30]. In Sect. 4 resorting to the NNRPIM several well-known 2D and 3D elastoplastic problems are solved. This work ends with the conclusions and remarks in Sect. 5.

2 Radial Natural Neighbours Interpolators

The NNRPIM uses the Voronoï diagrams and the Delaunay triangulation which are useful mathematical tools in the determination of the natural neighbours for each node belonging to the global nodal set. In order to determine the natural neighbours [32] for each node belonging to the global nodal set, $\mathbf{N} = 1, \ldots, N$, the NNRPIM uses the Voronoï diagrams and the Delaunay triangulation [27]. This theory is applicable to a $\mathscr{D}$-dimensional space. The Voronoï diagram of $\mathbf{N}$ is the partition of the domain defined by N sub-regions V_I, closed and convex. Each sub-region V_I is associated to the node I, n_I, in a way that any point in the interior of the V_I is closer to n_I than any other node n_J, where $n_J \in \mathbf{N}$ $(J \neq I)$, Fig. 1(a). The sub-regions V_k are the "Voronoï cells" which form the Voronoï diagram, $k = 1, \ldots, N$.

2.1 Nodal Connectivity

The nodal connectivity is imposed by the overlapping of the influence-cells [27], similar to the influence domain concept, which is obtained from the Voronoï cells. The cell formed by n nodes that contributes to the interpolation of the interest point $\mathbf{x}_I$ is called "influence-cell". Since it is simpler to represent, only the determination of the 2D influence-cell is presented, however this concept is applicable to a $\mathscr{D}$-dimensional space. The two distinct types of influence-cells are summarized in Fig. 1(b) and are defined as,

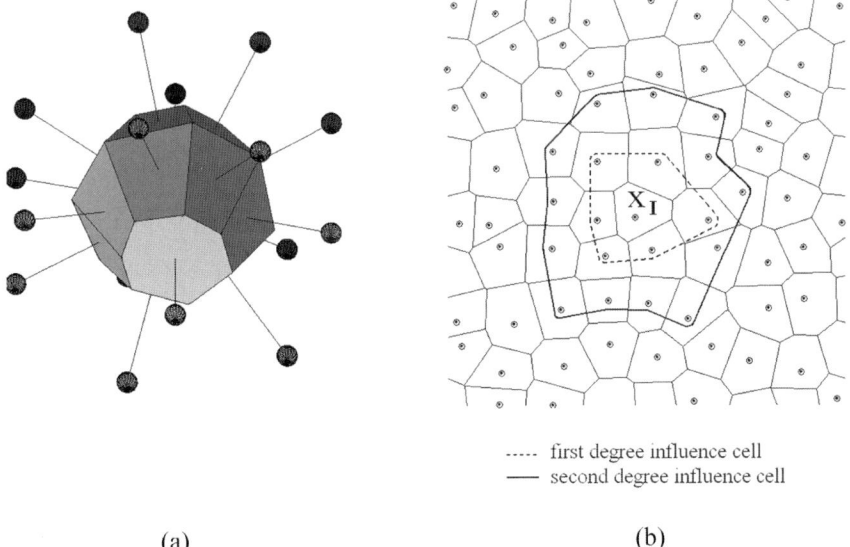

----- first degree influence cell
——— second degree influence cell

(a)   (b)

Fig. 1 (**a**) 3-Dimensional Voronoï cell. (**b**) Influence-cells

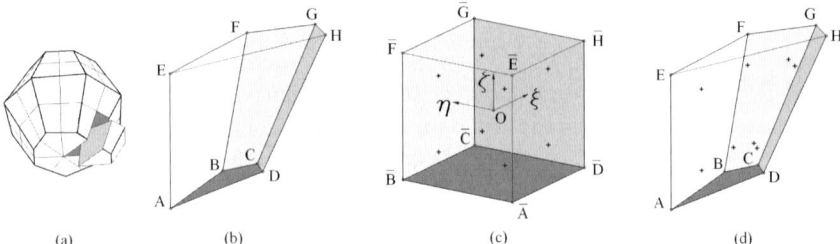

Fig. 2 (**a**) Voronoï cell and respective tetrahedrons. (**b**) Initial tetrahedron. (**c**) Initial tetrahedron isoparameterization and determination of the quadrature integration points. (**d**) Quadrature integration points in Cartesians coordinates

First degree influence-cell: A point of interest, $\mathbf{x}_I$, searches for its neighbour nodes, following the Natural Neighbour Voronoï construction. Thus the first degree influence-cell is composed by these first natural neighbours.

Second degree influence-cell: A point of interest, $\mathbf{x}_I$, searches for its neighbour nodes, in the same manner as in the first degree influence-cell. Then, based on a previous construction of the Voronoï diagram for the node mesh, the natural neighbours of the first natural neighbours of $\mathbf{x}_I$ are added to the influence-cell.

In this work, due the better behaviour, only second degree influence-cells are considered.

2.2 Numerical Integration

In an initial phase, after the domain discretization in a regular or an irregular nodal mesh, the Voronoï cells of each node are constructed. These cells can be considered as a background mesh for integration purpose, being determined the influence-cell for each one of these integration points. Using the Voronoï tessellation and the Delaunay triangulation small areas or volumes, respectively for the 2-Dimensional case or for the 3-Dimensional case, are established, Fig. 2(a). These areas or volumes, Fig. 2(b), can be isoparameterized and the Gauss-Legendre quadrature scheme applied, Fig. 2(c). In this work the Gauss-Legendre quadrature scheme used was the 1×1 in 2-Dimensional cases and the $1 \times 1 \times 1$ in 3-Dimensional cases. These two integrations schemes are sufficient for the used NNRPIM formulation, [27].

2.3 Interpolation Functions

This work uses the Radial Point Interpolators, [18, 19]. Consider a function $\mathbf{u}(\mathbf{x})$ defined in the domain Ω, which is discretized by a set of $\mathbf{N}$ nodes. It is assumed that only the nodes within the influence-cell of the point of interest $\mathbf{x}_I$ have effect on

$\mathbf{u}(\mathbf{x}_I)$. The value of function $\mathbf{u}(\mathbf{x}_I)$ at the point of interest $\mathbf{x}_I$ is obtained by,

$$\mathbf{u}(\mathbf{x}_I) = \sum_{i=1}^{n} R_i(\mathbf{x}_I)\, a_i(\mathbf{x}_I) \tag{1}$$

Being $R_i(\mathbf{x}_I)$ the Radial Basis Function (RBF), n the number of nodes inside the influence-cell of node I and $a_i(\mathbf{x}_I)$ the non-constant coefficient of $R_i(\mathbf{x}_I)$. The difference of the interpolation function presented in Eq. (1) with the one in [18,19] is the absence of the polynomial basis, which was proved to be dispensable in [27] for the NNRPIM formulation, reducing in this manner the computational effort of the interpolation function construction. The RBFs are well studied and developed in [18,19], in these functions the variable is the Euclidean norm r_{Ii} of the interest point $\mathbf{x}_I$ and the natural neighbour node $\mathbf{x}_i$, $r_{Ii} = \sqrt{(x_I - x_i)^2 + (y_I - y_i)^2 + (z_I - z_i)^2}$. In this work the Multiquadric RBF is used,

$$R(r_{Ii}) = \left(r_{Ii}^2 + c^2\right)^p \tag{2}$$

where c and p are two shape parameters that need to be optimized once the variation of these parameters affect the performance of the NNRPIM. This is the major disadvantage of meshless methods that use the RBFs. In the work presented in [27] it was found that when $c \to 0$ and $p \to 1$ higher accuracy is obtained, however these shape parameters cannot be $c = 0$ or $p = 1$ once it would produce a singular geometric matrix. Thus it was adopted $c = 0.0001$ and $p = 0.9999$.

The interpolation can be presented in the matricial form,

$$\mathbf{u}_s = \mathbf{R}\,\mathbf{a} = \begin{bmatrix} R(r_{11}) & R(r_{12}) & \dots & R(r_{1n}) \\ R(r_{21}) & R(r_{22}) & \dots & R(r_{2n}) \\ \vdots & \vdots & \ddots & \vdots \\ R(r_{n1}) & R(r_{n2}) & \dots & R(r_{nn}) \end{bmatrix} \begin{Bmatrix} a(\mathbf{x}_1) \\ a(\mathbf{x}_2) \\ \vdots \\ a(\mathbf{x}_n) \end{Bmatrix} \tag{3}$$

by solving Eq. (3),

$$\mathbf{a} = \mathbf{R}^{-1}\,\mathbf{u}_s \tag{4}$$

and substituting in Eq. (1) the interpolation function, $\varphi(\mathbf{x})$, is obtained,

$$\mathbf{u}(\mathbf{x}_I) = \mathbf{R}(\mathbf{x}_I)\,\mathbf{R}^{-1}\,\mathbf{u}_s = \varphi(\mathbf{x})\,\mathbf{u}_s \tag{5}$$

The partial derivative of $\varphi(\mathbf{x})$ in order to a variable ξ is defined as,

$$\varphi_{,\xi}(\mathbf{x}_I) = \mathbf{R}_{,\xi}^T(\mathbf{x}_I)\,\mathbf{R}^{-1} \tag{6}$$

The partial derivatives of the RBF in order to variable ξ,

$$\mathbf{R}_{,\xi}(r_{ij}) = 2p\left(r_{ij}^2 + c^2\right)^{p-1}(\xi_j - \xi_i) \tag{7}$$

The obtained interpolation functions, due to the chosen shape parameters [27], possesses the delta Kronecker property,

$$\varphi_i(\mathbf{x}_j) = \delta_{ij} = \begin{cases} 1 \ (i = j) \\ 0 \ (i \neq j) \end{cases} \quad i, j = 1, \dots, n \tag{8}$$

and satisfy the partition of unity,

$$\sum_{i=1}^{n} \varphi_i(\mathbf{x}_i) = 1 \tag{9}$$

An inconvenient property of the RPIMs interpolation functions is the lack of compatibility. This property is achieved in the RPIMs using the conforming RPIM [33], however this same study concluded that the RPIM is much more simple and efficient than the conforming RPIM.

2.4 Galerkin Weak Form

Consider Ω the solid domain. The equilibrium equations can be expressed by,

$$\nabla \sigma + b = 0 \quad in \ \Omega \tag{10}$$

being ∇ the gradient, σ the stress tensor and b the body forces. The natural and essential boundary condition are,

$$\begin{cases} \sigma \mathbf{n} = \bar{\mathbf{t}} \text{ on natural boundary} \\ \mathbf{u} = \bar{\mathbf{u}} \ \ \text{on essential boundary} \end{cases} \tag{11}$$

the prescribed displacement on the essential boundary Γ_u is defined as $\bar{\mathbf{u}}$, $\bar{\mathbf{t}}$ is the tension applied in the natural boundary Γ_t and $\mathbf{n}$ is the normal to the problem domain boundary. Considering Eq. (10) the Galerkin weak form can be presented as,

$$\mathscr{L} = \int_{\Omega} \delta \varepsilon^T \sigma \, d\Omega - \int_{\Omega} \delta \mathbf{u}^T \mathbf{b} \, d\Omega - \int_{\Gamma_t} \delta \mathbf{u}^T \mathbf{t} \, d\Gamma = 0 \tag{12}$$

where ε is defined as,

$$\varepsilon = \mathbf{L} \mathbf{u} \tag{13}$$

and the stress tensor,

$$\sigma = \mathbf{c} \varepsilon = \mathbf{c} \mathbf{L} \mathbf{u} \tag{14}$$

And considering the equation,

$$\mathbf{u} = \sum_{I}^{n} \varphi_I \mathbf{u}_I \tag{15}$$

being n the number of nodes inside the influence-cell, the following expression can be written,

$$\mathbf{Lu} = \mathbf{L}\sum_I^n \varphi_I \, \mathbf{u}_I = \sum_I^n \mathbf{L}\varphi_I \, \mathbf{u}_I = \sum_I^n \mathbf{B}_I \, \mathbf{u}_I \tag{16}$$

where $\mathbf{L}$ the differential operator, $\mathbf{B}$ the deformation matrix and $\mathbf{c}$ the material matrix,

$$\mathbf{B}_I = \begin{bmatrix} \dfrac{\partial \varphi_I}{\partial x} & 0 & 0 \\ 0 & \dfrac{\partial \varphi_I}{\partial y} & 0 \\ 0 & 0 & \dfrac{\partial \varphi_I}{\partial z} \\ \dfrac{\partial \varphi_I}{\partial y} & \dfrac{\partial \varphi_I}{\partial x} & 0 \\ 0 & \dfrac{\partial \varphi_I}{\partial z} & \dfrac{\partial \varphi_I}{\partial y} \\ \dfrac{\partial \varphi_I}{\partial z} & 0 & \dfrac{\partial \varphi_I}{\partial x} \end{bmatrix} \tag{17}$$

$$\mathbf{c}^{-1} = \begin{bmatrix} 1/E & -\upsilon/E & -\upsilon/E & 0 & 0 & 0 \\ -\upsilon/E & 1/E & -\upsilon/E & 0 & 0 & 0 \\ -\upsilon/E & -\upsilon/E & 1/E & 0 & 0 & 0 \\ 0 & 0 & 0 & 1/G & 0 & 0 \\ 0 & 0 & 0 & 0 & 1/G & 0 \\ 0 & 0 & 0 & 0 & 0 & 1/G \end{bmatrix} \tag{18}$$

Equation (12) can be developed and presented in a matricial form,

$$\mathscr{L} = \delta \mathbf{U}^T \left[\mathbf{K} \mathbf{U} - \mathbf{F} \right] = 0 \tag{19}$$

where,

$$\mathbf{K} = \int_\Omega \mathbf{B}_I^T \, \mathbf{c} \, \mathbf{B}_J \, d\Omega \quad \text{and} \quad \mathbf{F} = \int_\Omega \phi_I^T \, \mathbf{b} \, d\Omega + \int_{\Gamma_t} \phi_I^T \, \mathbf{\bar{t}} \, d\Gamma \tag{20}$$

Once the RPIM interpolation function possess the delta Kronecker property the essential boundary conditions can be directly imposed in the stiffness matrix $\mathbf{K}$, as in the FEM.

3 Small Strain Elastoplastic Formulation

3.1 Elastoplastic Constitutive Model

The small-strain formulation can be summarized as,

1. Additive decomposition of rate strain into elastic and plastic parts:

$$\dot{\varepsilon} = (\dot{\varepsilon}_e + \dot{\varepsilon}_p) \tag{21}$$

2. Stress rate and elastic strain relation:

$$\dot{\sigma} = \mathbf{c} : \dot{\varepsilon}_e = \mathbf{c} : (\dot{\varepsilon} - \dot{\varepsilon}_p) \tag{22}$$

3. The Plastic Prandtl-Reuss flow rule, where $\dot{\lambda}$ is the plastic rate multiplier and **a** is the flow vector:

$$\dot{\varepsilon}_p = \dot{\lambda}\frac{\partial F}{\partial \sigma} = \dot{\lambda}\,\mathbf{a} \tag{23}$$

4. The Von Mises [34] yield condition $F(\sigma,k)$, only depending on the magnitude of the deviatoric stress tensor, **S**, and of a hardening parameter k:

$$F(\sigma,k) = \frac{3}{2}\mathbf{S}_{ij}\mathbf{S}_{ij} - \sigma_Y^2(k) = 0 \tag{24}$$

5. The hardening functional, depending on the hardening rule [30]:

$$A = \dot{\lambda}^{-1}\frac{\partial \sigma_Y}{\partial k}\,k \tag{25}$$

6. The plastic rate parameter:

$$\dot{\lambda} = \frac{\mathbf{a}^T : \mathbf{c} : \dot{\varepsilon}}{A + \mathbf{a}^T : \mathbf{c} : \mathbf{a}} \tag{26}$$

7. Stress rate and strain rate relation:

$$\dot{\sigma} = \mathbf{c}^{ep} : \dot{\varepsilon} \tag{27}$$

8. Continuum elasto-plastic tangent modulus:

$$\mathbf{c}^{ep} = \mathbf{c} - \frac{(\mathbf{c} : \mathbf{a}^T) \otimes (\mathbf{c} : \dot{\varepsilon})}{A + \mathbf{a}^T : \mathbf{c} : \mathbf{a}} \tag{28}$$

3.2 Stress Return Algorithm

In this work the material behaviour is modelled in the form of an incremental relation between the incremental stress vector and the strain increment. In order to force the stress to return to the yield surface the "forward-Euler" procedure [31] presented in Fig. 3 is considered.

A trial stress σ_i^{trial} is obtained and it passes beyond the actualized yield surface $f(\sigma)$. The trial stress is divided in two parts, $(1 - R)\sigma_i^{trial}$ inside the yield surface and $R\sigma_i^{trial}$ outside the yield surface, where R is the reduction factor $R = \frac{\bar{\sigma}_i^{trial} - \sigma_Y}{\bar{\sigma}_i^{trial} - \bar{\sigma}_{i-1}}$. Being $\bar{\sigma}$ the σ effective stress and σ_Y the actualized yield stress. The flow vector **a** is defined and it is visible that it is normal to the yield surface at the intersection of σ_i^{trial} with the yield surface. The part of the stress that passes beyond the yield surface it is projected into the flow vector, originating $\Delta\sigma_a^p$, which is subtracted to $R\sigma_i^{trial}$, returning the stress to the yield surface. However it is visible and understandable that in fact the yield surface has not been reached yet, thus the process is repeated until $\Delta \cong 0$.

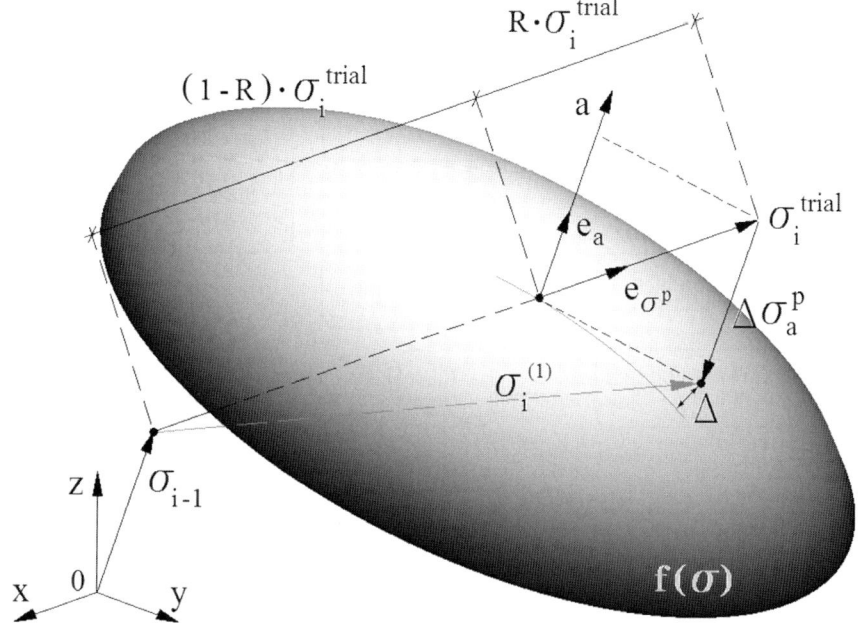

Fig. 3 Incremental stress changes at a point in an elasto-plastic continuum at initial yield

3.3 Non-linear Solution Algorithm

In this work the non-linear solution is obtained resorting to the Newton-Rapson initial stiffness method [30]. In this algorithm the stiffness matrix is calculated only once in the first iteration of the first increment. The solution algorithm, used in the NNRPIM code created by the authors, is presented in Fig. 4.

4 Numerical Examples

4.1 Cantilever Beam

A clamped beam with the geometric and material characteristics presented in Fig. 5(a) is analysed with the NNRPIM, considering it as a 2-Dimensional Plane Stress problem.

The problem domain is discretized in two distinct meshes a regular mesh of 561 nodes, Fig. 5(b), and a irregular mesh of the same mesh density, Fig. 5(c). The obtained results are compared with the FEM [35] and with the EFGM [36]. The vertical displacement of point A in relation to the tip force applied is presented in Fig. 6. It is perceptible by the figure that the NNRPIM results are closer to the solution

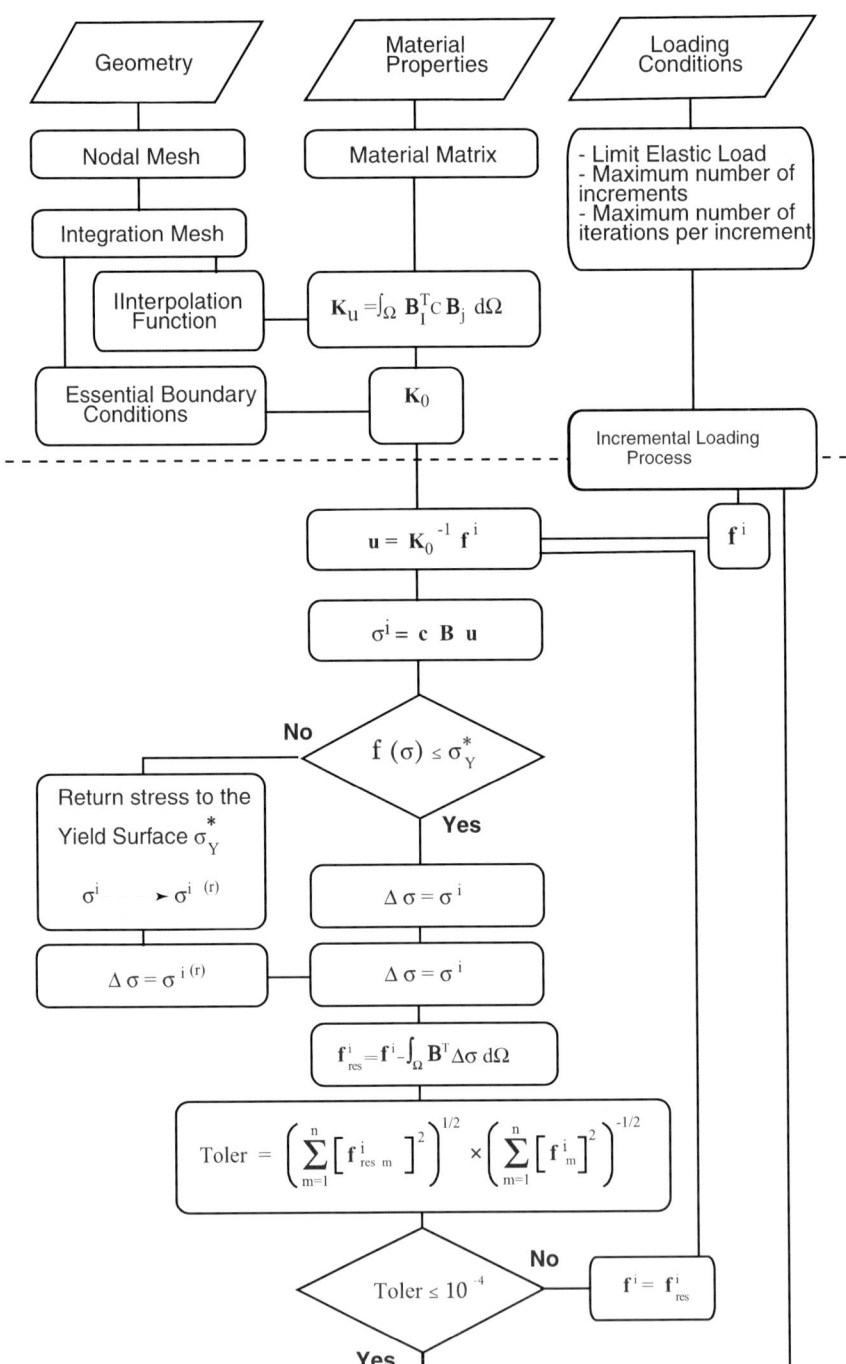

Fig. 4 NNRPIM non-linear solution algorithm

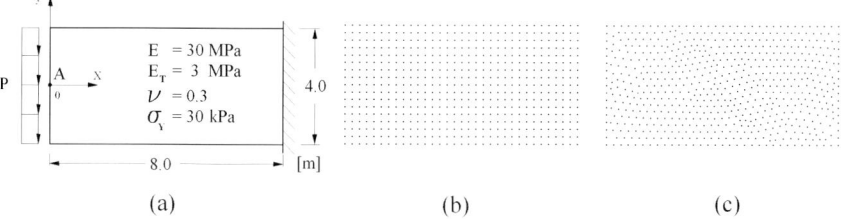

Fig. 5 (**a**) Cantilever Beam. (**b**) Regular Mesh. (**c**) Irregular Mesh

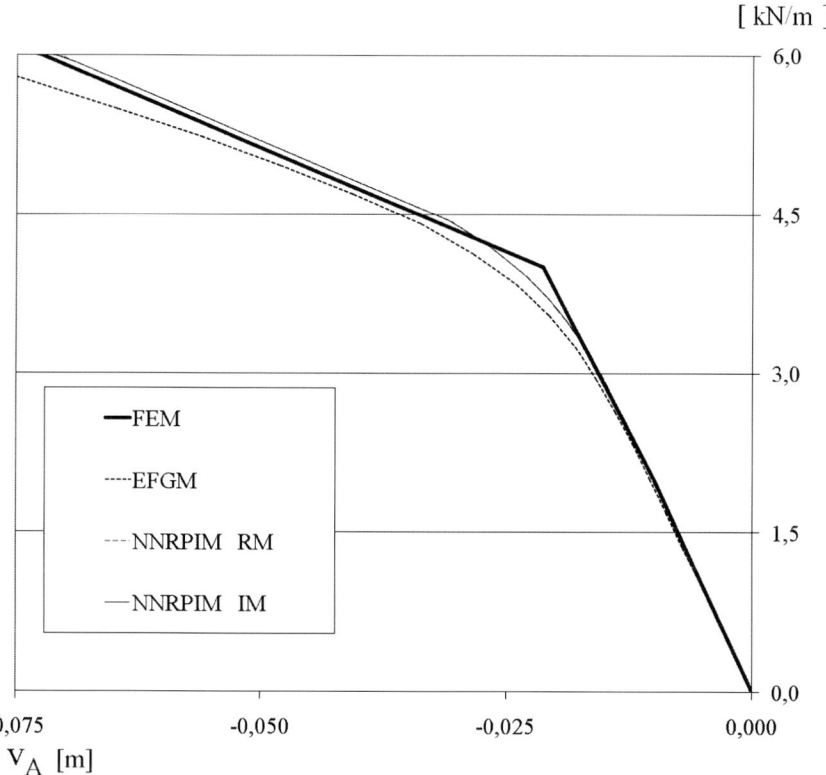

Fig. 6 Load/Displacement diagram for the vertical displacement in point A

obtained with the FEM than the EFGM solution. It is visible that the analysis with the irregular mesh (NNRPIM-IM) produces very similar results to the analysis with the regular mesh (NNRPIM-RM).

In Fig. 7 it is presented the normal stress σ_{xx} distribution along the essential boundary for increasing load levels. Both analysis, with the regular and the irregular mesh, adjust well, even for high load levels. The difference between the EFGM results and the NNRPIM results can be explained by the difficulty of the imposition of the boundary conditions in the EFGM. The boundary conditions in the EFGM are

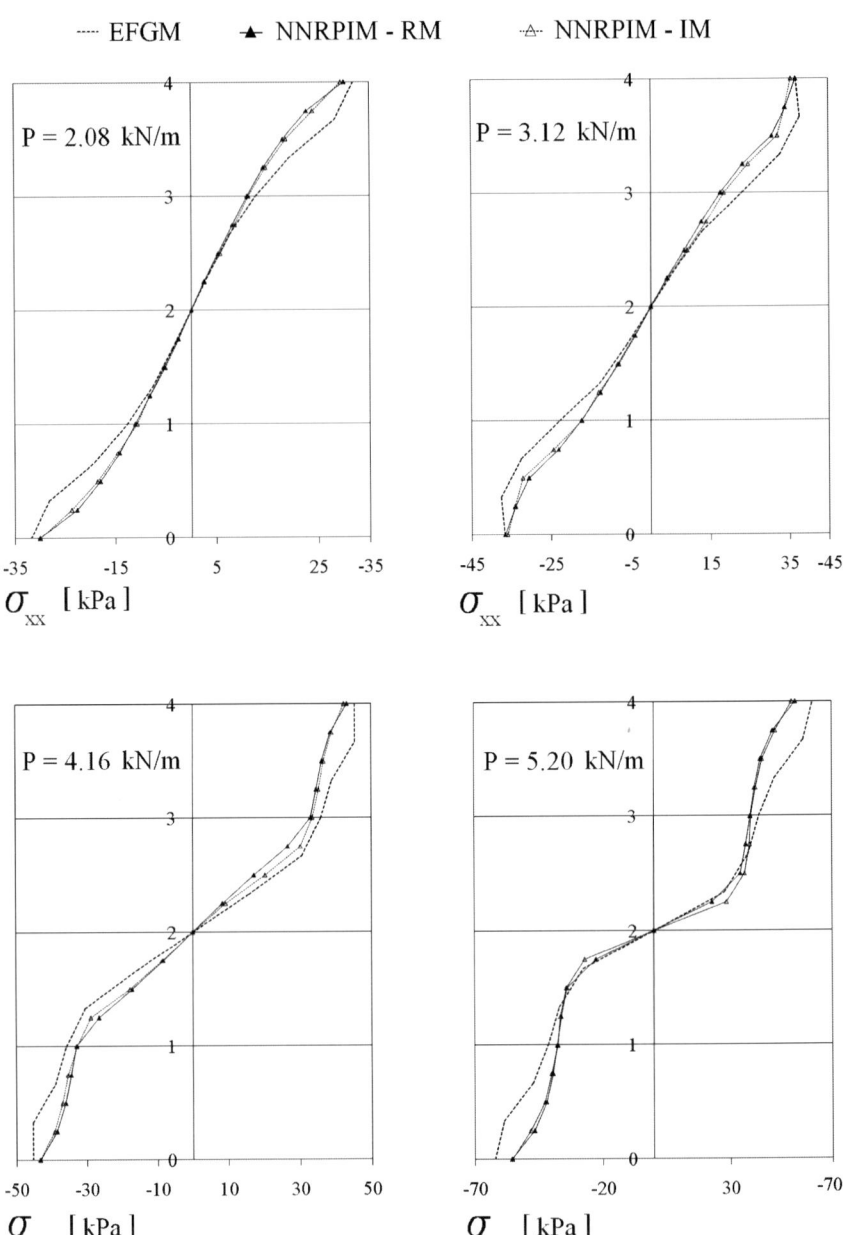

Fig. 7 Normal stress σ_{xx} distribution along the essential boundary for increasing load levels

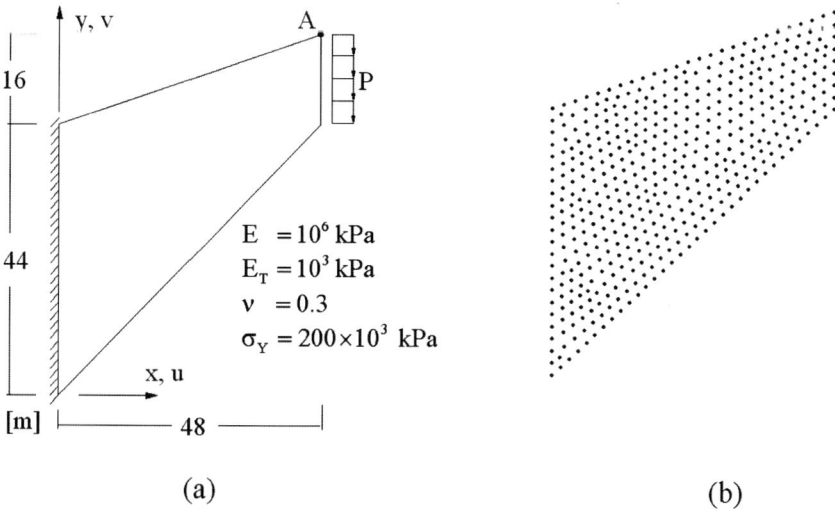

Fig. 8 (a) Cook Membrane. (b) Irregular 462 node Mesh

imposed by the penalty method [36], which is not a totally accurate way of imposing these conditions in approximation methods.

4.2 Cook Membrane

The Cook membrane presented in Fig. 8(a) is studied. The irregular mesh of 462 nodes on Fig. 8(b) is used in the NNRPIM analysis. The 2-Dimensional Plane Stress deformation theory is considered.

The NNRPIM solution is compared with the FEM results and the EFGM results [36]. In Fig. 9 it is presented the displacement on point A for an increasing load. As it is visible the NNRPIM adjust very well to the results obtained with the FEM analysis.

In Figs. 10 and 11 are presented respectively the normal stress σ_{xx} and the shear stress τ_{xy} distribution along the membrane essential boundary, for increasing load levels.

It is visible on both figures that the NNRPIM has a high correlation with the FEM, much better than the EFGM. Once again the reason is the imposition of the essential boundary conditions, in the NNRPIM these conditions are imposed with the direct imposition method, as the FEM, and in the EFGM the used imposition method is the Penalty Method. The good adjustment between the NNRPIM formulation and the FEM is maintained even for high load levels.

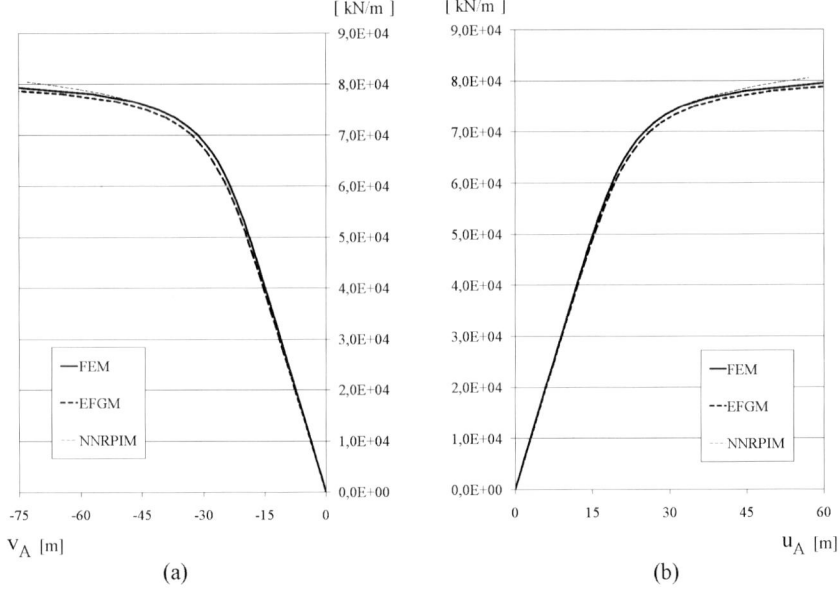

Fig. 9 (**a**) v_A and (**b**) u_A displacement components in relation to the applied load

4.3 Infinite Plate with a Circular Hole

Consider an infinite plate with a central hole subjected to a traction pressure in the extremities. Due the symmetry of the problem only a quarter is analysed. The geometrical and material properties of the problem are presented in Fig. 12(a). The problem is discretized in a irregular mesh of 660 nodes, Fig. 12(b), and analysed with the NNRPIM. The results are compared with the FEM and the EFGM.

The obtained results are compared with the nine node finite element and with the EFGM [37]. In Fig. 13 are presented the results regarding to the displacements of the interest points of the plate, marked in Fig. 12(a), in relation to the applied load.

In Fig. 14 it is presented the normal stress σ_{xx} distribution along the essential boundary defined by $\overline{AB}$ for increasing load levels.

As it is visible by the figure the NNRPIM adjust well to the FEM and the EFGM solutions. In this example there is no perturbation in the essential boundary in the EFGM. This is explained by the fact that the result of the EFGM is regarding a complete study of this problem, no symmetry was considered. As so the line $\overline{AB}$ is not in fact an essential boundary in the EFGM analysis [37]. Notice that the NNRPIM stress field in the boundary is as smooth as the EFGM stress field in an inner zone of the domain.

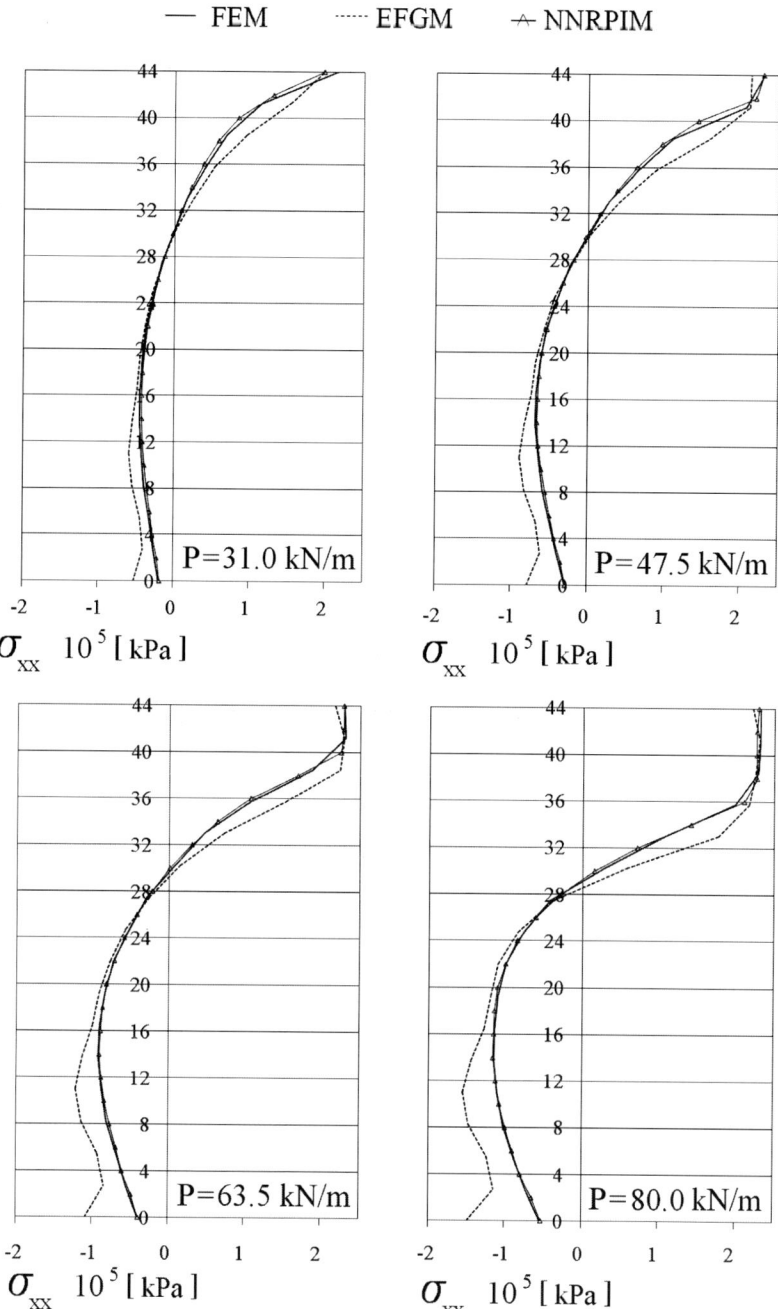

Fig. 10 Normal stress σ_{xx} distribution along the membrane essential boundary for increasing load levels

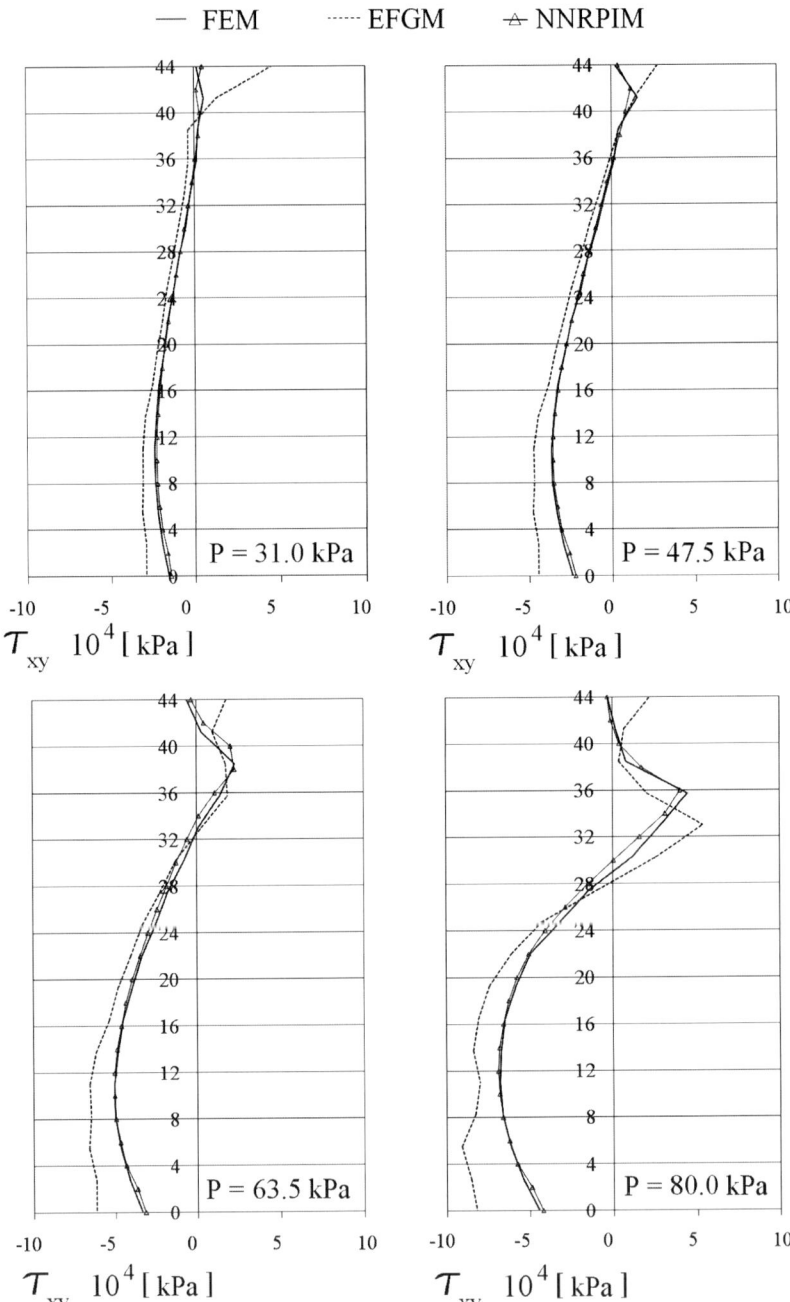

Fig. 11 Shear stress τ_{xy} distribution along the membrane essential boundary for increasing load levels

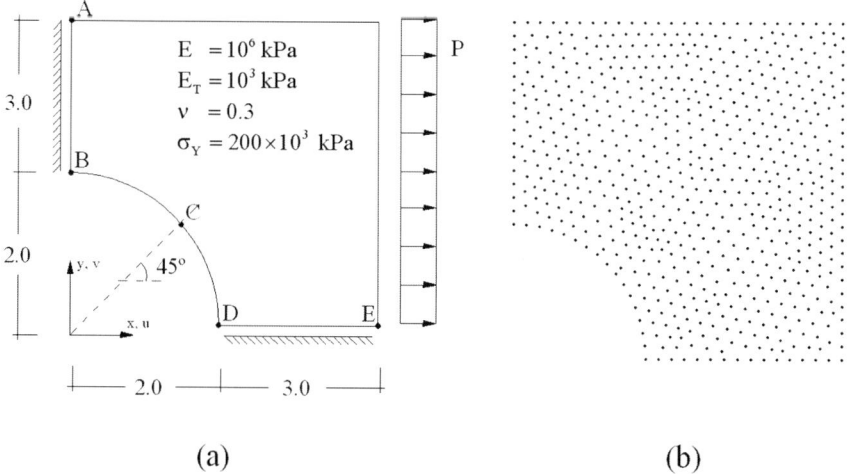

Fig. 12 (a) Quarter of an infinite plate with a circular hole. (b) Irregular 660 node Mesh

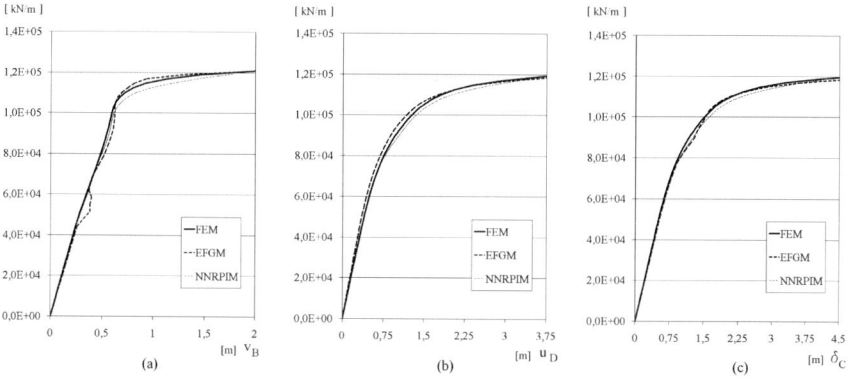

Fig. 13 (a) v_B, (b) δ_C and (c) u_D displacement components in relation to the applied load

4.4 Infinite Cylinder

An infinite thick pipe subjected to an internal pressure is studied. Due the symmetry and the infinite length of the problem only a quarter is analysed. The geometric characteristics are presented in Fig. 15(a). The material properties are: $E = 21MPa$, $E_T = 21kPa$, $v = 0.3$ and $\sigma_Y = 24kPa$. The problem is analysed considering the 2-Dimensional Plane Strain theory (NNRPIM-2D) and the 3-Dimensional Strain Theory (NNRPIM-3D). In the 3-Dimensional analysis is considered, along the zz axis, a thick of 50 mm and the restriction of displacement w in the faces at $z = 0\,mm$ and $z = 50\,mm$. The 2-Dimensional mesh used is presented in Fig. 15(b). In the 3-Dimensional analysis two different meshes are used, a regular mesh (RM) of 2205 nodes and a irregular mesh (IM) of the same node density, Fig. 15(c).

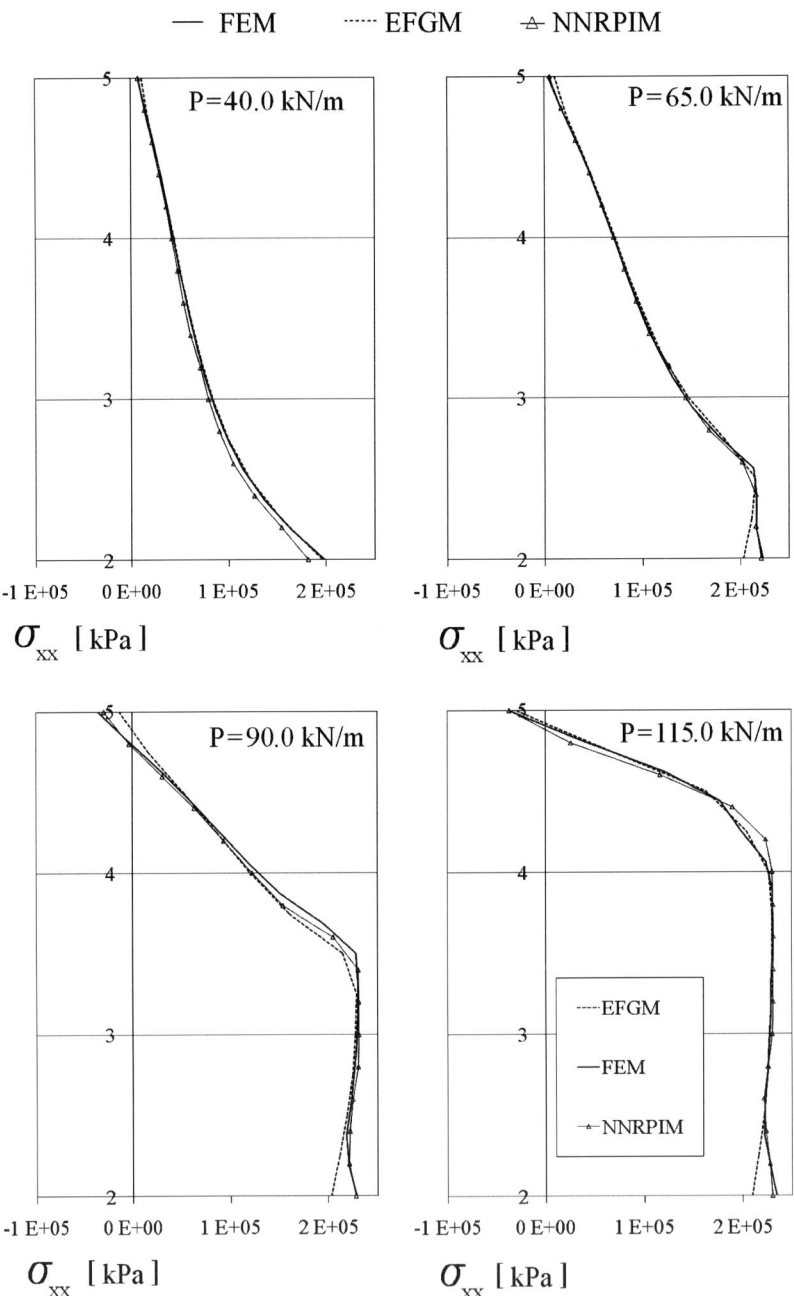

Fig. 14 Normal stress σ_{xx} distribution along essential boundary define by $\overline{AB}$ for increasing load levels

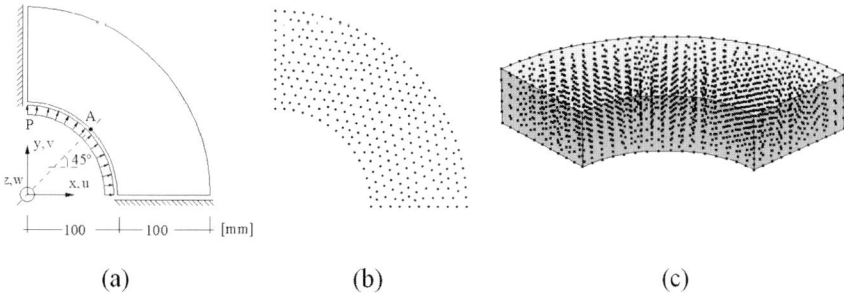

Fig. 15 (**a**) Cylinder geometric characteristics. (**b**) 2-Dimensional mesh with 315 nodes and (**c**) 3-Dimensional regular mesh with 2,205 nodes

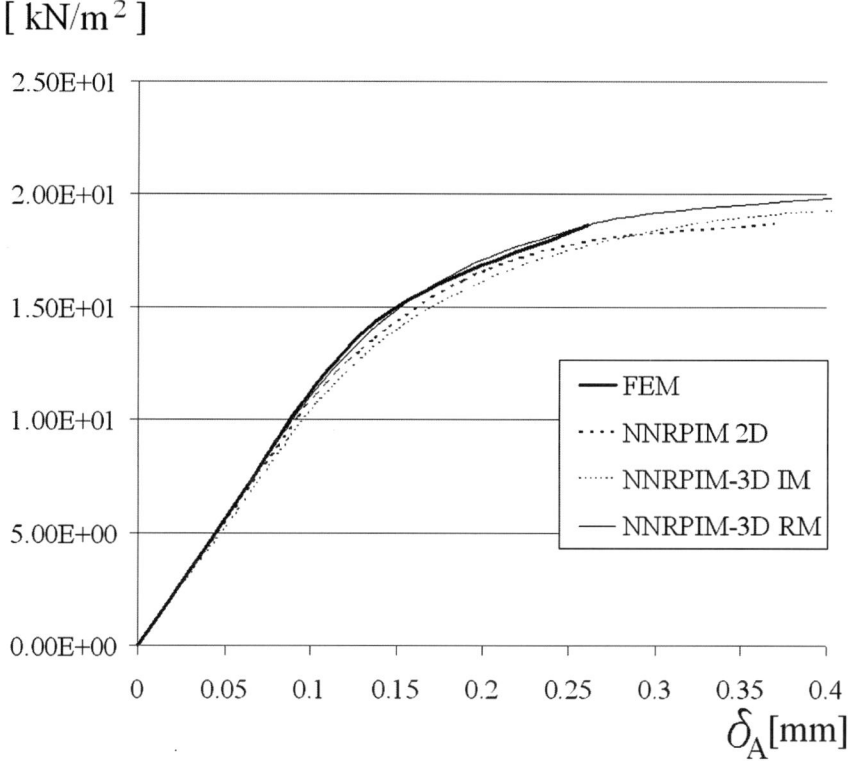

Fig. 16 (**a**) Load/displacement δ_A diagram

The displacement on point A, δ_A, for increasing loads is shown in Fig. 16. The NNRPIM results are compared with the FEM solution presented in [4]. It is visible the good correlation between all the NNRPIM formulations and the FEM solution. Once more there is not a significant difference between the results obtained with a regular mesh and the irregular mesh.

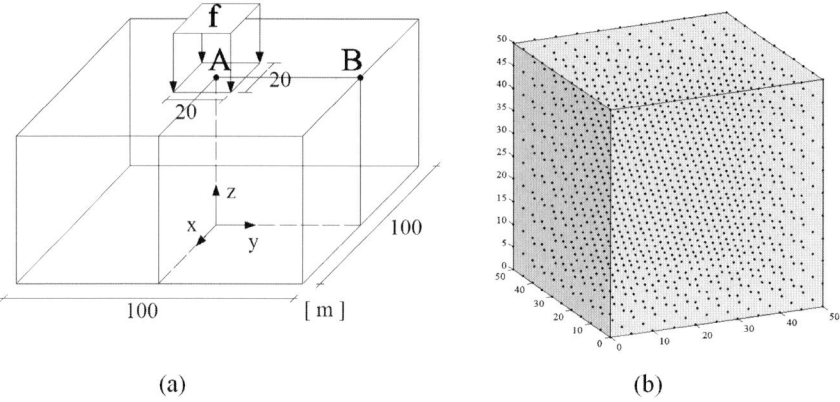

Fig. 17 (a) Block under pressure. (b) Regular mesh of 1,331 nodes

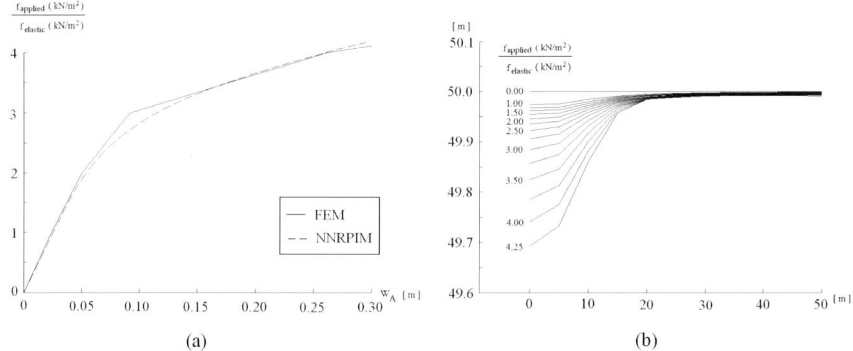

Fig. 18 (a) Load/displacement w_A diagram. (b) Load/displacement w diagram for line $\overline{AB}$

4.5 Block Under Concentrated Pressure

Consider the block under a centre concentrated pressure, as Fig. 17(a) indicates. Due to the symmetry only a quarter of the block is studied. The material properties are: $E = 210\,\text{MPa}$, $v = 0.3$, $\sigma_Y = 250\,\text{kPa}$ and the hardening parameter $A = 250$. The problem is discretized in a $11 \times 11 \times 11$ nodes regular mesh, Fig. 17(b).

The zz component of the displacement on point A, w_A, for increasing loads is shown in Fig. 18(a). It is perceptible the good correlation between the NNRPIM formulation and the FEM solution (8-nodes brick finite element) [38]. In Fig. 18(b) it is presented the evolution of the zz component of the displacement, w, on line $\overline{AB}$ for increasing load levels.

Figure 19 shows the effective stress map for different load levels. It is visible that the stress distribution is smooth.

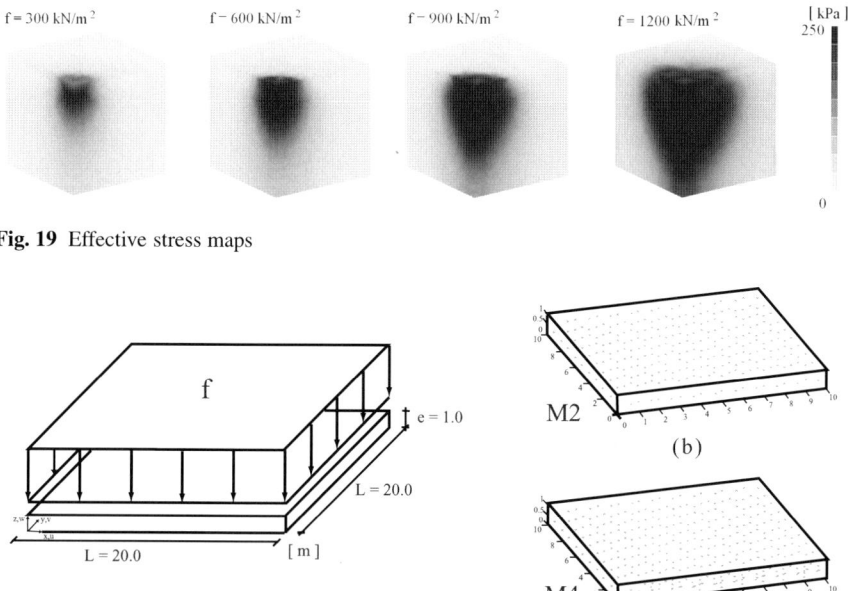

Fig. 19 Effective stress maps

Fig. 20 (a) Studied square plate. (b) Regular meshes with 363 nodes and (c) regular mesh with 605 nodes

4.6 Square Plate Under Transversal Load

A 3-Dimensional square plate under a distributed uniform transverse load is considered, Fig. 20(a). The material properties are: $E = 206\,\text{MPa}$, $E_T = 2.06\,\text{MPa}$, $v = 0.3$ and $\sigma_Y = 206\,\text{kPa}$. Due to the existing symmetry only one quarter of the plate is analysed. In this example two distinct cases are studied. The simple supported plate in all edges (SSSS) and the clamped plate in all edges (CCCC). In the SSSS NNRPIM analysis two different meshes are used, a 3-Dimensional regular mesh with three levels (M2) and a 3-Dimensional regular mesh with five levels (M4), Figs. 20(b, c).

In Fig. 21(a) it is presented the SSSS load/central displacement w_c diagram. The results are compared with the FEM solution [30] for plates with the First Order Shear Deformation Theory (FOSDT). As it is visible by Fig. 21(a) the analysis that uses more nodes along the thickness of the plate (M4) produces a solution which approaches more the FEM solution. As so, only the M4 mesh is used in the CCCC analysis.

In Fig. 21(b) it is presented the CCCC load/central displacement w_c diagram. As it is visible the NNRPIM formulation approach the FEM solution for plates, with the FOSDT, proposed by Barbosa [39] and Figueiras [40]. In Figs. 22(a, b) are respectively presented the effective stress maps for different load levels for the SSSS plate and the CCCC plate.

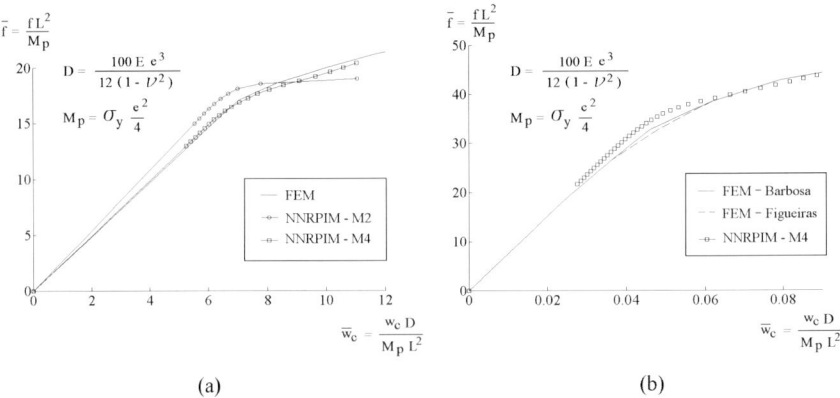

Fig. 21 Normalized load/vertical central displacement w_c diagram for the (a) SSSS plate and the (b) CCCC plate

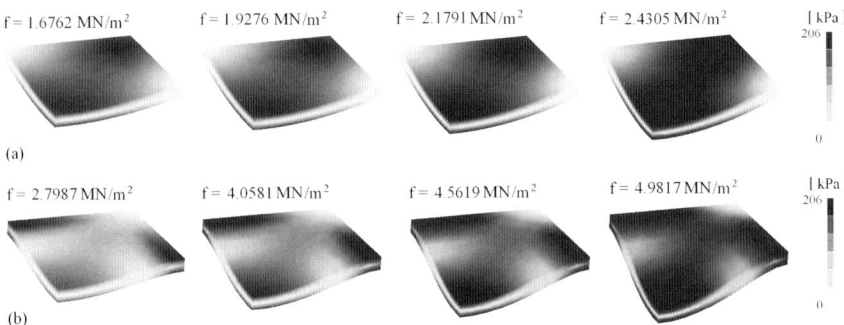

Fig. 22 Effective stress map for (**a**) the SSSS plate and (**b**) the CCCC plate

5 Conclusions

In this work the NNRPIM was extended to the elastoplastic analysis, considering a small strain formulation. The Newton-Raphson method was used for the solution of the nonlinear system of equations and the Von Mises yield criterion was considered. From the results obtained with the presented benchmark examples it can be concluded,

1. The non-linear solution algorithm was successfully implemented, as the algorithm of returning stress to the yield surface.
2. With the NNRPIM the imposition of the boundary conditions is straight-forward, reducing the computational time comparing with the approximation methods, where that does not happen. The NNRPIM results near the boundary are more accurate than the EFGM.
3. The mesh irregularity does not significantly affect the final results and the convex boundaries of the problem, due the NNRPIM node connectivity scheme, does

not represent a setback, as in other meshless methods that for instances use the influence domain concept.

4. The NNRPIM analysis produces a displacement field and a stress field smooth, accurate and very close to the FEM solution.

Acknowledgments The authors truly acknowledge the funding provide by Ministério da Ciência, Tecnologia e Ensino Superior – Fundação para a Ciência e a Tecnologia (Portugal), under grant SFRH/BD/31121/2006, and by FEDER/FSE, under grant PTDC/EME-PME/81229/2006.

References

1. T. Belytschko, Y. Krongauz, D. Organ, M. Fleming, and P. Krysl. Meshless methods: An overview and recent developments. *Computer Methods in Applied Mechanics and Engineering*, 139(1):3–47, 1999.
2. Y.T. Gu. Meshfree methods and their comparisons. *International Journal of Computational Methods*, 2(4):477–515, 2005.
3. K.J. Bathe. *Finite Element Procedures*. Prentice-Hall: Englewood Cliffs, NJ, 1996.
4. O.C. Zienkiewicz. *The Finite Element Method*. 4th ed. McGraw-Hill: London, 1989.
5. P. Lancaster and K. Salkauskas. Surfaces generation by moving least squares methods. *Mathematics of Computation*, 37:141–158, 1981.
6. B. Nayroles, G. Touzot, and P. Villon. Generalizing the finite element method: Diffuse approximation and diffuse elements. *Computational Mechanics*, 10:307–318, 1992.
7. T. Belytschko, Y.Y. Lu, and L. Gu. Element-free galerkin method. *International Journal for Numerical Methods in Engineering*, 37:229–256, 1994.
8. J. Dolbow and T. Belytschko. An introduction to programming the meshless element free galerkin method. *Archives in Computational Mechanics*, 5(3):207–241, 1998.
9. Y. Lu, T. Belytschko, and L. Gu. A new implementation of the element free galerkin method. *Computer Methods in Applied Mechanics and Engineering*, 113:397–414, 1994.
10. J.J. Monaghan. Smoothed particle hydrodynamics: Theory and applications to non-spherical stars. *Monthly Notices of the Astronomical Society*, 181:375–389, 1977.
11. W.K. Liu, S. Jun, and Y.F. Zhang. Reproducing kernel particle methods. *International Journal for Numerical Methods in Fluids*, 20(6):1081–1106, 1995.
12. S.N. Atluri and T. Zhu. A new meshless local petrov-galerkin (mlpg) approach in computational mechanics. *Computational Mechanics*, 22(2):117–127, 1998.
13. E. Oñate, S. Idelsohn, O.C. Zienkiewicz, and R.L. Taylor. A finite point method in computational mechanics – Applications to convective transport and fluid flow. *International Journal for Numerical Methods in Engineering*, 39:3839–3866, 1996.
14. K.J. Bathe and S. De. Towards an efficient meshless computational technique: The method of finite spheres. *Engineering Computations*, 18:170–192, 2001.
15. G.R. Liu and Y.T. Gu. A point interpolation method for two-dimensional solids. *International Journal for Numerical Methods in Engineering*, 50:937–951, 2001.
16. J.G. Wang, G.R. Liu, and Y.G. Wu. A point interpolation method for simulating dissipation process of consolidation. *Computer Methods in Applied Mechanics and Engineering*, 190:5907–5922, 2001.
17. G.R. Liu. A point assembly method for stress analysis for two-dimensional solids. *International Journal of Solid and Structures*, 39:261–276, 2002.
18. J.G. Wang and G.R. Liu. A point interpolation meshless method based on radial basis functions. *International Journal for Numerical Methods in Engineering*, 54:1623–1648, 2002.
19. J.G. Wang and G.R. Liu. On the optimal shape parameters of radial basis functions used for 2-d meshless methods. *Computer Methods in Applied Mechanics and Engineering*, 191:2611–2630, 2002.

20. L. Traversoni. Natural neighbour finite elements. *International Conference on Hydraulic Engineering Software, Hydrosoft Proceedings, Computational Mechanics Publications*, 2:291–297, 1994.
21. N. Sukumar, B. Moran, A.Yu Semenov, and V.V. Belikov. Natural neighbour galerkin methods. *International Journal for Numerical Methods in Engineering*, 50(1):1–27, 2001.
22. J. Braun and M. Sambridge. A numerical method for solving partial differential equations on highly irregular evolving grids. *Nature*, 376:655–660, 1995.
23. N. Sukumar, B. Moran, and T. Belytschko. The natural element method in solid mechanics. *International Journal for Numerical Methods in Engineering*, 43(5):839–887, 1998.
24. E. Cueto, M. Doblaré, and L. Gracia. Imposing essential boundary conditions in the natural element method by means of density-scaled -shapes. *International Journal for Numerical Methods in Engineering*, 49(4):519–546, 2000.
25. E. Cueto, N. Sukumar, B. Calvo, J. Cegoñino, and M. Doblaré. Overview and recent advances in the natural neighbour galerkin method. *Archives of Computational Methods in Engineering*, 10(4):307–387, 2003.
26. R. Sergio, S. Idelsohn, E. Oñate, N. Calvo, and F. Del Pin. The meshless finite element method. *International Journal for Numerical Methods in Engineering*, 58(6):893–912, 2003.
27. L.M.J.S. Dinis, R.M.N. Jorge, and J. Belinha. Analysis of 3d solids using the natural neighbour radial point interpolation method. *Computer Methods in Applied Mechanics and Engineering*, 196:2009–2028, 2007.
28. G.M. Voronoï. Nouvelles applications des paramètres continus à la théorie des formes quadratiques. *Deuxième Mémoire: Recherches sur les parallélloèdres primitifs, Journal für die reine und angewandte Mathematik*, 134:198–287, 1908.
29. B. Delaunay. Sur la sphére vide. a la memoire de georges voronoï. *Izv. Akad. Nauk SSSR, Otdelenie Matematicheskih i Estestvennyh Nauk*, 7:793–800, 1934.
30. D. Owen and E. Hinton. *Finite Element in Plasticity*. Pineridge Press: Swansea, 1980.
31. M.A. Crisfield. *Non-Linear Finite Element Analysis of Solids and Structures*. Wiley: Baffins Lane, Chichester, 1991.
32. R. Sibson. A vector identity for the dirichlet tesselation. *Mathematical Proceedings of the Cambridge Philosophical Society*, 87:151–155, 1980.
33. G.R. Liu, Y.T. Gu, and K.Y. Dai. Assessment and applications of interpolation methods for computational mechanics. *International Journal for Numerical Methods in Engineering*, 59:1373–1379, 2004.
34. R. Von Mises. Nachrichten der Kgl. Gesellschaft der Wissenschaften Göttingen. Klasse, pp. 582–592, 1913.
35. M. Brünig. Nonlinear analysis and elasto-plastic behavior of anisotropic structures. *Finite Elements in Analysis and Design*, 20:155–177, 1995.
36. J. Belinha and L.M.J.S. Dinis. Elasto-plastic analysis of anisotropic problems considering the element free galerkin method. *Revista Internacional de Métodos Numéricos para Cálculo y Diseño en Ingeniería.*, 22(2):87–117, 2006.
37. J. Belinha. *Elasto-Plastic Analysis Considering the Element Free Galerkin Method*. M.Sc. thesis, Faculdade de Engenharia da Universidade do Porto, 2004.
38. R.J. Alves de Sousa, R.N. Natal Jorge, R.A. Fontes Valente, and J.M.A. César de Sá. A new volumetric and shear locking-free 3d enhanced strain element. *Engineering Computations*, 20(7):896–925, 2003.
39. J.A.T. Barbosa. *Análise Não Linear Por Elementos Finitos de Placas e Cascas Reforçadas*. Tese de Doutoramento, Faculdade de Engenharia da Universidade do Porto, 1992.
40. J.A. Figueiras. *Ultimate Load Analysis of Anisotropic and Reinforced Concrete Plates and Shells*. University of Wales, Ph.D. thesis, C/Ph/72/83, Swansea, 1983.

Static and Damage Analyses of Shear Deformable Laminated Composite Plates Using the Radial Point Interpolation Method

Armel Djeukou(✉) and Otto von Estorff

Abstract The radial point interpolation method is employed to analyse the static deflection and damage of thin and thick laminated composite plates using a high order shear deformable theory. The problem domain is represented by regularly distributed nodes. A variational formulation is used to derive the discrete system equations which are based on the third order plate theory suggested by Reddy. The essential boundary conditions are imposed separately, as in the FEM, by means of the penalty method since the RPIM shape functions possess the Kronecker delta function property. The Gauss quadrature scheme is used to perform the integration over the cells and the layers numerically. A Hashin-type failure criterion and a gradual unloading is used with an incremental loading approach for the modelling of damage initiation and evolution. In order to demonstrate the validity and accuracy of the current method, several examples are investigated and the results are compared to those available in the literature. Moreover, side-to-thickness ratios as well as the influence of the number of nodes are discussed.

Keywords: Composite Materials · Radial Point Interpolation Method · Damage analysis

1 Introduction

The use of composite structures has steadily increased during the past decades, and in many applications, particularly within the aeronautical field, the locking design of a component is the result of an intensive simulation of its mechanical behaviour. In

A. Djeukou
Airbus Deutschland GmbH, Airbus Allee 1, D-28199 Bremen,
e-mail: armel.djeukou@tu-harburg.de

O. von Estorff
Institute of Modelling and Computation, Hamburg University of Technology, Denickestrasse 17, D-21073 Hamburg, Germany

A.J.M. Ferreira et al. (eds.) *Progress on Meshless Methods, Computational Methods in Applied Sciences.*

most cases the finite element method (FEM) is used to perform these investigations, since it is a robust methodology that may be seen as a kind of standard if it comes to static and dynamic, linear or non-linear stress analysis of composite structures. However, the FEM exhibits the typical shortcomings of all numerical methods based on meshes or elements which are connected by nodes: High element deformation ratios are not allowed. Consequently, in some analyses meshing and re-meshing are necessary, burdensome, and often lead to difficulties when dealing with complex problems (e.g., large deformation problems, dynamic crack problems, etc.).

To overcome this handicap, a new class methodologies, namely so called "mesh-free methods" has been developed and achieved remarkable progress over the past few years. Examples of these methods can be found in [1–6, 9, 10]. All of these meshless methods have in common that they do not require an element discretization of the problem domain. The approximation functions are constructed entirely in terms of a set of nodes, and no element or connectivity of the nodes is needed. In view of this, meshfree methods have a very good potential to become a powerful new generation of numerical methods in the future.

Currently, the moving least-squares (MLS) approximation is widely used in meshfree methodologies in order to construct shape functions. However this type of shape functions does not possess the Kronecker delta function properties and the approximation is computationally expensive because of the complexity in algorithms for computing shape functions and its derivatives.

To construct shape functions with the Kronecker delta function properties, point interpolation methods (PIM) based on polynomial basis functions have been proposed by Liu [7]. These formulations also turned out to be more accurate and easier to implement. The Kronecker delta function properties make it easy to directly enforce essential boundary conditions as in the FEM. However the PIM system matrix, namely the method moment matrix, can be singular in some cases. This issue is overcome by using radial basis functions, which leads, e.g., to the radial point interpolation method (RPIM) as suggested in [8–10].

For the analysis of laminated composite plates, the classical laminated plate theory (CPT) and the first order shear deformation theory (FSDT) are commonly used. Since the transverse shear deformation is not taken into account in these formulations, the CPT can only lead to good results in the case of thin plates. For the FSDT, a shear correction factor is needed to compensate that omission. Unfortunately, this factor depends on the material coefficients, the geometry, the stacking scheme, and the boundary conditions, which cannot be easily determined for practical problems.

In this contribution, the radial point interpolation method (RPIM) using the third order shear deformation theory (TSDT) given by Reddy [11] is applied for the formulation of composite laminates. The introduction of cubic variation of the displacement in the TSDT formulation avoids the need of shear correction coefficients. This theory can represent the kinematics in a better way and it provides more accurate inter-laminar stress distributions.

In order to predict the onset of damage and to model damage evolution in composites, a damage model consisting of a linear elastic undamaged response of the material, a damage initiation criterion, and a damage evolution law are required. The damage initiation criterion used in this contribution is an anisotropic damage model based on the work of Hashin [16, 17] introducing different criteria for the matrix, the fibre failure, and the delamination. The damage evolution is characterised by the degradation of the material stiffness. To demonstrate the validity of the present formulation, numerical examples for the static and damage analyses of composite laminates with different layups and boundary conditions are presented. The Loads in the damage analyses are applied incrementally. The problem domain is discretized using regularly distributed nodes. The solutions are verified by means of results given in the literature.

2 Radial Point Interpolation Method (RPIM)

Consider a field function $u(\mathbf{x})$ defined in the problem domain Ω discretized by a set of field nodes. Then the RPIM interpolation can be written as

$$u(\mathbf{x}) = \sum_{i=1}^{n} R_i(\mathbf{x})a_i + \sum_{j=1}^{m} p_j(\mathbf{x})b_j = \mathbf{R}^T(\mathbf{x})\mathbf{a} + \mathbf{p}^T(\mathbf{x})\mathbf{b}, \tag{1}$$

where $R_i(\mathbf{x})$ is a radial basis function (RBF), n is the number of RBFs, $p_j(\mathbf{x})$ is a monomial in the space coordinates $\mathbf{x}^T = [x, y]$, and m is the number of polynomial basis functions.

When $m = 0$, pure RBFs are used. Otherwise, the RBF is augmented with m polynomial basis functions. The coefficients a_i and b_j are constants yet to be determined.

In the radial basis function $R_i(\mathbf{x})$, the variable is only the distance between the point of interest $\mathbf{x}$ and a node at $\mathbf{x}_i$, i.e. for 2-D problems

$$r = \sqrt{(x - x_i)^2 + (y - y_i)^2}. \tag{2}$$

There exist a number of different types of radial basis functions. Four often used RBFs, namely the multi-quadrics (MQ) function, the Gaussian (Exp) function, the thin plate spline (TPS) function, and the Logarithmic radial basis function, are listed in Table 1. In utilizing RBFs, several shape parameters need to be determined in order to obtain a good performance. In general, for given types of problems these parameters can be determined by numerical examinations. For examples in the MQ-RBF, there are two shape parameters: α_c and q, to be determined by the analyst. When $q = \pm 0.5$, it is the standard MQ-RBF. Wang and Liu [14] left the parameter q open to any real variable, and found that $q = 0.98$ or 1.03 led to proper results for most problems studied in the case of two-dimensional solids. d_c is set to be 2.

In order to determine a_i and b_j in Eq. (1), a support domain is formed for the point of interest at $\mathbf{x}$, and n field nodes are included in the support domain. Coefficients a_i and b_j in Eq. (1) can be determined by enforcing Eq. (1) to be satisfied at these n nodes surrounding the point of interest at $\mathbf{x}$. This leads to n linear equations, one for

Table 1 Typical radial basis functions with dimensionless shape parameters

	Name	Expression	Shape parameters
1	Multi-quadrics (MQ)	$R_i(x,y) = \left(r_i^2 + (\alpha_c d_c)^2\right)^q$	$\alpha_c \geq 0, q$
2	Gaussian (EXP)	$R_i(x,y) = \exp\left[-\alpha_c \left(\frac{r_i}{d_c}\right)^2\right]$	α_c
3	Thin Plate Spline (TPS)	$R_i(x,y) = r_i^{\eta}$	η
4	Logarithmic	$R_i(x,y) = r_i^{\eta} \log r_i$	η

d_c is a characteristic length that relates to the nodal spacing in the local support domain of the point of interest $\mathbf{x}$, and is usually the average nodal spacing for all the nodes in the local support domain.

each node. The matrix form of these equations can be expressed as

$$\mathbf{U}_s = \mathbf{R}_0 \mathbf{a} + \mathbf{P}_m \mathbf{b}, \tag{3}$$

where the vector of function values $\mathbf{U}_s$ is defined by

$$\mathbf{U}_s = \left\{ u_1 \ u_2 \ \dots \ u_n \right\}^T, \tag{4}$$

$\mathbf{R}_0$ is the moment matrix of the RBFs, given by

$$\mathbf{R}_0 = \begin{bmatrix} R_1(r_1) & R_2(r_1) & \cdots & R_n(r_1) \\ R_1(r_2) & R_2(r_2) & \cdots & R_n(r_2) \\ \cdots & \cdots & \cdots & \cdots \\ R_1(r_n) & R_2(r_n) & \cdots & R_n(r_n) \end{bmatrix}_{(n \times n)}, \tag{5}$$

and the polynomial moment matrix is defined by

$$\mathbf{P}_m^T = \begin{bmatrix} 1 & 1 & \cdots & 1 \\ x_1 & x_2 & \cdots & x_n \\ y_1 & y_2 & \cdots & y_n \\ \vdots & \vdots & \ddots & \vdots \\ p_m(\mathbf{x}_1) & p_m(\mathbf{x}_2) & \cdots & p_m(\mathbf{x}_n) \end{bmatrix}_{(m \times n)}. \tag{6}$$

Finally, the vector of coefficients for RBFs is

$$\mathbf{a}^T = \left\{ a_1 \ a_2 \ \cdots \ a_n \right\}, \tag{7}$$

and the vector of coefficients for polynomial is given by

$$\mathbf{b}^T = \left\{ b_1 \ b_2 \ \cdots \ b_n \right\}. \tag{8}$$

In Eq. (5), the distance r_k, occurring in $R_i(r_k)$, is obtained by

$$r_k = \sqrt{(x_k - x_i)^2 + (y_k - y_i)^2}. \tag{9}$$

At this point it should be noted, that there are $n+m$ variables in Eq. (3). The additional m equations can be added using the following m constraint conditions

$$\sum_{i=1}^{n} p_j(\mathbf{x}_i)a_i = \mathbf{P}_m^T \mathbf{a} = 0, \quad j = 1, 2, \ldots, m. \tag{10}$$

Combining Eqs. (3) and (10) yields a set of equations, here given in the matrix notation,

$$\tilde{\mathbf{U}}_s = \begin{bmatrix} \mathbf{U}_s \\ \mathbf{0} \end{bmatrix} = \underbrace{\begin{bmatrix} \mathbf{R}_0 & \mathbf{P}_m \\ \mathbf{P}_m^T & \mathbf{0} \end{bmatrix}}_{\mathbf{G}} \begin{bmatrix} \mathbf{a} \\ \mathbf{b} \end{bmatrix} = \mathbf{G}\mathbf{a}_0, \tag{11}$$

where

$$\mathbf{a}_0^T = \left\{ a_1 \; a_2 \; \cdots \; a_n \; b_1 \; b_2 \; \cdots \; b_m \right\}, \tag{12}$$

$$\tilde{\mathbf{U}}_s = \left\{ u_1 \; u_2 \; \cdots \; u_n \; 0 \; 0 \; \cdots \; 0 \right\}. \tag{13}$$

Since the matrix $\mathbf{R}_o$ is symmetric, the matrix $\mathbf{G}$ will also be symmetric.
  Solving Eq. (11) yields

$$\mathbf{a}_0 = \left\{ \begin{array}{c} \mathbf{a} \\ \mathbf{b} \end{array} \right\} = \mathbf{G}^{-1}\tilde{\mathbf{U}}_s. \tag{14}$$

Re-writing Eq. (1), such that

$$u(\mathbf{x}) = \mathbf{R}^T(\mathbf{x})\mathbf{a} + \mathbf{p}^T(\mathbf{x})\mathbf{b} = \left\{ \mathbf{R}^T(\mathbf{x}) \; \mathbf{p}^T(\mathbf{x}) \right\} \left\{ \begin{array}{c} \mathbf{a} \\ \mathbf{b} \end{array} \right\}, \tag{15}$$

and using Eq. (14), one obtains

$$u(\mathbf{x}) = \left\{ \mathbf{R}^T(\mathbf{x}) \; \mathbf{p}^T(\mathbf{x}) \right\} \mathbf{G}^{-1}\tilde{\mathbf{U}}_s = \tilde{\Phi}^T(\mathbf{x})\tilde{\mathbf{U}}_s, \tag{16}$$

where the RPIM shape functions can be expressed as

$$\begin{aligned} \tilde{\Phi}^T(\mathbf{x}) &= \left\{ \mathbf{R}^T(\mathbf{x}) \; \mathbf{p}^T(\mathbf{x}) \right\} \mathbf{G}^{-1} \\ &= \left\{ \phi_1(\mathbf{x}) \; \phi_2(\mathbf{x}) \; \cdots \; \phi_n(\mathbf{x}) \; \phi_{n+1}(\mathbf{x}) \; \cdots \; \phi_{n+m}(\mathbf{x}) \right\}. \end{aligned} \tag{17}$$

Finally, the RPIM shape functions corresponding to the nodal displacements vector $\Phi(\mathbf{x})$ are obtained as

$$\Phi^T(\mathbf{x}) = \left\{ \phi_1(\mathbf{x}) \; \phi_2(\mathbf{x}) \; \cdots \; \phi_n(\mathbf{x}) \right\}. \tag{18}$$

Equation (16) can be re-written such that

$$u(\mathbf{x}) = \Phi^T(\mathbf{x})\mathbf{U}_s = \sum_{i=1}^{n} \phi_i u_i, \tag{19}$$

and the derivatives of $u(\mathbf{x})$ are easily obtained as

$$u_l(\mathbf{x}) = \mathbf{\Phi}_l^T(\mathbf{x})\mathbf{U}_s, \tag{20}$$

where l denotes either the coordinates x or y. A comma designates a partial differentiation with respect to the indicated spatial coordinate that follows.

3 Third Order Shear Deformation Theory

As a representative shear deformable laminate, a laminated plate of $a \times b \times c$ as shown in Fig. 1 is considered. The displacements of the plate in the (x,y,z) directions are denoted as (u,v,w), respectively.

Based on the third order deformation theory given by Reddy [11], the displacement field within one layer is assumed to be

$$\left\{ \begin{matrix} u \\ v \\ w \end{matrix} \right\} = \begin{bmatrix} 1 & 0 & -\gamma z^3 \frac{\partial}{\partial x} & z - \gamma z^3 & 0 \\ 0 & 1 & -\gamma z^3 \frac{\partial}{\partial y} & 0 & z - \gamma z^3 \\ 0 & 0 & 1 & 0 & 0 \end{bmatrix} \left\{ \begin{matrix} u_0 \\ v_0 \\ w_0 \\ \varphi_x \\ \varphi_y \end{matrix} \right\}, \quad \text{or } \mathbf{u} = \mathbf{H}\mathbf{u}_0, \tag{21}$$

where $\gamma = 4/(3h^2), h$ is the thickness of the laminate and (u_0, v_0, w_0) are the displacements of a point on the neutral-plane in the (x,y,z) direction, respectively. (φ_x, φ_y) are the rotations about the (x,y) axis. They are defined as

$$\mathbf{u}_0 = \left\{ \begin{matrix} u_0 \\ v_0 \\ w_0 \\ \varphi_x \\ \varphi_y \end{matrix} \right\} = \sum_{I=1}^{n} \begin{bmatrix} \phi_{uI} & 0 & 0 & 0 & 0 \\ 0 & \phi_{vI} & 0 & 0 & 0 \\ 0 & 0 & \phi_{wI} & 0 & 0 \\ 0 & 0 & 0 & \phi_{xI} & 0 \\ 0 & 0 & 0 & 0 & \phi_{yI} \end{bmatrix} \left\{ \begin{matrix} u_I \\ v_I \\ w_I \\ \varphi_{xI} \\ \varphi_{yI} \end{matrix} \right\} = \sum_{I=1}^{n} \mathbf{\Phi}_I \mathbf{u}_I \tag{22}$$

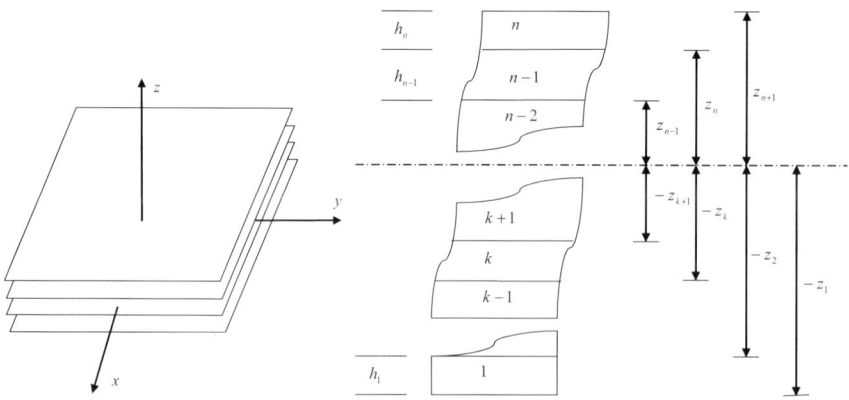

Fig. 1 A typical laminate plate and its coordinate system

where n is the number of nodes in the support domain of a point of interest x and ϕ_{uI}, ϕ_{vI}, ϕ_{wI}, ϕ_{xI}, and ϕ_{yI} are different shape functions which may be independent of each other. It should be noted, that in this work, the shape functions in all directions are set to be equal. The linear strains are given by

$$
\begin{Bmatrix} \varepsilon_{xx} \\ \varepsilon_{yy} \\ \varepsilon_{xy} \\ \varepsilon_{xz} \\ \varepsilon_{yz} \end{Bmatrix} =
\begin{bmatrix}
\frac{\partial}{\partial x} & 0 & -\gamma z^3 \frac{\partial^2}{\partial x^2} & (z - \gamma z^3)\frac{\partial}{\partial x} & 0 \\
0 & \frac{\partial}{\partial y} & -\gamma z^3 \frac{\partial^2}{\partial y^2} & 0 & (z - \gamma z^3)\frac{\partial}{\partial y} \\
\frac{\partial}{\partial y} & \frac{\partial}{\partial y} & -2\gamma z^3 \frac{\partial^2}{\partial x \partial y} & (z - \gamma z^3)\frac{\partial}{\partial y} & (z - \gamma z^3)\frac{\partial}{\partial x} \\
0 & 0 & (1 - \beta z^2)\frac{\partial}{\partial x} & (1 - \beta z^2) & 0 \\
0 & 0 & (1 - \beta z^2)\frac{\partial}{\partial y} & 0 & (1 - \beta z^2)
\end{bmatrix}
\begin{Bmatrix} u_0 \\ v_0 \\ w_0 \\ \varphi_x \\ \varphi_y \end{Bmatrix}, \quad \text{or } \boldsymbol{\varepsilon_p} = \mathbf{L} \mathbf{u_0}
$$

(23)

where $\beta = 3\gamma$.

Since most laminates are typically rather thin, a plane state of stress can be assumed. For an orthotropic laminate, the strain-stress relations can be denoted in the form of

$$
\boldsymbol{\sigma_p} = \mathbf{D} \boldsymbol{\varepsilon_p}, \mathbf{D} =
\begin{bmatrix}
\bar{Q}_{11} & \bar{Q}_{12} & \bar{Q}_{16} & 0 & 0 \\
\bar{Q}_{12} & \bar{Q}_{22} & \bar{Q}_{26} & 0 & 0 \\
\bar{Q}_{16} & \bar{Q}_{26} & \bar{Q}_{66} & 0 & 0 \\
0 & 0 & 0 & \bar{Q}_{44} & \bar{Q}_{45} \\
0 & 0 & 0 & \bar{Q}_{45} & \bar{Q}_{55}
\end{bmatrix},
$$

(24)

written in the global coordinate system of the whole plate, and the values $\bar{Q}_{ij}$ are derived as

$$
\begin{aligned}
\bar{Q}_{11} &= Q_{11} \cos^4 \alpha + 2(Q_{12} + 2Q_{66}) \sin^2 \alpha \cos^2 \alpha + Q_{22} \sin^4 \alpha, \\
\bar{Q}_{12} &= (Q_{11} + Q_{22} - 4Q_{66}) \sin^2 \alpha \cos^2 \alpha + Q_{12}(\sin^4 \alpha + \cos^4 \alpha), \\
\bar{Q}_{16} &= (Q_{11} - Q_{12} - 2Q_{66}) \sin \alpha \cos^3 \alpha + (Q_{12} - Q_{22} + 2Q_{66}) \sin^3 \alpha \cos \alpha, \\
\bar{Q}_{22} &= Q_{11} \sin^4 \alpha + 2(Q_{12} + 2Q_{66}) \sin^2 \alpha \cos^2 \alpha + Q_{22} \cos^4 \alpha, \\
\bar{Q}_{26} &= (Q_{11} - Q_{12} - 2Q_{66}) \sin^3 \alpha \cos \alpha + (Q_{12} - Q_{22} + 2Q_{66}) \sin \alpha \cos^3 \alpha, \quad (25) \\
\bar{Q}_{66} &= (Q_{11} + Q_{22} - 2Q_{12} - 2Q_{66}) \sin^2 \alpha \cos^2 \alpha + Q_{66}(\sin^4 \alpha + \cos^4 \alpha), \\
\bar{Q}_{44} &= Q_{44} \cos^2 \alpha + Q_{55} \sin^2 \alpha, \\
\bar{Q}_{45} &= (Q_{55} - Q_{44}) \cos \alpha \sin \alpha, \\
\bar{Q}_{55} &= Q_{55} \cos^2 \alpha + Q_{44} \sin^2 \alpha.
\end{aligned}
$$

All the Q_{ij} values are defined in the material coordinate of the laminate, where α is the angle of the fibre orientation of the ply, i.e., the ply angle. The Q_{ij} values are given by

$$
Q_{11} = \frac{E_1}{1 - v_{12} v_{21}}, \quad Q_{12} = \frac{v_{12} E_2}{1 - v_{12} v_{21}}, \quad Q_{22} = \frac{E_2}{1 - v_{12} v_{21}},
$$
$$
Q_{66} = G_{12}, \quad Q_{44} = G_{13}, \quad Q_{55} = G_{23}, \quad v_{21} E_1 = v_{12} E_2
$$

(26)

in which (E_i, G_{ij}, v_{ij}) are Young's modulus, shear modulus, and Poisson's ratio, respectively. Subscript 1 denotes the principle material (or fibre) direction.

4 Composite Damage Modelling

Damage in fibre laminates can be distinguished between damage in the prepreg (matrix and fibre) and delamination (Fig. 2).

A laminate will show stress redistribution after first ply failure and additional loading of the laminate will still be possible. Therefore, for a correct description of a laminate structure also the effects of damage evolution have to be described. Damage evolution influences the material properties of the damaged ply. For a modeling on a meso-level, the degradation of the properties at a ply level is needed. In this paper, a local material matrix is assigned to each node of a layer in the problem domain. In the case of damage at a node location after a load step, the respective material properties (assigned to the node) corresponding to the failure mode are set to be zero for the next load increment. This could be considered as a gradual unloading, since undamaged nodes continue to carry the load.

$$\mathbf{C_{n,1}} = \begin{bmatrix} Q_{11} & 0 & 0 & 0 & 0 \\ & 0 & 0 & 0 & 0 \\ & & 0 & 0 & 0 \\ & & & Q_{44} & 0 \\ sym & & & & Q_{55} \end{bmatrix} \quad \text{for matrix failure} \tag{27}$$

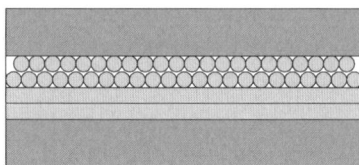

(a) Undamaged composite

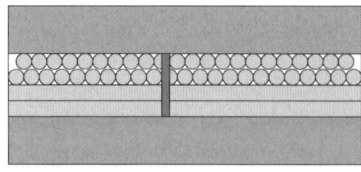

(b) Fibre and matrix failure   (c) Delamination

Fig. 2 Damage in composites

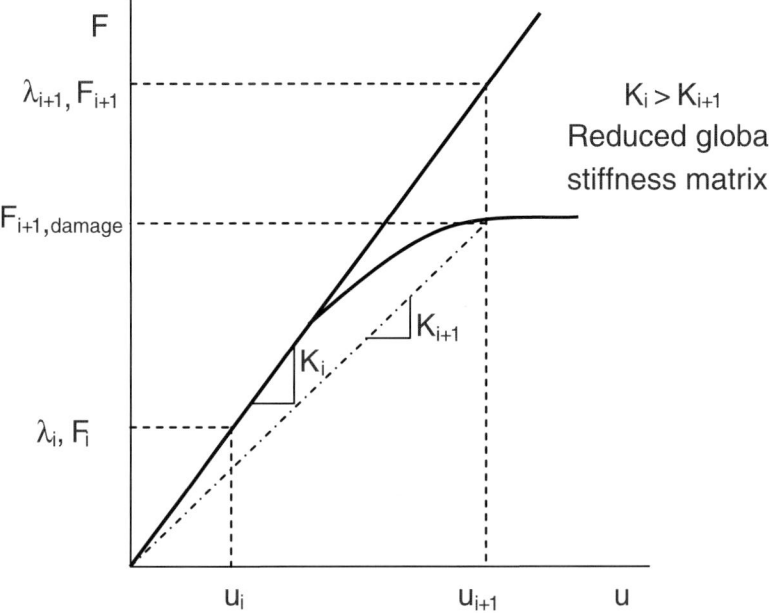

Fig. 3 Load displacement curve in a damage analysis

$$\mathbf{C_{n,1}} = \begin{bmatrix} 0 & Q_{12} & 0 & 0 & 0 \\ & Q_{22} & 0 & 0 & 0 \\ & & Q_{66} & 0 & 0 \\ & & & Q_{44} & 0 \\ sym & & & & Q_{55} \end{bmatrix} \quad \text{for fibre failure} \qquad (28)$$

$$\mathbf{C_{n,1}} = \begin{bmatrix} Q_{11} & Q_{12} & 0 & 0 & 0 \\ & Q_{22} & 0 & 0 & 0 \\ & & Q_{66} & 0 & 0 \\ & & & 0 & 0 \\ sym & & & & 0 \end{bmatrix} \quad \text{for delamination} \qquad (29)$$

with $\mathbf{C_{n,1}}$ as the layer local stiffness matrix, n as the node number and l as the layer number. Q_{ij} are defined in (26).

For a one dimensional consideration, Fig. 3 shows the load-displacement curve.

The failure criterion used here is a Hashin-type failure criterium with different failure criteria for matrix, fibre failure and delamination according to [16, 17].

For matrix failure the following criteria are used

$$f_m = \sqrt{\left(\frac{\sigma_{yy}}{\sigma_{yTM}}\right)^2 + \left(\frac{\tau_{xy}}{\tau_{xyTM}}\right)^2 + \left(\frac{\tau_{yz}}{\tau_{yzM}}\right)^2} \quad \sigma_{yy} \geq 0$$

$$f_m = \sqrt{\left(\frac{\sigma_{yy}}{\sigma_{yCM}}\right)^2 + \left(\frac{\tau_{xy}}{\tau_{xyM}}\right)^2 + \left(\frac{\tau_{yz}}{\tau_{yzM}}\right)^2} \quad \sigma_{yy} \leq 0$$

$$\qquad (30)$$

where σ_{yTM} and σ_{yCM} are the failure stresses perpendicular to the fibre direction in tension and compression, respectively. The failure stresses for shear are τ_{xyM} and τ_{yzM}. Failure occurs when f_m exceeds its threshold value 1.

The failure criterion for fibre failure is given by

$$
\begin{aligned}
f_f &= \sqrt{\left(\frac{\sigma_{xx}}{\sigma_{xTM}}\right)^2 + \left(\frac{\tau_{xy}}{\tau_{xyM}}\right)^2 + \left(\frac{\tau_{yz}}{\tau_{yzM}}\right)^2} \qquad \sigma_{xx} \geq 0 \\
f_f &= \sqrt{\left(\frac{\sigma_{xx}}{\sigma_{xCM}}\right)^2} \qquad \sigma_{xx} \leq 0
\end{aligned}
\tag{31}
$$

Here σ_{xTM} and σ_{xCM} are the failure stresses in fibre direction in tension and compression, respectively. Similar as before, failure occurs when f_f exceeds its threshold value 1.

The initiation of delamination is generally due to matrix failure between the prepregs. In this model, the failure criterion for delamination is given by

$$
f_d = \sqrt{\left(\frac{\tau_{xz}}{\tau_{xzM}}\right)^2 + \left(\frac{\tau_{yz}}{\tau_{yzM}}\right)^2}
\tag{32}
$$

Here τ_{xzM} and τ_{yzM} are the failure stresses in the xz and the yz directions. Failure occurs when f_d exceeds its threshold value 1.

5 Discrete System of Equations

The dynamic equations of a laminated plate can be derived by applying the principle of virtual work to an elastic structure under dynamic loading. The principle requires that the work of discrete forces, body forces, and surface tractions due to an infinitesimal virtual displacement should be equal to the sum of strain energy and dissipated strain energy variations, i.e.,

$$
\int_{\Omega} \delta \mathbf{u}^T (-\rho \ddot{\mathbf{u}}) d\Omega + \int_{\Omega} \delta \mathbf{u}^T \mathbf{b} d\Omega + \int_{\Gamma_t} \delta \mathbf{u}^T \bar{\mathbf{t}} d\Gamma = \int_{\Omega} \delta \boldsymbol{\varepsilon}^T \mathbf{C} \dot{\boldsymbol{\varepsilon}} d\Omega + \int_{\Omega} \delta \boldsymbol{\varepsilon}^T \boldsymbol{\sigma} d\Omega \tag{33}
$$

where $\int_{\Omega} \delta \mathbf{u}^T \mathbf{b} d\Omega$ is the virtual work of non-inertial body forces, $\int_{\Omega} \delta \mathbf{u}^T (-\rho \ddot{\mathbf{u}}) d\Omega$ represents the virtual work of the inertial body forces, and $\int_{\Gamma_t} \delta \mathbf{u}^T \bar{\mathbf{t}} d\Gamma$ is the virtual work of the surface tractions. The term $\int_{\Omega} \delta \boldsymbol{\varepsilon}^T \boldsymbol{\sigma} d\Omega$ denotes the variation of the specific strain energy and $\int_{\Omega} \delta \boldsymbol{\varepsilon}^T \mathbf{C} \dot{\boldsymbol{\varepsilon}} d\Omega$ is the dissipated specific strain energy during the virtual motion.

Substituting Eq. (22) into Eq. (33), the well known equation of motion can be obtained as

$$\mathbf{M}\ddot{\mathbf{U}} + \mathbf{P}\dot{\mathbf{U}} + \mathbf{K}\mathbf{U} = \mathbf{F}, \tag{34}$$

where $\mathbf{M}$, $\mathbf{P}$ and $\mathbf{K}$ are the mass, damping and stiffness matrix, respectively. $\mathbf{F}$ is the dynamic global force vector.

For the static analysis, as conducted in this paper, $\mathbf{M}$ and $\mathbf{P}$ are set to be zero. Equation (34) then becomes

$$\mathbf{K}\mathbf{U} = \mathbf{F}. \tag{35}$$

The stiffness matrix and force vector are formed by assembling the matrices and vectors associated with the nodes I and J, and with the layer k, as given by

$$\mathbf{K} = \sum_{k=1}^{P}\sum_{I=1}^{N}\sum_{J=1}^{N}\mathbf{K}_{kIJ}; \quad \mathbf{K}_{kIJ} = \int_{\Omega}\int_{z_k}^{z_{k+1}}\mathbf{\Phi}_I^T\mathbf{L}^T\mathbf{DL}\mathbf{\Phi}_J dz d\Omega = \int_{\Omega}\int_{z_k}^{z_{k+1}}(\mathbf{B}_I^T)_{5\times5}\mathbf{D}(\mathbf{B}_J)_{5\times5}dz d\Omega \tag{36}$$

$$\mathbf{F} = \sum_{k=1}^{P}\sum_{I=1}^{N}\mathbf{F}_{kI}; \quad \mathbf{F}_{kI} = \mathbf{F}^b + \mathbf{F}^t = \int_{\Omega}\int_{z_k}^{z_{k+1}}\mathbf{A}_I^T\mathbf{b}dz d\Omega + \int_{\Gamma_t}\int_{z_k}^{z_{k+1}}\mathbf{A}_I^T\mathbf{t}dz d\Gamma \tag{37}$$

$$\mathbf{A}_I = \mathbf{H}\mathbf{\Phi}_I = \begin{bmatrix} \phi_{uI} & 0 & -\gamma z^3\phi_{wI,x} & (z-\gamma z^3)\phi_{xI} & 0 \\ 0 & \phi_{vI} & -\gamma z^3\phi_{wI,y} & 0 & (z-\gamma z^3)\phi_{yI} \\ 0 & 0 & \phi_{wI} & 0 & 0 \end{bmatrix} \tag{38}$$

$$\mathbf{B}_I = \mathbf{L}\mathbf{\Phi}_I = \begin{bmatrix} \phi_{uI,x} & 0 & -\gamma z^3\phi_{wI,xx} & (z-\gamma z^3)\phi_{xI,x} & 0 \\ 0 & \phi_{vI,y} & -\gamma z^3\phi_{wI,yy} & 0 & (z-\gamma z^3)\phi_{yI,y} \\ \phi_{uI,y} & \phi_{vI,x} & -2\gamma z^3\phi_{wI,xy} & (z-\gamma z^3)\phi_{xI,y} & (z-\gamma z^3)\phi_{yI,x} \\ 0 & 0 & (1-\beta z^2)\phi_{wI,x} & (1-\beta z^2)\phi_{xI} & 0 \\ 0 & 0 & (1-\beta z^2)\phi_{wI,y} & 0 & (1-\beta z^2)\phi_{yI} \end{bmatrix} \tag{39}$$

In Eq. (36), P is the number of layers and N the numbers of nodes. In this work, the force vector is assumed to be applied at $z = 0$.

$\mathbf{U}$ in Eq. (35) represents the global nodal parameters. The corresponding nodal displacements are obtained by using Eq. (22). Finally, the penalty method is employed for enforcing the essential boundary condition.

6 Numerical Examples

6.1 Static Deflection Analysis

To examine the efficiency and accuracy of the present formulations, static deflection analyses are studied. All examples consider laminated composites with length $a = b = 10.0\,\mathrm{mm}$, subjected to transverse loads. The material properties for each layer are

$$\begin{aligned} E_1 &= 25.0E_2, \quad G_{12} = G_{13} = 0.5E_2, \\ G_{23} &= 0.2E_2, \quad \nu_{12} = \nu_{13} = 0.25 \end{aligned} \tag{40}$$

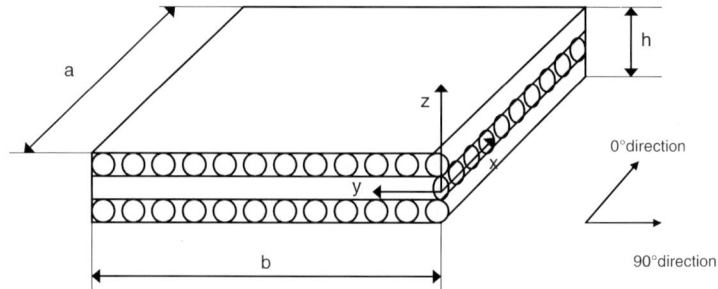

Fig. 4 A $[0°/90°/0°]$ symmetric laminate

Fig. 5 Panel loading for static
analysis

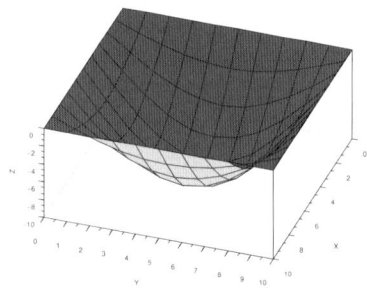

Two symmetric laminates, $[0°/90°/90°/0°]$ and $[0°/90°/0°]$, as well as one non-symmetric laminate $[0°/90°]$ are considered with the origin of the coordinate system being located at the lower right corner of the mid-plane, see Fig. 4.

A sinusoidal distribution of the transverse load for static analysis (Fig. 5) is defined as

$$f(x,y) = f_0 \cos \frac{\pi \left(x - \frac{a}{2}\right)}{a} \sin \frac{\pi y}{b}. \tag{41}$$

In all cases, a quadratic influence domain is used with an edge length of three times the average nodal spacing. Integration of the weak form is carried out with a 4×4 gauss quadrature order over each cell and 2×2 gauss quadrature order over each layer.

Different kinds of boundary conditions are used. They are summarized in the following table (Table 2), where S means "simply supported", C refers to "fully clamped", and F stands for "free".

Several cases are studied here. If not specified otherwise, regular rectangle background meshes (20×20) are used for the integration of the stiffness matrix or the load vectors, and regularly distributed field nodes (21×21) are employed for the approximation of the field variables. The results for deflections are presented in normalized form as

$$\bar{w} = w_0(a/2, b/2)\frac{h^3 E_2}{b^4 f_0} \times 10^2. \tag{42}$$

Figure 6 shows a typical out-of-plane displacement distribution of the plate.

Table 2 Applied boundary conditions

Name	$x = 0.0$	$x = a$	$y = 0.0$	$y = b$
1 SSSS	$v_0 = w_0 = \phi_y = 0$	$v_0 = w_0 = \phi_y = 0$	$u_0 = w_0 = \phi_x = 0$	$u_0 = w_0 = \phi_x = 0$
2 CCCC	$u_0 = v_0 = w_0$ $= \phi_x = \phi_y = 0$	$u_0 = v_0 = w_0$ $= \phi_x = \phi_y = 0$	$u_0 = v_0 = w_0$ $= \phi_x = \phi_y = 0$	$u_0 = v_0 = w_0$ $= \phi_x = \phi_y = 0$
3 SSCC	$v_0 = w_0 = \phi_y = 0$	$v_0 = w_0 = \phi_y = 0$	$u_0 = v_0 = w_0$ $= \phi_x = \phi_y =$	$u_0 = v_0 = w_0$ $= \phi_x = \phi_y = 0$
4 SSFF	$v_0 = w_0 = \phi_y = 0$	$v_0 = w_0 = \phi_y = 0$	–	–

Fig. 6 $[0°/90°/0°]$ typical out-of-plane displacements distribution

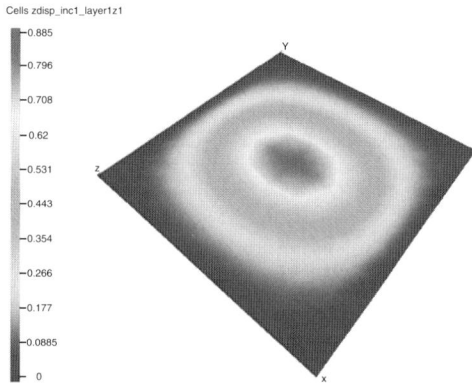

Table 3 Non-dimensionalized maximum deflections $\bar{w}$ in simply supported symmetric cross-ply $[0°/90°/0°]$ square laminates under sinusoidally distributed transverse load using different nodal densities $(a/h = 10)$

Theory, source	$\bar{w}$						$\bar{w}$ [11]
	9×9	11×11	13×13	16×16	19×19	21×21	
CPT [13]	0.4297	0.4299	0.4303	0.4305	0.4307	0.4308	0.4313
TSDT [13]	0.7117	0.7121	0.7124	0.7124	0.7124	0.7125	0.7125
TSDT Present	0.7273	0.7243	0.7219	0.7198	0.7170	0.7162	0.7125

Six different nodal densities are used to demonstrate the convergence of the method applied on a cross-ply symmetric laminate $[0°/90°/0°]$. Simply supported boundary conditions SSSS are assumed. The results are presented in Table 3 together with the FEM solutions obtained by Reddy [11] and results of the classical laminate theory as well as the EFG third order theory given by Dai [13].

It can be seen that a very good convergence is achieved. The accuracy increases with an increasing number of the field nodes.

The next case is a cross-ply symmetric square laminate $[0°/90°/90°/0°]$ with four layers of equal thickness. It is subjected to the sinusoidally distributed transverse load defined above. Four kinds of side-to-thickness ratios are used for the simulations. The results are listed in Table 4. Simply supported boundary conditions SSSS are applied. The solutions by the FEM of Reddy [11], the EFG of Dai [13], and the

Table 4 Non-dimensionalized maximum deflections $\bar{w}$ in simply supported symmetric cross-ply $[0°/90°/90°/0°]$ square laminates under sinusoidally distributed transverse load

Theory, source		a/h			
		4	10	20	100
CPT [11]	0.431				
FSDT [11]		1.710	0.663	0.491	0.434
TSDT [11]		1.894	0.715	0.506	0.434
TSDT [13]		1.8930	0.7147	0.5060	0.4342
TSDT, N = 11 [15]		1.8848	0.7115	0.5039	0.4496
ELS [12]		1.954	0.743	0.517	0.438
TSDT Present		1.9034	0.7169	0.5070	0.4337

Table 5 Non-dimensionalized maximum deflections $\bar{w}$ in simply supported symmetric cross-ply $[0°/90°/0°]$ square laminates under sinusoidally distributed transverse load

Theory, source		a/h		
		4	10	100
CPT [11]	0.4313			
TSDT [11]		1.9218	0.7125	0.4342
TSDT [13]		1.9210	0.7124	0.4341
TSDT Present		1.9475	0.7162	0.4357

multi-quadratic radial basis method (MRBM) of Ferreira [15] are also given in the table together with those by the 3-D elasticity theory (ELS [12]).

It can be observed that the third order theory gives more accurate results compared to the first order deformation theory. The present method gives results quite similar to those of the FEM and EFG. The difference between the TSDT and the ELS gets smaller with increasing ratio a/h, whereas the differences between the TSDT remain very little. This means, that the present method is rather accurate not only for thin laminates but also for thick ones.

A cross-ply laminate with three layers $[0°/90°/0°]$ is also examined using different side-to-thickness ratios. The results are presented in Table 5. They are compared with FEM solutions given by Reddy [11] and the EFG results of Dai [13]. It can be seen that the present method also shows good agreements with both methods.

In the presented cases, the calculations are carried out for symmetric laminates with the same boundary conditions. For non-symmetric laminates, however, there exists coupling between bending and extension, which complicates the simulation. Therefore, a two-layer non-symmetric cross-ply laminate is also considered, see Table 6. The results are once more compared to the FEM [11] and EFG [13]. Good agreements can also be reached for SSSS and SSFF boundary conditions. For SSCC a slight difference is observed for the thick laminate. The difference decreases the thinner the laminate gets.

Table 6 Non-dimensionalized maximum deflections $\bar{w}$ of non-symmetric cross-ply $[0°/90°]$ square laminates under sinusoidally distributed transverse load with various boundary conditions

b/h	Theory, source	SSSS	SSCC	SSFF
5	Analytical [11]	1.667	1.088	2.624
	TSDT [11]	1.667	1.068	2.647
	TSDT [13]	1.636	1.058	2.584
	TSDT Present	1.710	1.435	2.695
10	Analytical [11]	1.216	0.617	1.992
	TSDT [11]	1.214	0.605	2.002
	TSDT [13]	1.205	0.596	1.954
	TSDT Present	1.228	0.786	2.015

Table 7 Limit values for the damage analysis

τ_{xTM} (MPa)	τ_{xCM} (MPa)	τ_{yTM} (MPa)	τ_{yCM} (MPa)	τ_{xyM} (MPa)	τ_{xzM} (MPa)	τ_{yzM} (MPa)
2,430	2,000	50	160	50	50	50

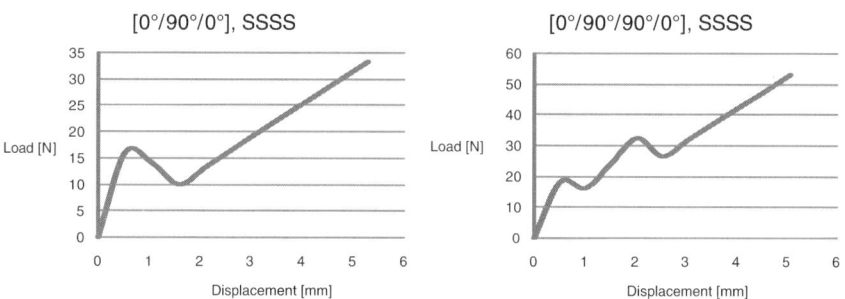

Fig. 7 Calculated load – displacement curves

6.2 Damage Analysis

For a representative damage analysis, the same plate as for the preceding investigations is used. Its edge length to thickness ratio is $a/h = 50$. A concentrated load is applied in the middle of the plate, and simply supported (SSSS) as well as clamped boundary conditions (CCCC) are assumed. Two different laminate layups, namely $[0°/90°/0°]$ and $[0°/90°/90°/0°]$, are considered. The node density was chosen to be (11×11). For the material degradation analysis, the limit values for matrix and fibre failure are given in Table 7.

The load displacement curves obtained for $[0°/90°/0°]$ and $[0°/90°/90°/0°]$ laminates with SSSS boundary conditions are shown in Fig. 7.

It can be noted that the global stiffness matrix changes, i.e. the stiffness is reduced, as soon as failure occurs. The increments following the failure occurrence is then computed with the reduced stiffness matrix. One load peak is observed for

Table 8 Out-of-plane displacements for linear (simple static) and nonlinear (with possible damage) analysis

	$[0°/90°/0°]$, linear, load $= 33.3$ N (mm)	$[0°/90°/0°]$, nonlinear, load $= 33.3$ N (mm)	$[0°/90°/90°/0°]$, linear, load $= 53.3$ N (mm)	$[0°/90°/90°/0°]$, nonlinear, load $= 53.3$ N (mm)
SSSS	0.89	5.30	1.35	5.06
CCCC	0.38	2.27	0.56	2.12

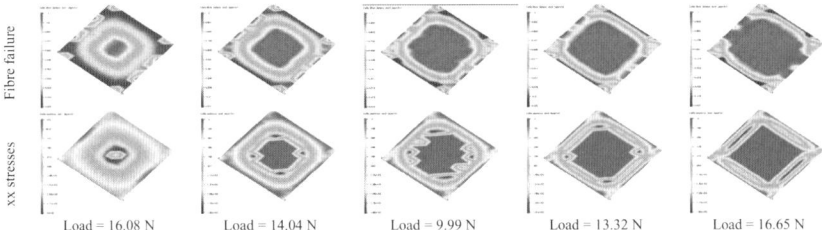

Load = 16.08 N   Load = 14.04 N   Load = 9.99 N   Load = 13.32 N   Load = 16.65 N

Fig. 8 Fibre failure index and xx stresses in the first layer of a $[0°/90°/0°]$ laminate

$[0°/90°/0°]$ laminate and two for $[0°/90°/90°/0°]$ laminate. These peaks represent the 90° layers failure onset.

The out of plane displacements can be compared to those of the static deflection analysis with the same parameters, geometry and loading, see Table 8.

In these examples, the reduction of the global stiffness matrix implies a large displacement for both simply supported and clamped boundary conditions as expected. The effects of clamped boundary conditions on the out-of-plane displacements can also be noted. Clamped boundary conditions imply a stiffening of the panel.

A substantial amount of data has been investigated. For reason of conciseness, however, only results by a $[0°/90°/0°]$ laminate with simply supported boundary conditions are presented here.

The gradual degradation of the material of the first layer can be observed. By increasing the load, fibre failure occurs in the middle of the plate. Fibres that fail cannot carry any additional loading. This may be seen in the stress map of the same plates given in Fig. 8. As expected, stress concentration around the damaged area can be observed.

Figures 9 and 10 show the matrix failure index in the first layer and the delamination index between the first and the second layers. In contrary to the matrix failure that appears in the middle of the plate first, delamination begins near the middle of the plate.

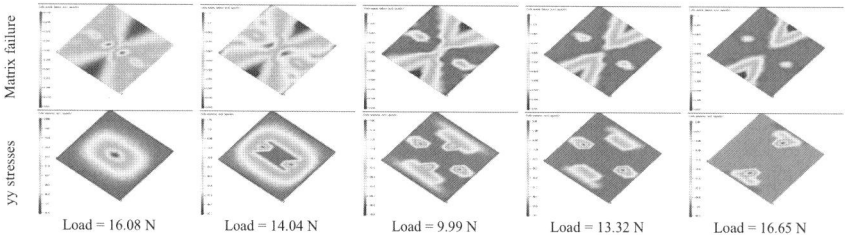

Fig. 9 Matrix failure index and yy stresses in the second layer of a $[0°/90°/0°]$ laminate

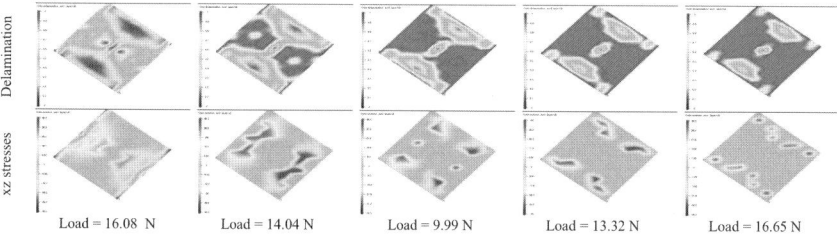

Fig. 10 Delamination and xz stresses between the first and second layers of a $[0°/90°/0°]$ laminate

7 Conclusions

In this paper, the Reddy higher-order shear deformation theory of laminated elastic shells for static deflection and damage analyses has been implemented representing the field variables by a set of regularly distributed nodes. In fact, a meshless procedure, namely the radial point interpolation method, has been chosen to investigate the mechanical behaviour of the laminated systems. It may be expected that the higher order theory leads to more accurate results than the classical or first order shear deformation theory and no shear correction factor is needed.

The essential boundary conditions have been imposed by the penalty method. The stiffness matrix was assembled using a 4×4 gauss quadrature over the cells and a 2×2 gauss quadrature over the layers.

Two representative numerical examples have been given: The first one is dealing with laminates of different side-to-thickness ratios, boundary conditions, and ply angles. To demonstrate the applicability and the accuracy of the new approach, the obtained results have been compared to the solutions of the three commonly used laminate theories. Moreover, comparisons with FEM results as well as with analytical and other meshless methods available from literatures have been conducted. In all cases, very good agreements could be achieved showing the convergence and the accuracy of the new approach. In the second example, the new model has been used for a damage analysis of different laminate layups. The formulation is based on Hashin type failure criteria. Also here the new approach led to results which seem to be quite reliable. A detailed verification is planned in the near future.

References

1. X.L. Chen, G.R. Liu, S.P. Lim, An element free Galerkin method for the free vibration analysis of composite laminates of complicated shape. Composite Structures 59 (2003) 279–289.
2. Y.Y. Lu, T. Belytschko, L. Gu, A new implementation of the element free Galerkin method. Computational Methods in Applied Mechanics and Engineering 113 (1994) 397–414.
3. S.N. Atluri, T. Zhu, A new meshless local Petrov-Galerkin (MLPG) approach in computational mechanics. Computational Mechanics 22 (1998) 117–127.
4. J. Dolbow, T. Belytschko, An introduction to programming the meshless element-free Galerkin method. Archives of Computational Methods in Engineering 5 (1998) 207–241.
5. T. Belytscko, Y.Y. Lu, L. Gu, Crack propagation by element-free Galerkin methods. Engineering Fracture Mechanics 32 (1995) 2547–2570.
6. P. Krysl, T. Belytscko, Analysis of thin plates by the element-free Galerkin method. Computational Mechanics 17 (1995) 26–35.
7. G.R. Liu, Y.T. Gu, A point interpolation method for two-dimensional solids. International Journal for Numerical Methods in Engineering 50 (2001) 937–951.
8. G.R. Liu, Y.T. Gu, A local radial point interpolation method (LR-PIM) for free vibration analysis of 2-D solids. Journal of Sound and Vibration 246 (2001-1) 29–46.
9. G.R. Liu, Mesh free methods: moving beyond the finite element method, CRC Press, New York, 2002.
10. G.R. Liu, Y.T. Gu, An introduction to meshfree methods and their programming, Springer, New York, 2005.
11. J.N. Reddy, Mechanics of laminated composite plates - theory and analysis, CRC Press, New York, 1997.
12. N.J. Pegano, S.J. Hatfield, Elastic behaviour of multilayered bi-directional composites. American Institute of Aeronautics and Astronautics Journal 10 (1972) 931–933.
13. K.Y. Dai, G.R. Liu, K.M. Lim, X.L. Chen, A mesh-free method for static and free vibration analysis of shear deformable laminated composite plates. Journal of Sound and Vibration 269 (2004) 633–652.
14. J.G. Wang, G.R. Liu, Radial point interpolation method for no-yielding surface models, Proceedings of the first M.I.T. conference on computational fluid and solid mechanics, 12–14 June 2001, 538–540.
15. A.J.M. Ferreira, C.M.C. Roque, R.M.N. Jorge, Static and free vibration analysis of composite shells by radial basis functions. Engineering Analysis with Boundary Elements 30 (2006) 719–733.
16. Z. Hashin, A. Rotem, A fatigue criterion for fiber-reinforced materials. Journal of Composite Materials 7 (1973) 448–464.
17. Z. Hashin, Failure criteria for unidirectional fiber composites. Journal of Applied Mechanics 47 (1980) 329–334.

Analysis of Tensile Structures with the Element Free Galerkin Method

Bruno Figueiredo(✉) and Vitor M.A. Leitão

Abstract In this work an implementation of a meshless method, the element free Galerkin method proposed by Belytschko et al. [1], for the analysis of three-dimensional laminar (thin) anisotropic structures is presented. By using the mapping technique proposed by Noguchi [5] the geometry of arbitrary curved surfaces is expanded in the two-dimensional space and the bases of convected co-ordinate system are utilized for expressing the strain and stress components in the virtual work principle. The nodes are generated in this two-dimensional space and the convected co-ordinates are used in the moving least-squares approximation of the displacement field. Generally shaped three-dimensional tensile structures require geometrically non-linear analysis. In the work described herein these effects are formulated in terms of the total Lagrangian method.

Keywords: Membrane structures · non-linear geometrical analysis · total Lagrangian method · EFGM

1 Introduction

The EFGM, originally developed by Belytschko et al. [1, 2], relies on nodes rather than on 'elements. The elements are replaced by sets of nodes within domains of influence of a given node. The variable of interest (the displacements, in our case) is approximated at a given point by finding the nodes within its domain of influence and by constructing the corresponding approximation function. These functions are found by means of the Moving Least-Squares concept which is basically a Least-Squares procedure that, as the name says, is not "statically" defined, it moves as the domains of influence for each given node move.

B. Figueiredo and V.M.A. Leitão

DECivil/ICIST, Instituto Superior Técnico, TU Lisbon, Portugal, e-mail: brunof@civil.ist.utl.pt, e-mail: vitor@civil.ist.utl.pt

A.J.M. Ferrcira et al. (eds.) *Progress on Meshless Methods, Computational Methods in Applied Sciences.*

The MLS way of constructing the approximation functions does not require the definition of elements but makes it harder (than with traditional mesh methods such as the FEM) to impose the essential boundary conditions as the approximation functions no longer verify the Kronecker delta property.

Another aspect of the EFG formulation that requires a more careful analysis is that of the numerical integrations needed to assemble the systems of equations. In the EFGM this is normally carried out by resorting to a background integration grid [3].

The analysis of general membrane structures on the basis of methods without elements still in an incipient phase when compared with the application with other types of structural problems, nominated, shells, plates, problems these for which the results with the EFGM are sufficiently positive.

The objectives of this study are to further enhance the EFGM for the analysis of general three-dimensional tensile anisotropic structures.

The outline of the work is as follows. In Section 2, the moving least-squares method is briefly reviewed. The EFGM formulation for the analysis of membrane structures is described in Section 3 whereas Section 4 deals with its numerical implementation. In Section 5 three geometrically non-linear membrane problems are illustrated to validate the proposed method. Finally, conclusions are presented in Section 6.

2 MLS Approximation and EFGM

As referred above, the Element Free Galerkin method relies on Moving Least-Squares to construct the approximation functions. This technique will now be briefly described for the one-dimensional case as shown in Fig. 1.

A component of displacement $u(x)$ is approximated by a basis of functions (usually, but not necessarily, of the polynomial type), as follows, where n is the number of terms of the basis:

$$u^h(x) = \sum_{j=1}^{n} p_j(x)a_j(x) = p^T(x)a(x) \tag{1}$$

Coefficients $a(x)$ are determined by minimizing the following weighted functional:

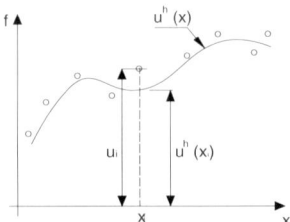

Fig. 1 Schematic representation of the moving least squares concept

$$J = \sum_{I=1}^{n} w(x - x_I)[u_L^h(x_I, x) - u_I]^2 = \sum_{I=1}^{n} w(x - x_I)[p^T(x_I)a(x) - u_I]^2 \qquad (2)$$

$$\frac{dJ}{da} = 0 \; a(x) = A(x)^{-1}B(x)u \qquad (3)$$

$$A(x) = \sum_{i}^{n} w(x - x_I)p(x_I)p^T(x_I) \qquad (4)$$

$$B(x) = \sum_{i}^{n} w(x - x_I)p^T(x_I) \qquad (5)$$

where u_I is the (unknown) nodal value of u at a sampling point x_I and the $w(x - x_I)$ is a weight function, $w(r)$, which satisfies $w(\rho) = 0$, $dw/dr(\rho) = 0$ and $d^2w/dr^2(\rho) = 0$ (ρ is a radius of the domain of influence at x):

Substituting Eq. (2) into Eq. (1), $u^h(x)$ can be represented by the nodal value u_I, as:

$$u^h(x) = p^T(x)A^{-1}(x)B(x)u = \phi^T(x)u = \sum_{I=1}^{n} \phi_I(x)u_I \qquad (6)$$

where $\phi_I(x)$ is a component of vector ϕ which corresponds to a shape function in FEM. Since the approximation is based on the moving least-squares concept, which is not an interpolation, the displacement u^h at a node x_I usually does not coincide with the nodal value u^h.

3 Membrane Structures Analysis by Using EFGM

Membrane structures are characterized by being very thin and by being acted upon with loads that may only be of the tensile/compressive type, that is, no bending is present.

For membranes of general shape and, especially, if they are subjected to transversal loads, it is essential to take into account the geometric stiffness of the membrane, that is, to consider geometrically non-linear effects. In this section the total Lagrangian method for tensile structures analysis with element free Galerkin method is briefly described.

By using the mapping technique the geometry of arbitrary curved surface is expanded in the two dimensional space, and the bases of convected co-ordinate system are utilized for the expression of strain and stress components involved in the virtual work principle.

Only nodal data are generated on this two-dimensional space and the convected co-ordinates are used in the moving least squares interpolation method for the approximation of the displacement field.

3.1 Definition of Curved Surface

There are several ways to approximate the surface geometry of spatial structures. To be consistent with the approximation of the displacement fields, Krysl et al. [4] use the moving least-squares interpolation for the surface approximation of general shells and investigate its properties in detail.

In the numerical tests presented in this paper, the curved surface, which has C^1 continuity, can be expressed as an explicit form of an analytical function.

For the general curved surface, the Cartesian co-ordinates of the appropriate number of sampling points should be given in advance and the spline function or the Lagrange polynomial function must be defined.

It should be noted that in case the structure is not composed of smooth surfaces only and C^1 continuity is not satisfied, the proposed method cannot be applied directly and special care is needed for the C^0 continuity condition at the intersections.

In the proposed formulation, which follows closely the work of Noguchi [5], the three-dimensional geometry of 3D membranes is expanded into the two-dimensional space by mapping from the Cartesian co-ordinate system to the convected co-ordinate system, as shown in Fig. 2.

The position vector X on the initial configuration of the mid-surface can be expressed as the explicit function, $f(r^1, r^2)$ when Lagrange polynomial functions are used:

$$X = f(r^1, r^2) = \sum_{J=1}^{l} \sum_{I=1}^{k} H_I(r_1) H_J(r_2) X(r_I^1, r_J^2) \tag{7}$$

$$H_I(r^1) = \frac{(r^1 - r_1^1)(r^1 - r_2^1)...(r^1 - r_{I-1}^1)(r^1 - r_{I+1}^1)...(r^1 - r_k^1)}{(r_I^1 - r_1^1)(r_I^1 - r_2^1)...(r_I^1 - r_{I-1}^1)(r_I^1 - r_{I+1}^1)...(r_I^1 - r_k^1)} \tag{8}$$

$$H_J(r^2) = \frac{(r^2 - r_1^2)(r^2 - r_2^2)...(r^2 - r_{J-1}^2)(r^2 - r_{J+1}^2)...(r^2 - r_l^2)}{(r_J^2 - r_1^2)(r_J^2 - r_2^2)...(r_J^2 - r_{J-1}^2)(r_J^2 - r_{J+1}^2)...(r_J^2 - r_l^2)} \tag{9}$$

To define the curved surface, $k \times l$ sampling points are needed. If the range of convected coordinate is $-1 \leq r^i \leq 1$ ($i = 1, 2$) and each sampling point is distributed uniformly in both r^1 and r^2 directions on the convected co-ordinate system, $H_I(r^1)$ and $H_J(r^2)$ coincide with the Lagrangian shape functions used in finite elements.

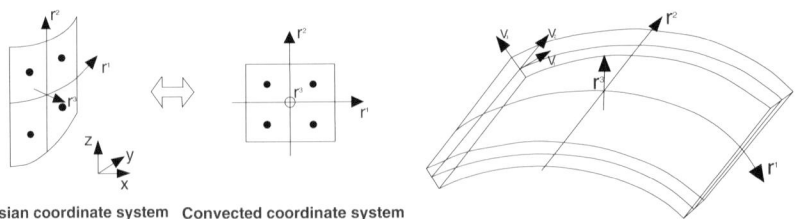

Cartesian coordinate system   Convected coordinate system

Fig. 2 (*Left*) Mapping of curved surface; (*Right*) three-dimensional shape

Fig. 3 Generated nodes and quadrature points on $r^1 - r^2$ space in the convected co-ordinate system

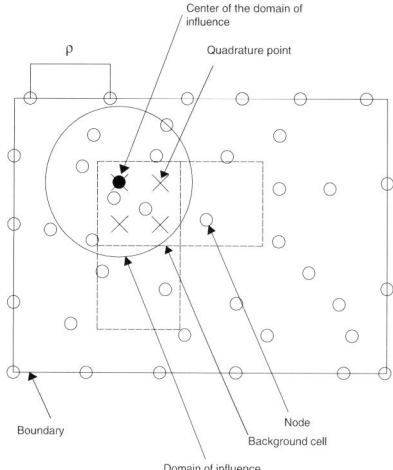

Only the nodal data is generated in this two-dimensional space and the convected co-ordinates $r^1 - r^2$ (Fig. 3) are used, in the moving least-squares technique, for the approximation of the displacement field. The stresses and the strains in the virtual work statement are defined in this coordinate system (Noguchi [5]).

3.2 Total Lagrangian Method

As seen before, geometrically non-linear analysis is required. The total Lagrangian method is adopted in the formulation. The bases of convected co-ordinate system are utilized for the expression of the Green-Lagrange strain tensor, the second Piola-Kirchhoff stress tensor and the fourth-order constitutive tensor (Noguchi [5]). Green-Lagrange strain tensor may be written as follows:

$$ {}^t_0 E = \frac{1}{2} ({}^t g_i {}^t g_j - G_i G_j) G_i \otimes G_j \tag{10} $$

where G_i and G^i are covariant and contravariant base vectors at time 0, respectively, and g_i is the covariant base vector at time t.

According to the definition of base vector in the convected co-ordinate system:

$$ G_i = \frac{\partial X}{\partial r^i} = \sum_{I=1}^{K} \frac{\partial N_I(r^1, r^2)}{\partial r^i} X_I, \quad i = 1, 2 \tag{11} $$

$$ g_i = \frac{\partial^t X}{\partial r^i} = \frac{\partial (X + {}^t u)}{\partial r^i} = G_i + \frac{\partial^t u}{\partial r^i}, \quad i = 1, 2 \tag{12} $$

Substituting Eqs. (11, 12) into Eq. (10), the Green-Lagrange strain tensor and the incremental Green-Lagrange strain from time t to $t' = t + \Delta t$ are given as follows:

$$
{}_0^t E = \frac{1}{2} \left(G_i \frac{\partial^t u}{\partial r^j} + G_j \frac{\partial^t u}{\partial r^i} + \frac{\partial^t u}{\partial r^i} \frac{\partial^t u}{\partial r^j} \right) G_i \otimes G_j
\tag{13}
$$

$$
{}_0 E = {}_0^{t'} E - {}_0^t E = \frac{1}{2} \left({}^t g_i \frac{\partial u}{\partial r^j} + {}^t g_j \frac{\partial u}{\partial r^i} + \frac{\partial u}{\partial r^i} \frac{\partial u}{\partial r^j} \right) G_i \otimes G_j = (e_{ij} + \eta_{ij}) G^i \otimes G^j
\tag{14}
$$

where e is the linear and η the non-linear part of ${}_0 E$, respectively.

As for the constitutive equation, the constant fourth-order elastic constitutive tensor, ${}_0 C$, is assumed in the relation between ${}_0^t S$ and ${}_0^t E$.

$$
{}_0^t S^{ij} = {}_0 C^{ijkl} {}_0^t E_{kl}
\tag{15}
$$

Taking material time derivative, we have equivalently the following constitutive equations:

$$
{}_0^t \dot{S}^{ij} = {}_0 C^{ijkl} {}_0^t \dot{e}_{kl}
\tag{16}
$$

Elastic constants are usually defined as tensor components decomposed by the orthonormal base vectors in the local Cartesian co-ordinate system (Fig. 2), such as $V_i (i = 1, 2, 3)$. Denoting these constitutive tensor components by ${}_0 \bar{C}^{mnop}$, the constitutive tensor components are obtained by using the transformation:

$$
{}_0 C^{ijkl} = (G^i \cdot V_m)(G^j \cdot V_n)(G^k \cdot V_o)(G^l \cdot V_p) {}_0 \bar{C}^{mnop}
\tag{17}
$$

$$
G^i = \frac{G_j \times G_k}{[G_1 G_2 G_3]} \qquad (i, j, k) = (1, 2, 3), (2, 3, 1), (3, 1, 2)
\tag{18}
$$

where [abc] shows a scalar triple product and is defined as $a \times b \cdot c$. The director unit normal V_3 is obtained by outer product of the covariant base vectors G_1 and G_2, as:

$$
V_3 = \frac{G_1(r^1, r^2, 0) \times G_2(r^1, r^2, 0)}{|G_1(r^1, r^2, 0) \times G_2(r^1, r^2, 0)|}
\tag{19}
$$

where G_i is obtained from Eq. (11). V_1 may be determined by the projection of e_1, the first unit base vector in the global Cartesian co-ordinate system. Hence:

$$
V_2 = \frac{V_3 \times e_1}{|V_3 \times e_1|}, \quad V_1 = V_2 \times V_3
\tag{20}
$$

Finally, the equation of virtual work can be written in the following way:

$$\int_V {}^t_0S : \delta e\, dV = \int_{r_1}\int_{r_2}\int_{r_3} {}^t_0S^{ij}\delta e_{ij}[G_1G_2G_3]dr^1dr^2dr^3$$

$$= \int_V {}^t_0\dot{S} : \delta e dV + \int_V {}^t_0S : (\delta\eta)^\bullet dV$$

$$\int_{r_1}\int_{r_2}\int_{r_3} {}_0C^{ijkl}\dot{e}_{kl}\delta e_{ij}[G_1G_2G_3]dr^1dr^2dr^3 + \int_{r_1}\int_{r_2}\int_{r_3} {}^t_0S^{ij}(\delta\eta_{ij})^\bullet[G_1G_2G_3]dr^1dr^2dr^3$$

$$dV = [G_1G_2G_3]dr^1dr^2dr^3$$

$$(21)$$

4 Numerical Implementation

Brief details are now given on the implementation of the above described formulation in the MATLAB environment.

4.1 Displacement Field Approach

Spatial cartesian coordinates are written as a function of curvilinear coordinates by using the Lagrange polynomials. For analysing spatial membrane structures, only translational degrees of freedom are taken into account; rotational degrees of freedom can be neglected. The displacement vector may be written as follows:

$$u^h(x) = \sum_I^n \phi_I(x)u_I \quad u^h = \begin{Bmatrix} u^h \\ v^h \\ w^h \end{Bmatrix} \quad \phi_I = \begin{bmatrix} \phi_I^u & 0 & 0 \\ 0 & \phi_I^v & 0 \\ 0 & 0 & \phi_I^w \end{bmatrix} \quad u_I = \begin{bmatrix} u_I \\ v_I \\ w_I \end{bmatrix} \quad (22)$$

4.2 Tangent Stiffness Matrix

The tangent stiffness matrix may be written as a sum of two parts: elastic linear part $[K_E]$ and geometrical non linear part $[K_G]$.

$$[K] = [K_E] + [K_G] \tag{23}$$

$$[K_E] = \int_{r^1}\int_{r^2} ([B_0] + [A]\cdot[G])^T [C]([B_0] + [A]\cdot[G])\cdot t\cdot[G_1G_2G_3]dr^1dr^2 \tag{24}$$

$$[K_G] = \int_{r^1}\int_{r^2} [G]^T\cdot[\tau]\cdot[G]\cdot t\cdot[G_1G_2G_3]dr^1dr^2 \tag{25}$$

In the previous equations, t is the membrane thickness, $[G_1 G_2 G_3]$ denotes a scalar triple product and is defined as $G_1 \times G_2 \cdot G_3$. Matrix τ represents the initial stresses and it is written as follows:

$$\tau = \begin{bmatrix} \sigma_x I_3 & \tau_{xy} I_3 & \tau_{xz} I_3 \\ \tau_{yx} I_3 & \sigma_y I_3 & \tau_{yz} I_3 \\ \tau_{zx} I_3 & \tau_{zy} I_3 & \sigma_z I_3 \end{bmatrix} \tag{26}$$

where I_3 is the third-order identity matrix.

4.3 Nodal Forces

The global nodal forces vector stores the contribution of the forces distributed in the volume and the forces applied at the surfaces. The forces b distributed in the volume V may be computed as follows:

$$f_I = \int_V \phi_I b \, dV \tag{27}$$

The contribution of the surface forces T_t is:

$$f_I = \int_{T_t} \phi_I^T \bar{t} \, d\Gamma \tag{28}$$

The internal forces in the membrane are:

$$\{Q\} = \int_{r^1} \int_{r^2} ([B_0] + [A] \cdot [G])^T \{\sigma\} \cdot t \cdot [G_1 G_2 G_3] dr^1 dr^2 \tag{29}$$

4.4 Essential Boundary Conditions

In the cases studied in this work, the penalty method is adopted to impose the essential boundary conditions. Comparing to the Lagrange multiplier method, the penalty technique is easier to implement and it is somehow faster as no additional degrees of freedom are required. The penalty matrix and the penalty vector are assembled in the stiffness matrix and the external force vector, respectively, as follows.

$$K_{IJ}^\alpha = \int_{\Gamma_u} \phi_I^T \alpha \phi_J d\Gamma \tag{30}$$

$$F_{IJ}^\alpha = \int_{\Gamma_u} \phi_I^T \alpha \bar{u} d\Gamma \tag{31}$$

where:

$$\phi_I = \begin{bmatrix} \phi_I & 0 & 0 \\ 0 & \phi_I & 0 \\ 0 & 0 & \phi_I \end{bmatrix} \quad \phi_J = \begin{bmatrix} \phi_J & 0 & 0 \\ 0 & \phi_J & 0 \\ 0 & 0 & \phi_J \end{bmatrix} \quad \alpha = \begin{bmatrix} \alpha & 0 & 0 \\ 0 & \alpha & 0 \\ 0 & 0 & \alpha \end{bmatrix} \quad (32)$$

Finally, the non linear system of equations may be written as follows.

$$\left([K] + [K]^{\alpha} \right) \cdot \{u\} = \{F\} + \{F^{\alpha}\} + \{Q(u)\} \quad (33)$$

5 Numerical Tests

5.1 Plane Membrane

The membrane analyzed in this example (Fig. 4) is a plane membrane (a real two-dimensional membrane) subject to an initial extension in two directions. Both, isotropic and orthotropic (0°, 45° and 90°) elastic behaviours are considered. The material properties correspond to that of a polyester film whose thickness is 1 mm. Membrane analysis was carried out by using the proposed formulation and by the general-purpose finite element method program **ADINA**.

Two different meshes, namely grids of 11×11 and 21×21 equally spaced nodes, are used (Fig. 5).

Linear basis functions for the MLS approximation, third-order polynomial weight function and domain of influence of radius ρ were considered. The radius ρ is normalized by the minimum distance of two neighboring nodes denoted by c. The background cell grid, needed for integration purposes, is a square surrounded by the four neighboring nodes. Three-by-Three Gauss integration is carried out in each cell.

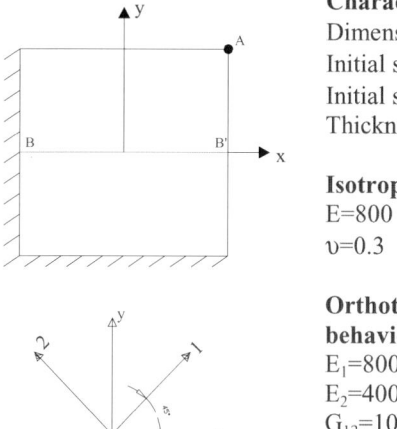

Characteristics:
Dimensions: 1×1 m
Initial strain ε_x: 0.01
Initial strain ε_y: 0.01
Thickness: 1 mm

Isotropic behaviour:
E=800 GPa
υ=0.3

Orthotropic behaviour:
E_1=800 MPa
E_2=400 MPa
G_{12}=100 MPa
υ_{12}=0.3

Fig. 4 Plane membrane

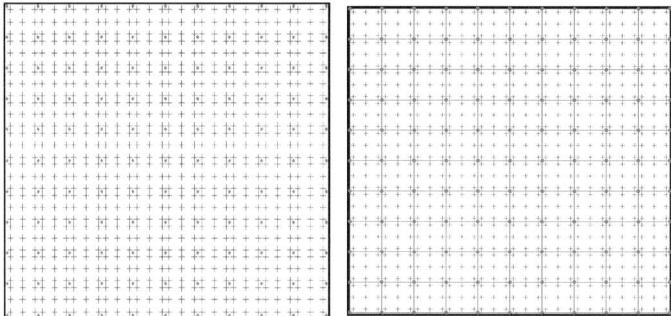

Fig. 5 Membrane discretization (*left*) EFGM (*right*) FEM

The boundary conditions were imposed by penalty method by using a penalty factor of 10^{10}.

For comparison purposes finite element discretizations using four-node quadrilateral with 3×3 Gauss integration were also conducted. The same arrangements of integration stations are use for both EFG and FEM thus making the comparison more realistic.

Both numerical implementations, ADINA and the present code, used the iterative Newton-Raphson method with 10 load increments in the solution of the non linear system of equations. The convergence criterion adopted is based on the variation of the displacements from the *i-th* to the *i-th+1* iterations with a tolerance of convergence of 10^{-10}.

In what follows the comparison between EFGM and FEM is made by using the same degrees of freedom, the same polynomial basis for the displacements and the same quadrature rule in each integration cell.

Figure 6 shows the variation of the displacement at Point A with the angle of orthotropy, the displacements and the principal stresses σ_I along $B - B'$ alignment.

From Fig. 6(a) it is possible to see that both components of the displacement at Point A are equal when the orthotropy is at an angle of $0°$, $45°$ or $90°$ with the axes of the global coordinate system. It is verified that for an angle between $0°$ and $45°$ the displacements at Point A along direction x diminishes at exactly the same ratio as the corresponding displacements in direction y increases. For an angle between $45°$ and $90°$, the inverse phenomenon is verified.

Comparing the displacements components for orthotropic situations along the alignment $B - B'$, it allows to conclude that only for the $45°$ case the displacements values are lesser than the observed ones for the isotropic situation.

The comparison between the stress plots obtained by the FEM and by the EFGM is quite good and shows the accuracy and suitability of the computational model developed in this work.

In Fig. 7 and Fig. 8 the displacement fields and the principal stresses distributions in the membrane are illustrated for the isotropic and orthotropic ($45°$) behavior situations, respectively. The plots were obtained from the EFGM and the FEM models

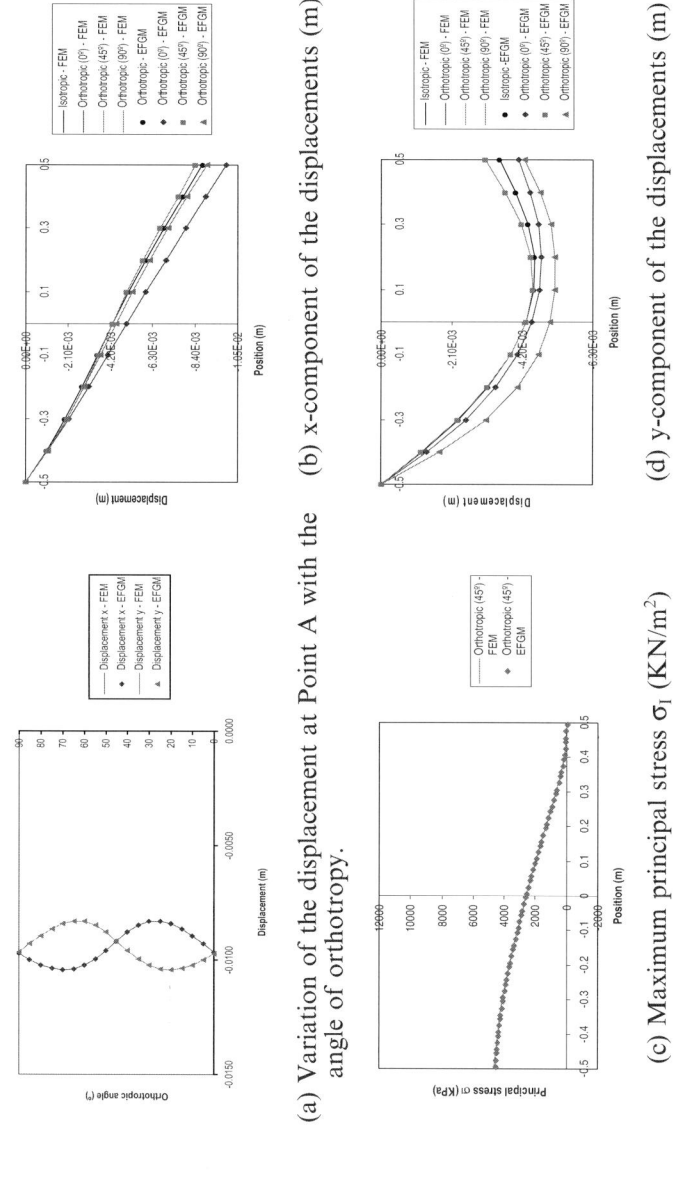

(a) Variation of the displacement at Point A with the angle of orthotropy.

(b) x-component of the displacements (m)

(c) Maximum principal stress σ_I (KN/m^2)

(d) y-component of the displacements (m)

Fig. 6 Plane membrane results

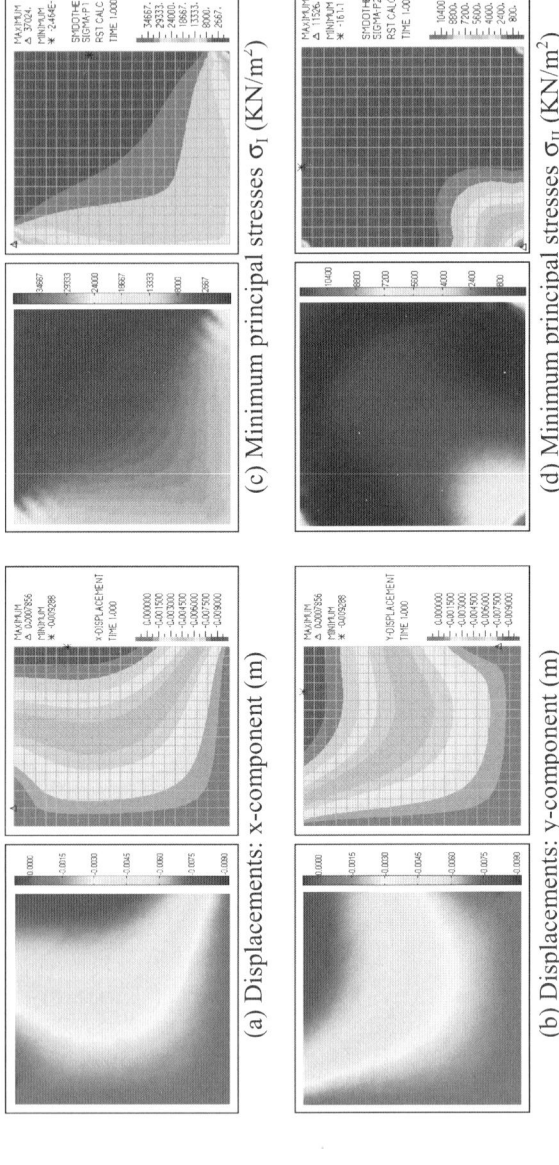

(a) Displacements: x-component (m)

(b) Displacements: y-component (m)

(c) Minimum principal stresses σ_{I} (KN/m²)

(d) Minimum principal stresses σ_{II} (KN/m²)

Fig. 7 Displacements and principal stresses – isotropic case (in each subfigure the EFGM result is the left hand side one)

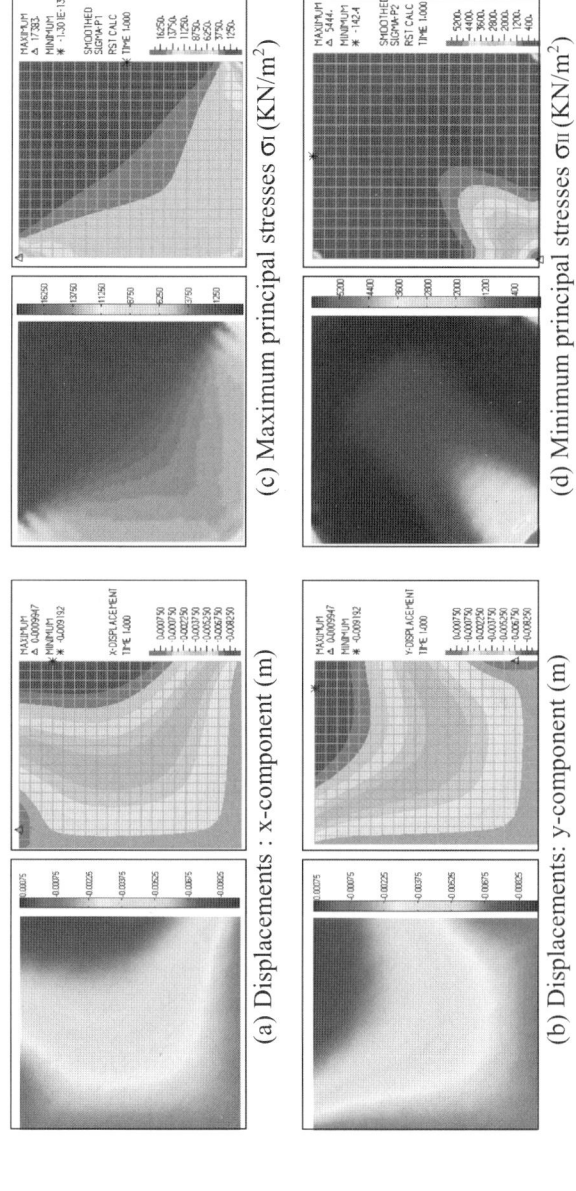

(a) Displacements : x-component (m)

(b) Displacements: y-component (m)

(c) Maximum principal stresses σ_I (KN/m²)

(d) Minimum principal stresses σ_{II} (KN/m²)

Fig. 8 Displacements and principal stresses – orthotropic case (45°)

by using meshes/distributions of 21×21 nodes/points with $2 + \varepsilon$ supports in the construction of the EFGM approximation functions.

By looking at the displacement fields shown below it is possible to see that in the case of isotropic behavior, the plot of the x-component of the displacements y is exactly the same as the plot of the y-component rotated by $90°$, due to symmetry of the membrane with respect to its diagonal line (at $45°$ with both axes). If the membrane has orthotropic behavior at an angle of $45°$, the displacements along x and y are very similar to the ones in the previous case, both in terms of numerical values and distribution.

The analysis of the stress components distributions shows that in both cases of material behavior, the maximum principal stress σ_I reaches values higher than that of the initial stress (8 MPa) in the neighborhood of the edges where essential boundary conditions are being imposed especially near the corner formed by the two edges and along the diagonal line of the membrane.

The membrane, in this region, is subjected to an increase of the stresses in the two main directions, an increase which is more evident in the isotropic situation. In the remaining region of the membrane the value of the maximum principal stress is lower than that of the initial stress, which indicates that relaxation of the state of initial tension has occurred.

The values of the minimum principal stresses σ_{II} are (nearly at all points) positive thus showing that wrinkling does not occur.

5.2 "Cable" Membrane

The membrane shown in Fig. 9 is yet another basic example but it serves the purpose of checking further the nonlinear capabilities of the model now for a three-dimensional membrane.

The membrane is subjected to, first, an initial strain, and then to a point load applied at the center of the membrane. The analysis was carried out in the same

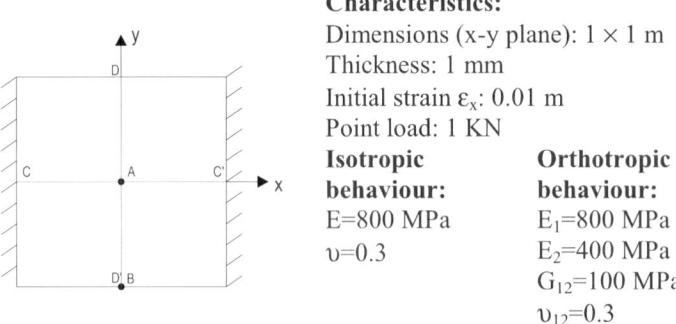

Characteristics:
Dimensions (x-y plane): 1×1 m
Thickness: 1 mm
Initial strain ε_x: 0.01 m
Point load: 1 KN

Isotropic behaviour:	**Orthotropic behaviour:**
E=800 MPa	E_1=800 MPa
υ=0.3	E_2=400 MPa
	G_{12}=100 MPa
	υ_{12}=0.3

Fig. 9 Cable membrane

way as for the previous example, that is, the same nodal arrangement, the same background integration, the same numerical method used for comparison.

In Fig. 10 the deformed configurations along alignments $C - C'$ and $D - D'$ are shown for both isotropic and orthotropic behavior.

The distribution of the vertical displacements along the alignments shows, in the vicinity of the point load, an approximately parabolic behavior. In this region the curvature of the parabola is approximately the same but the maximum value of the displacement depends on the degree of anisotropy.

For the two cases of orthotropic behavior considered, 45° and 90°, the displacements at the edge are, respectively, about 3.86 and 3.96 times the displacements obtained on the same location for the isotropic case.

The analysis of stresses (Fig. 11) shows that the maximum principal stress is higher than the value of the initial tension (8 MPa) in the region close to alignment C-C' and the diagonal line at 45°, for the 0° and 45° cases. The minimum principal stress σ_{II} is always positive (except for some local points) meaning that wrinkling does not occur.

5.3 Hyperbolic Paraboloid

The Hyperbolic Paraboloid of rigid boundaries is a demanding test for any formulation. The membrane tested has square shape (when projected on the xy plane) with length equal to 6.15 m in the plane (8.7 m is the real length). The height (the difference between the risen and the lowered corners) is 6.15 m (thus, the lateral slopes being 100%), and the membrane central point is at an intermediate height (3.075 m above the lowered corners). The isotropic and the 45° anisotropic cases were considered in the membrane analysis.

Both numerical models used a distribution of 11×11 points/nodes. Three-noded triangular elements were used in the FEM model (Fig. 12). The EFGM model uses a linear basis for the displacement field, a cubical spline as weight function and supports of size $1 + \varepsilon$. The essential boundary conditions are imposed by the penalty method where the penalty factor is 10^{10}.

The analysis of this type of membranes (real 3D ones) requires an initial step (prior to the application of any load) which is to find out what is the actual shape of the membrane. For engineers dealing with membranes this is known as the "form finding stage". The initial geometry to start the form finding procedure of membranes may be quite arbitrary. In the case under analysis here, the equilibrium geometry shape was represented by a bi-quadratic function considering 9 sampling points.

The material characteristics of the membrane are as follows: $E = 800$ MPa; $v = 0$. The thickness is 1 mm and the membrane is subjected to an initial strain of 0.01 in both directions on its plane.

The wind dynamic pressure was obtained by using two methodologies: the RSA [6] (the Portuguese structural code) and an approach proposed by Tabarrok [7]. The

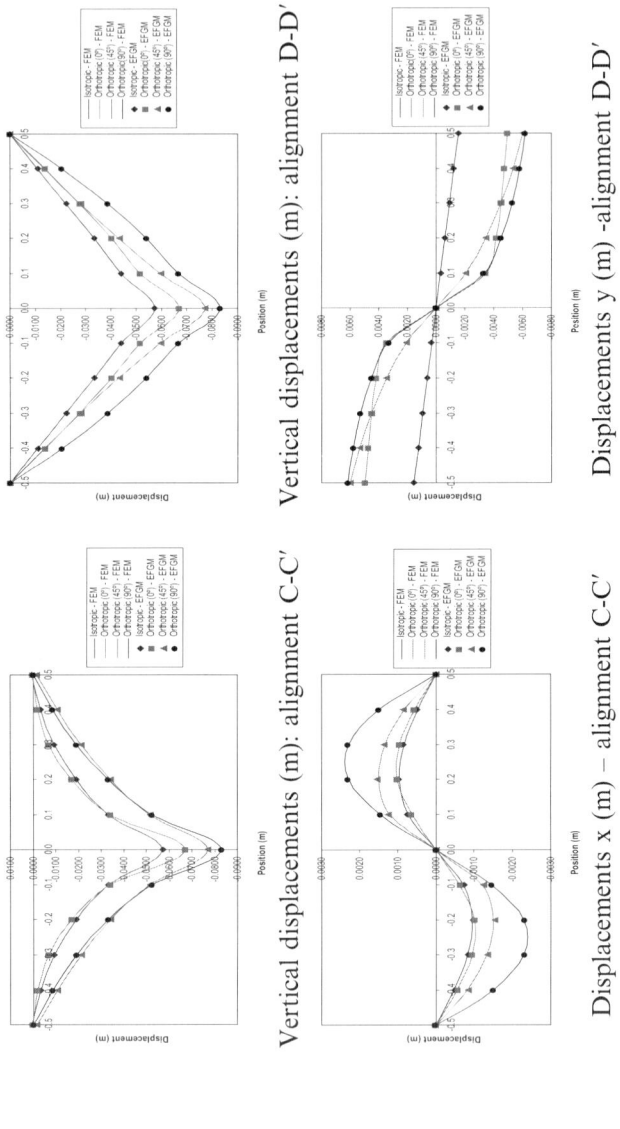

Fig. 10 "Cable" membrane results

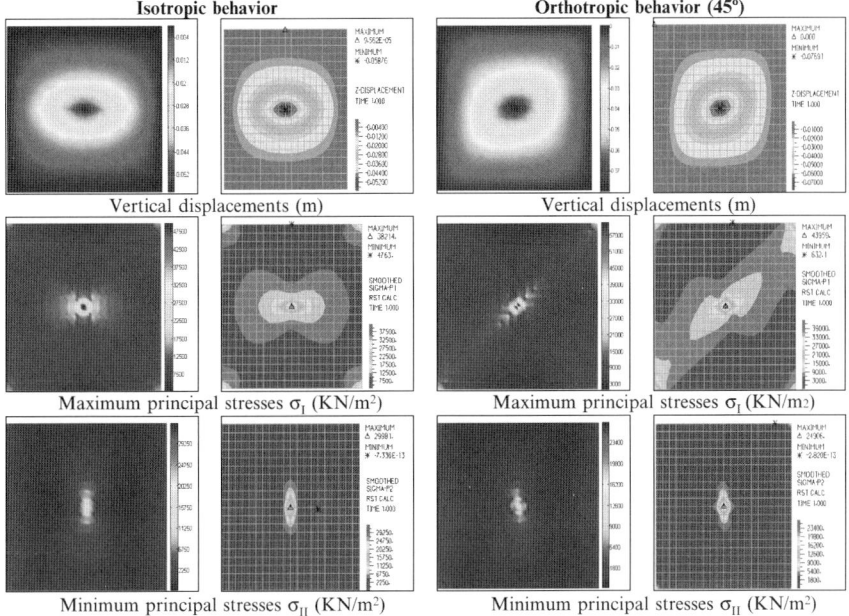

Isotropic behavior

Vertical displacements (m)

Maximum principal stresses σ_I (KN/m²)

Minimum principal stresses σ_{II} (KN/m²)

Orthotropic behavior (45°)

Vertical displacements (m)

Maximum principal stresses σ_I (KN/m₂)

Minimum principal stresses σ_{II} (KN/m²)

Fig. 11 Plots of the vertical component of the displacements and principal stresses

Fig. 12 Hyperbolic
Paraboloid – geometry and
mesh

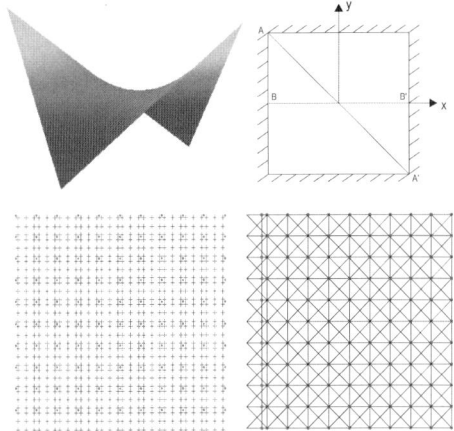

pressure coefficient was obtained for the pressure load indicated in Fig. 13. Tabarrok's model assumes that the coefficient of pressure on a surface is proportional to the cosine of the angle formed by the normal to the surface and the attack wind direction ($p = w\delta_p \cos(\alpha)$), where δ_p is constant in the whole structure.

Figure 14 shows the vertical displacements and the principal stresses.

The vertical displacement is highest in the neighborhood of the membrane's corner near the top diagonal line. The highest stressed fiber is also found in this region.

Fig. 13 Wind load

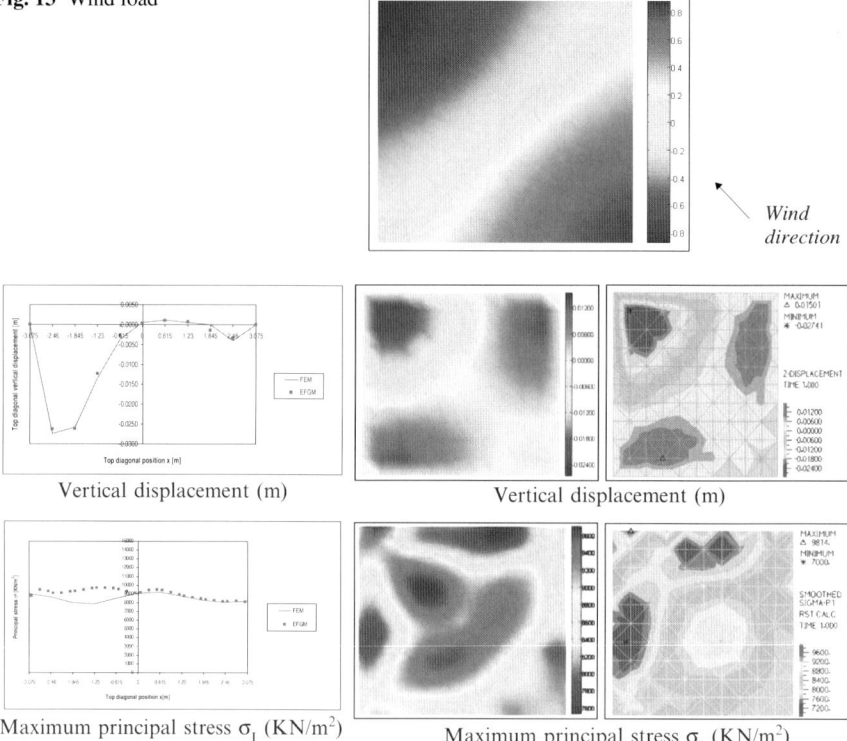

Vertical displacement (m)   Vertical displacement (m)

Maximum principal stress σ₁ (KN/m²)   Maximum principal stress σ₁ (KN/m²)

Fig. 14 Wind load – isotropic behaviour

The top diagonal line is subjected essentially to descending displacements, except for a small region situated between the center of the membrane center and the corner where the values are negligible. The vertical displacements vary between -2.74 and 1.5 cm.

The analysis of the maximum principal stresses shows that there are points where the stress is lower than 8 MPa (which is the initial tension) which indicates relaxation of the state of tension in the two principal directions. At other points σ_I is greater that 8 MPa which means that an increase of tension in the two principal directions has occurred.

Orthotropic behavior was also considered. Two cases were tested: orthotropy along the top diagonal ($\theta = 135°$) and along the bottom diagonal ($\theta = 45°$). In Fig. 15 the vertical displacements and maximum principal stress σ_I along the top diagonal line are presented for the 135° case.

One realizes that for the 135° orthotropic behavior the vertical displacements in the top diagonal line are lower than in the previous case between the top corner and the point situated at 1.23 m of the origin of the coordinates system. From this

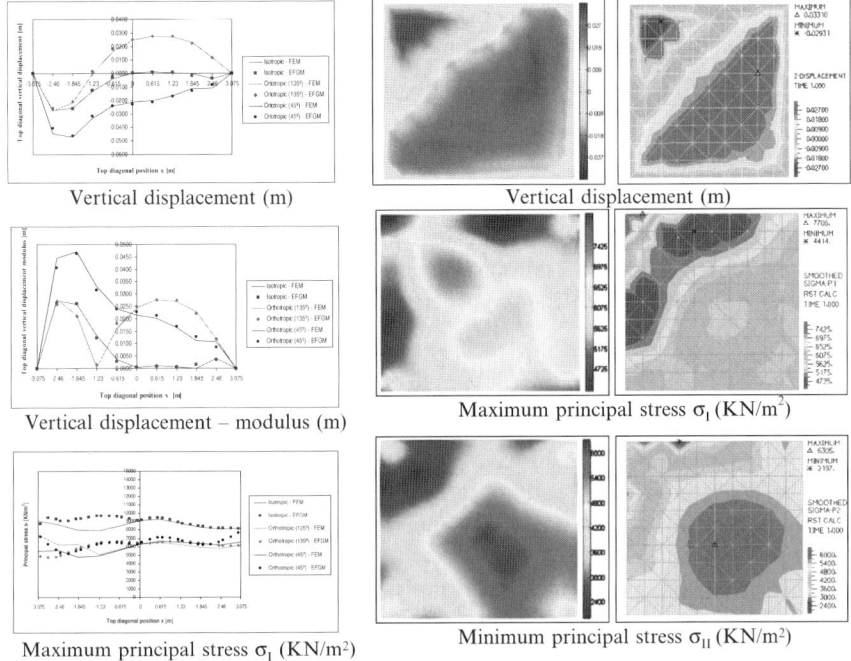

Vertical displacement (m)

Vertical displacement – modulus (m)

Maximum principal stress σ_I (KN/m²)

Vertical displacement (m)

Maximum principal stress σ_I (KN/m²)

Minimum principal stress σ_{II} (KN/m²)

Fig. 15 Wind load – orthotropic behaviour

point the displacements increase in parabolic manner reaching the maximum value of 0.276 m maximum at a distance of 0.615 m from the inbound corner.

When the 45° orthotropic case is considered, contrary to the other cases, the vertical displacements are decreasing in the whole top diagonal line. This behavior leads to higher vertical displacements in the region between the corner and the top diagonal line center, the maximum vertical displacement in this region being about 1.80 times the corresponding displacement for the isotropic behavior case. Then the displacements decrease up to an intermediate distribution between the orthotropic behavior (135°) and the isotropic behavior in the region situated between the top diagonal line center and inbound corner.

The isotropic behavior leads always to higher values than those for the non-isotropic case.

Comparing the two orthotropic situations, the situation where the top diagonal line is more rigid (orthotropic behavior, 135°) leads to the highest values of maximum principal stresses σ_I between the outbound corner and the membrane center.

On the other hand, the situation where the bottom diagonal line is more rigid (orthotropic behavior, 45°) leads to the highest values of maximum principal stresses σ_I from the centre to the corner.

The minimum principal stresses σ_{II} are positive in the whole membrane (except for a few local points) showing that wrinkling does not occur.

6 Conclusions

In the present study, an implementation of the EFGM meshless method for the analysis of three-dimensional anisotropic membranes is presented. The geometrically non-linear effects are taken into consideration by means of the total Lagrangian method.

The formulation and computational model develop in this work are validated by means of three examples of membrane structures. Both, the isotropic and the orthotropic behaviors are considered.

For comparison purposes all examples tested with the present formularization were also analyzed with a general-purpose finite elements code, ADINA.

The result of this comparison is quite positive in the sense that results of similar quality are obtained with these two different techniques with the extra advantage that, due to the absence of mesh, it is easier to model variations of the material characteristics (going from the isotropic to the orthotropic case) and of the loads (the wind loads, for example).

Acknowledgements The authors gratefully acknowledge the support of Fundação para a Ciência e a Tecnologia.

References

1. Belytschko, T.; Lu, Y. Y.; and Gu, L.; "Element free Galerkin methods"; *International Journal for Numerical Methods in Engineering*; 37, 229–256; 1994
2. Lu, Y. Y.; Belytschko, T.; and Gu, L.; "A new implementation of the element free Galerkin method"; *Computer Methods in Applied Mechanics and Engineering*; 113, 397–414; 1994
3. Dolbow J. and Belytschko T.; "An introduction to programming the meshless element free Galerkin method"; *Archives of Computational Methods in Engineering*; 43, 785–819; 1998
4. Krysl, P. and Belytschko, T.; "Analysis of thin shells by the element-free Galerkin method"; *International Journal of Solids and Structures*; 33, 3057–3080; 1996
5. Noguchi, H.; "Element free analysis of shell and spatial structures"; *International Journal for Numerical Methods in Engineering*; 47, 1215–1240; 2000
6. RSA-Regulamento de Segurança e Acções para Estruturas de Edifícios e Pontes, *Porto Editora*; 1983
7. Tabarrok, B. and Qin, Z.; "Dynamic analysis of tension structures"; *Computers & Structures*; 62, 467–474; 1997

Towards an Isogeometric Meshless Natural Element Method

David González, Elías Cueto(✉), and Manuel Doblaré

Abstract The problem of generalizing the Natural Element Method in terms of higher-order consistency and continuity is addressed here with several possible solutions. In this work we review some of them. First, a study of the possible benefits of enriching the interpolation using the Partition of Unity paradigm is considered. Different enrichments were tested leading to different reproducing properties. Another possible solution to the problem is the use of iterated Voronoi diagram interpolants, due to G. Farin. This last solution is done by means of the de Boor's algorithm, the same employed to obtain B-splines by linear combinations of linear interpolants in one dimension. We propose another form of B-spline-like interpolants that employs the de Boor's algorithm, but with a simpler structure. In order to obtain a smooth interpolant, a review of the Hiyoshi-Sugihara interpolant is also made. In this case, although high-order smoothness can be achieved, the consistency remains linear. By employing the proposed algorithms, however, this consistency can be improved to the desired degree.

This new class of meshless methods closely resembles the isogeometric analysis developed by [*T.J.R. Hughes et al. Isogeometric analysis: CAD, finite elements, NURBS, exact geometry, and mesh refinement. Computer Methods in Applied Mechanics and Engineering, 194:4135–4195, 2005*]. However, unlike B-splines, this new class of interpolants does not rely on an underlying tensor-product quadrilateral mesh. It is based upon the Delaunay triangulation of the cloud of knots and does not require any regularity on the connectivity.

In addition, the method thus generated conserves many of the attractive features of the Natural Element Method, such as strict interpolation on the boundary, and thus direct imposition of essential boundary conditions. After a theoretical description of the proposed method, some numerical examples are shown to test its performance in the context of Linear Elastostatics.

D. González, E. Cueto, and M. Doblaré
Group of Structural Mechanics and Materials Modelling, Aragón Institute of Engineering Research (*I3A*), University of Zaragoza, María de Luna, 5, E-50018 Zaragoza, Spain, e-mail: ecueto@unizar.es

A.J.M. Ferreira et al. (eds.) *Progress on Meshless Methods, Computational Methods in Applied Sciences.*
237

Keywords: Isometric Meshless Method · Natural Element Method · Voronoi Diagrams

1 Introduction

The Natural Element Method (NEM) is nowadays a well-established technique in the family of meshless methods that has attracted attention of many researchers. Both from the theoretical point of view [6, 9, 10, 23, 27, 28] and its application to different problems [1, 3, 14, 18], the Natural Element Method has demonstrated its success for a wide variety of problems, mainly those involving large strains, free surfaces, etc. Some of its interesting characteristics are the "exact" (i.e., within the order of the approximation) imposition of essential boundary conditions and the exact numerical integration (achieved by the use of Stabilized Conforming Nodal Integration, see [8, 13]).

In essence, the NEM is a Galerkin method based on the use of natural neighbour interpolation (either Thiessen [25], Laplace [5] or Sibson [19] interpolation schemes) instead of piece-wise polynomials, employed in the case of the FEM. The fact that, unlike most other meshless methods, NEM shape functions are strictly interpolant, clearly simplifies the issue of imposing essential boundary conditions, as mentioned before. In fact, many authors consider natural neighbour interpolation as the most suitable generalization of linear interpolation to two or higher dimensions, without an underlying quadrilateral or tensor-product mesh [12].

Precisely, the fact that the NEM possesses linear consistency and $\mathscr{C}^0$ continuity only is perhaps on the basis of its limited popularity, when compared to other meshless methods, which easily achieve higher-order consistency and even $\mathscr{C}^\infty$ continuity. Only one attempt has been made to overcome this difficulty, up to our knowledge, by applying a quadratic consistency and $\mathscr{C}^1$ interpolant based on natural neighbours that, however, does not seem to posses any further generalization [22]. This interpolant can be used, for instance, for solving fourth-order partial differential equations such as those arising from the theory of Kirchhoff plate bending.

In this chapter we review the available methods to overcome these NEM limitations. We review the use of PU enrichments and also a new generation of methods that go back to the foundations of B-splines and how by linear combinations of linear interpolants, higher-order curves can be obtained. B-spline curves can be obtained by means of the so-called de Boor's algorithm [12]. For the surface case, tensor product B-spline surfaces were initially proposed in [11]. An extensive review of this topic can be found in [12, Chapter 16]. Tensor product B-spline surfaces are, however, very rigid. For instance, no tensor product surface can have the connectivity of a double torus. This algorithm is here generalized, without the use of tensor products, to higher dimensions. This is done by employing different natural neighbour interpolation schemes [16].

2 Natural Neighbour Interpolation

There are several interpolation schemes based on the use of natural neighbours. Historically, the first one is due to Thiessen [25] and possesses constant consistency only. It has been employed to approximate pressures in Hellinger-Reissner-like variational principles [15], although it is obviously not suitable for second-order partial differential equations.

The most popular natural neighbour interpolant is probably the one due to Sibson [19]. Before describing this interpolant we introduce some basic geometrical entities that are needed for further developments.

2.1 Voronoi/Dirichlet Diagrams

Consider a model composed by a cloud of points $N = \{x_1, x_2, \ldots, x_m\} \subset \mathbb{R}^d$, for which there is a unique decomposition of the space into regions such that each point within these regions is closer to the node to which the region is associated than to any other in the cloud. This kind of space decomposition is called a Voronoi diagram (also Dirichlet tessellation) of the cloud of points and each Voronoi cell is formally defined as (see Fig. 1):

$$T_I = \{x \in \mathbb{R}^d : d(x, x_I) < d(x, x_J) \, \forall \, J \neq I\}, \tag{1}$$

where $d(\cdot, \cdot)$ is the Euclidean distance function.

The dual structure of the Voronoi diagram is the Delaunay triangulation,[1] obtained by connecting nodes that share a common $(d-1)$-dimensional facet. While the Voronoi structure is unique, the Delaunay triangulation is not, there being some so-called *degenerate* cases in which there are two or more possible Delaunay triangulations (consider, for example, the case of triangulating a square in 2D, as depicted in Fig. 1 (right)). Another way to define the Delaunay triangulation of a

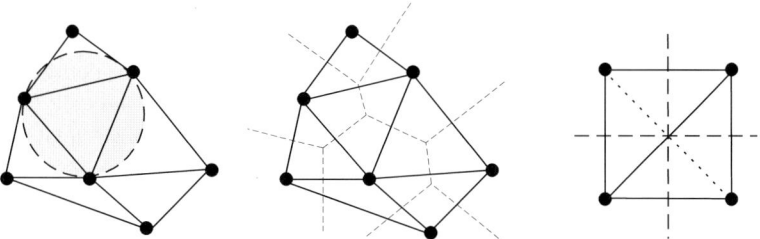

Fig. 1 Delaunay triangulation and Voronoi diagram of a cloud of points

---

[1] Even in three-dimensional spaces, it is common to refer to the Delaunay tetrahedralisation with the word *triangulation* in the vast majority of the literature.

set of nodes is by invoking the *empty circumcircle* property, which means that no node of the cloud lies within the circle covering a Delaunay triangle. Two nodes sharing a facet of their Voronoi cell are called *natural neighbours* and hence the name of the technique.

Equivalently, the second-order Voronoi diagram of the cloud is defined as

$$T_{IJ} = \{x \in \mathbb{R}^d : d(x, x_I) < d(x, x_J) < d(x, x_K) \; \forall \, J \neq I \neq K\}. \tag{2}$$

Based on these definitions, different natural neighbour interpolation schemes have been proposed. For a brief review of the most popular, see [21].

2.2 Thiessen Interpolation

The simplest of the natural neighbour-based interpolants is the so-called Thiessen's interpolant [25]. Its interpolating functions are defined as

$$\psi_I(x) = \begin{cases} 1 & \text{if } x \in T_I \\ 0 & \text{elsewhere} \end{cases}. \tag{3}$$

The Thiessen interpolant is a piece-wise constant function, defined over each Voronoi cell. It defines a method of interpolation often referred to as *nearest neighbour* interpolation, since a point is given a value defined by its nearest neighbour. Although it is obviously not valid for the solution of second-order partial differential equations, it can be used to interpolate the pressure in formulations arising from Hellinger-Reissner-like mixed variational principles [15].

2.3 Sibson's Interpolation

The most extended natural neighbour interpolation method is the Sibson interpolant [19, 20]. Consider the introduction of the point x in the cloud of nodes. Due to this introduction, the Voronoi diagram will be altered, affecting the Voronoi cells of the natural neighbours of x. Sibson [19] defined the natural neighbour coordinates of a point x with respect to one of its neighbours I as the ratio of the cell T_I that is transferred to T_x when adding x to the initial cloud of points to the total volume of T_x. In other words, if $\kappa(x)$ and $\kappa_I(x)$ are the Lebesgue measures of T_x and T_{xI} respectively, the natural neighbour coordinates of x with respect to the node I is defined as

$$\phi_I(x) = \frac{\kappa_I(x)}{\kappa(x)}. \tag{4}$$

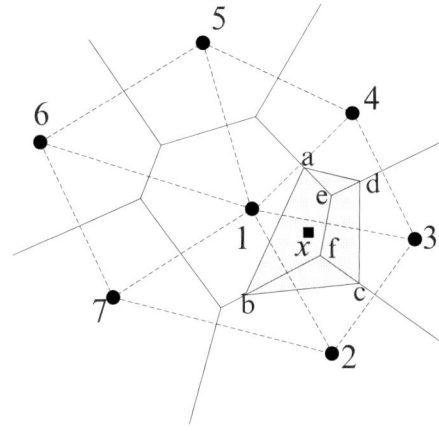

Fig. 2 Definition of the natural neighbour coordinates of a point *x*

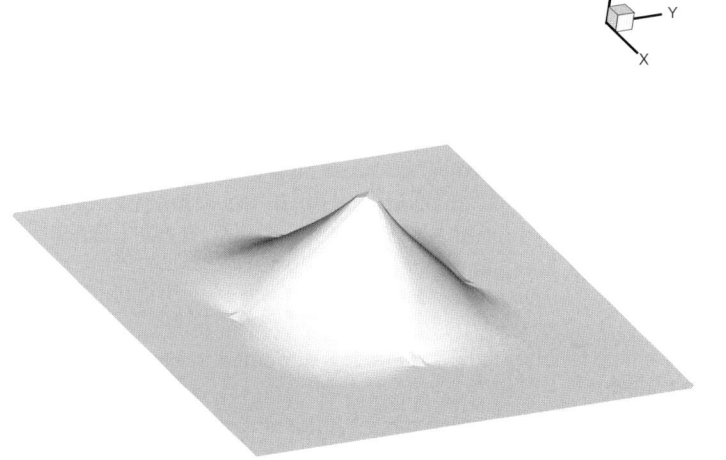

Fig. 3 Typical function $\phi(x)$

In Fig. 2 the shape function associated to node 1 at point *x* may be expressed as

$$\phi_1(x) = \frac{A_{abfe}}{A_{abcd}}. \tag{5}$$

Sibson's interpolation scheme possesses the usual reproducing properties for this class of problems, that is, it verifies the partition of unity property (constant consistency) and the linear consistency property. Thus, it is suitable for the solution of second-order PDE. Other interesting properties such as the Kronecker delta property [23] and linear interpolation on the boundary [9, 28] are also verified by the NEM.

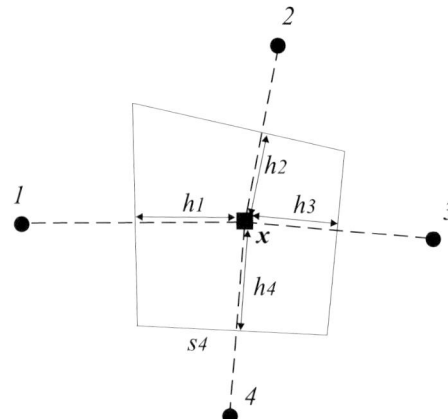

Fig. 4 Definition of
non-Sibsonian coordinates

2.4 Other Natural Neighbour-Based Interpolation Schemes

Other natural neighbour-based interpolation schemes exist, always with linear consistency and $\mathscr{C}^0$ continuity. The most popular one, due to its low computational
cost, is the so-called Laplace interpolation. This Laplace or *non-Sibsonian* interpolation has received considerable attention, since it involves magnitudes of one
order less of the space dimension (i.e., the calculation of areas in three-dimensional
problems, for instance, instead of volumes). If we define the cell intersection
$t_{IJ} = \{\boldsymbol{x} \in T_I \cap T_J, J \neq I\}$ (note that t_{IJ} may be an empty set) we can define the value

$$\alpha_J(\boldsymbol{x}) = \frac{|t_{\boldsymbol{x}J}|}{d(\boldsymbol{x},\boldsymbol{x}_J)/2}. \tag{6}$$

Thus, the shape function related to node 4 at point $\boldsymbol{x}$ in Fig. 4 is defined as

$$\phi_4^{ns}(\boldsymbol{x}) = \frac{\alpha_4(\boldsymbol{x})}{\sum_{J=1}^n \alpha_J(\boldsymbol{x})} = \frac{s_4(\boldsymbol{x})/h_4(\boldsymbol{x})}{\sum_{J=1}^n \left[s_J(\boldsymbol{x})/h_J(\boldsymbol{x}) \right]}, \tag{7}$$

where s_J represents the length of the Voronoi segment associated to node J and n
represents the number of natural neighbours of the point under consideration, $\boldsymbol{x}$.
$n = 4$ in this example. h_J represents the distance from the evaluation point to the
edge of the voronoi cell, see Fig. 4.

The interested reader should consult [24] for more details. Another approach
combines a Moving least squares approximation with a natural neighbour-based
support for the window function, leading to the so-called *pseudo*-NEM approach [2].

3 Smooth Natural Neighbour Interpolation

Recently, Hiyoshi and Sugihara have proposed a generalization of natural neighbour
coordinates of a set of points, see [17]. Prior to its introduction, it is necessary to
define the concept of Laguerre-Voronoi diagram (weighted Voronoi diagrams). Let

$B = \{b_1, b_2, \ldots, b_n\}$ be a set of closed balls in $\mathbb{R}^d$, i.e., $b_I = (x_I, \rho_I)$, where x_I is the center of the ball and ρ_I its radius. The *weighted distance* from a point to a ball will be thus defined as

$$\pi_I(x) = ||x - x_I|| - \rho_I. \tag{8}$$

The Laguerre-Voronoi diagram will thus be defined as the decomposition of the space into regions of the form:

$$V_I = \{x \in \mathbb{R}^d : \pi_I(x) < \pi_J(x) \ \forall \ J \neq I\}. \tag{9}$$

The Laguerre-Voronoi diagrams share many of they properties with standard Voronoi diagrams, but not all. For instance:

- The Laguerre-Voronoi cell associated to a given node does not necessarily exist.
- If it exists, it is a convex polytope.
- The Laguerre-Voronoi facet is perpendicular to the line joining two neihgbouring nodes.

See [17] and references therein for detailed proofs of these assertions.

The key aspect in defining this generalized system of coordinates is considering a set of nodes equipped with increasing ball radii. Note that the Laguerre-Voronoi diagram for $\rho_I = 0$, $\forall I$, coincides with the standard Voronoi diagram. If, starting from zero, all ball radii are increased, except the one associated to the evaluation point, therefore its Laguerre-Voronoi cell will consequently shrink until disappearing for sufficiently large ball radius.

In Fig. 5, a Laguerre-Voronoi cell is depicted for an evaluation point x surrounded by other four nodes, x_1 to x_n. The distance between the evaluation point and a

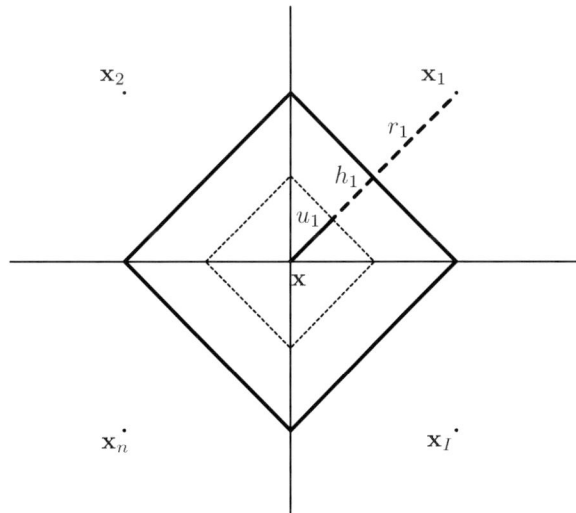

Fig. 5 Cells shrinking in Laguerre-Voronoi diagrams. The dashed-line cell represents the shrunken original cell (continuous line) when the weight $\rho \neq 0$

given node x_I is denoted as r_I, whereas the distance from the evaluation point to the Voronoi $(n-1)$-dimensional facet is here labelled as h_I. This same distance is labelled as u_I in a Laguerre-Voronoi diagram.

After this definitions, we can define the following coordinates over a shrinking cell associated to an evaluation point x:

$$\zeta_I^k(x) = r_I \int_0^{u_J} \zeta_I^{k-1}(x) du_J', \tag{10}$$

with

$$\zeta_I^0(x) = \frac{s_I}{u_I}, \tag{11}$$

s_I being the $(d-1)$-dimensional measure of the Laguerre-Voronoi facet and x_J one of the natural neighbours of node x_I. du_J' represents a dummy argument for the integration along the segments u_I, defined in Fig. 5. It is possible then to normalize this coordinates to give:

$$\psi_I^k(x) = \frac{\zeta_I^k(x)}{\sum_{J=1}^n \zeta_J^k(x)} \tag{12}$$

where the super-index k represents how many times the obtained shape function can be continuously differentiated [17]. In the above definition, the $\psi_I^0(x)$'s are well defined only when x does not coincide with any of the nodes x_I. In that case,

$$\psi_I^0(x_J) = \delta_{IJ}. \tag{13}$$

The coordinates defined above generalize the sequence Laplace $\rightarrow$ Sibson and provide an interesting means for interpolating data with smoother shape functions. Some remarkable properties of these so-called "generalized" coordinates are (see [17]):

- Exactness (or, equivalently, the so-called Kronecker's delta property):

$$\psi_I^k(x_J) = \delta_{IJ}.$$

- $\mathscr{C}^k$ continuity if x lies on the Delaunay sphere of the data sites (nodes). Continuity other than $\mathscr{C}^0$ at the nodes themselves has not been proved analytically, but computational experiments performed in [17] and also by the authors suggest its fulfillment. $\mathscr{C}^\infty$ continuity is obtained elsewhere.

In Fig. 6 the shape functions with $\mathscr{C}^1$ and $\mathscr{C}^2$ continuity are shown for a set of nodes placed on a regular lattice.

This set of interpolating functions has linear precision, at most. This is easily proven if we consider the limiting case of only three neighbouring nodes around a given evaluation point. It is impossible to define a quadratic surface given only three data.

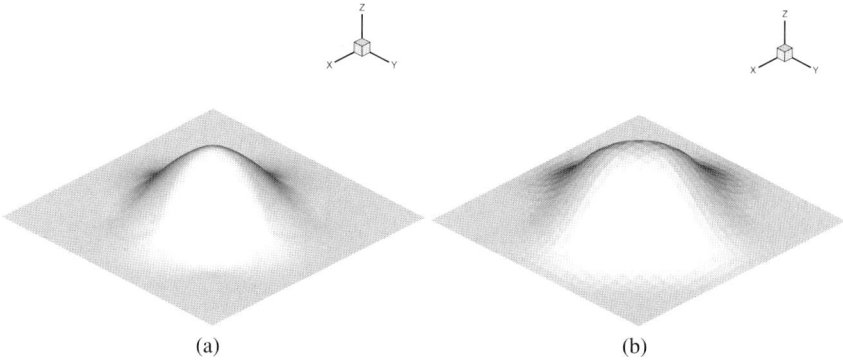

Fig. 6 Hiyoshi-Sugihara interpolants with $\mathscr{C}^1$ (**a**) and $\mathscr{C}^2$ (**b**) continuity, respectively

4 Higher-Order Interpolants

In order to generalize the Natural Element Method in terms of higher order consistency new schemes have been recently developed, similarly to the Hiyoshi-Sugihara interpolant [17] discussed above. In this section we review some of them.

4.1 Enriched Natural Neighbour Interpolants

The Natural neighbour approximation possesses linear completeness at most. In fact, there exists an interpolation scheme that reproduces a quadratic polynomial, but it requires the interpolation of the derivative of the essential field and a higher computational cost. Another possible solution to this problem may be to enrich the interpolant using the Partition of Unity paradigm. Some enriched approximations have been developed and analyzed in [15].

This philosophy was employed in order to overcome locking in problems with incompressibility restrictions, see [15]. It is well known that not all of the approximation schemes within a Hellinger-Reissner-like variational principle lead to stable methods. Only approximations schemes that verify the so-called LBB or inf-sup condition give stable results. On the other hand, it is also well known that mixed (Sibson-Sibson or Sibson-Thiessen) approximations in natural neighbour Galerkin methods do not always verify the LBB condition. In order to improve the stability of the resultant schemes, a enrichment of the interpolant by using the Partition of Unity paradigm was proposed. Several enrichments were tested, some of them leading to stable formulations, as proved in [15].

By employing the Partition of Unity method, we enriched the Sibson interpolant with different polynomial fields. In essence, the Partition of Unity method builds enriched approximations by adding functions to the trial space that approximate the

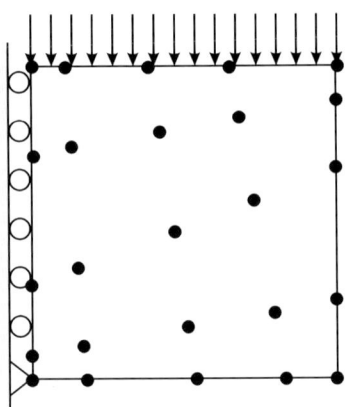

Fig. 7 Nodal arrangement for
the inf-sup test

solution accurately. If V_I is known to approximate well the solution locally, a global approximation space can be constructed as follows:

$$V := \sum_I \phi_I V_I = \{\sum_I \phi_I \mathbf{v}_I \mid \mathbf{v}_I \in V_I\} \subset H^1(\Omega) \tag{14}$$

where ϕ_I constitutes a partition of unity (i.e., $\sum_I \phi_I = 1$). In the PUM, the important aspect is that the global approximation space V inherits the approximation properties of the local spaces V_I and the smoothness of the partition of unity ϕ_I.

In order to verify the fulfillment of the LBB condition, Bathe [7] proposed a numerical test based on the use of a finite, small, set of meshes. In [15] we applied this test to a set of three different meshes for the problem in Fig. 7.

The evolution of the numerically predicted inf-sup value is depicted in Fig. 8. As can be noticed, the unenriched Sibson interpolant with discontinuous or continuous approximation for pressures does not lead to stable results. Among different possibilities, the enrichment with the set $V_I = \{1, x, y, xy\}$ gives rise to linear dependencies in the resultant system of equations, while the enrichment with the set $V_I = \{1, x^2, y^2\}$ shows spurious pressure modes. The resulting shape functions are plotted in Fig. 9.

Enrichment with the set $\{1, xy\}$ provides, however, a stable formulation. This enrichment closely resembles the MINI finite element by Arnold et al. [4] (linear triangle with cubic bubble for the displacements and discontinuous pressures).

4.2 Splines over Iterated Voronoi Diagrams

We briefly introduce the original concept of Splines over iterated Voronoi diagrams, an unpublished work due to G. Farin. In his work, a method which produces B-splines-like surfaces (or curves) in any dimension is presented. The de Boor's algorithm for B-Splines curves is based on repeated piecewise linear interpolation.

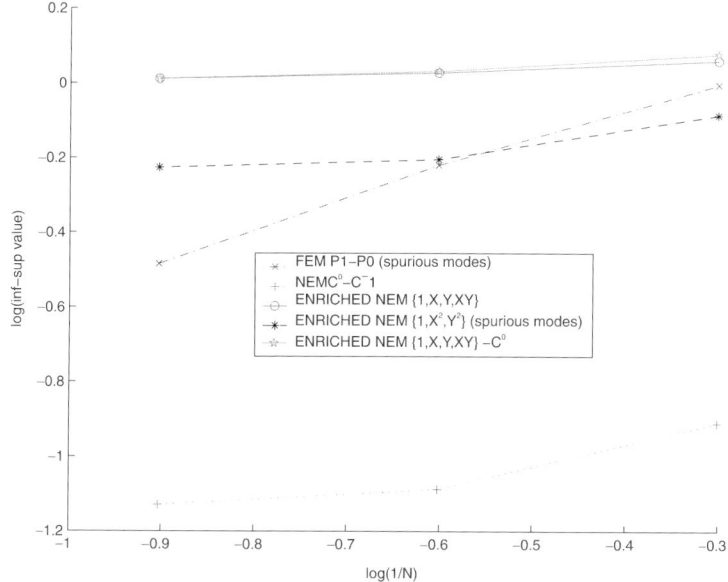

Fig. 8 Evolution of the inf-sup value for different enrichments

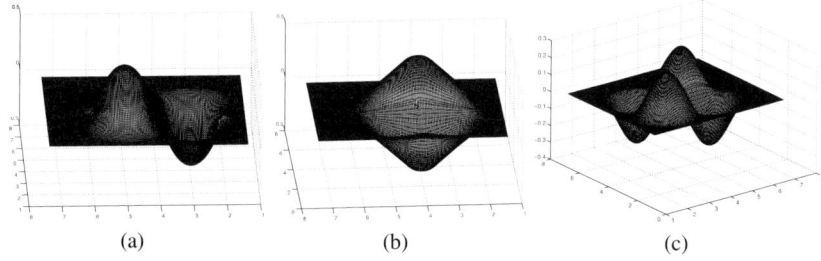

Fig. 9 Enriched Sibson interpolants: (**a**) ϕx, (**b**) ϕy and (**c**) ϕxy

In the context of surfaces, Farin proposed to replace the concept of piecewise linear interpolation by that of Sibson interpolation, described before.

4.2.1 Description and Evaluation

Let us consider a set of points $\mathbf{U}_0$, and denote by $\mathbf{U}_1 = \mathscr{V}(\mathbf{U}_0)$ the set of vertices of the Voronoi Diagram associated to $\mathbf{U}_0$. This recursive process can be extended to the *depth* of scheme, n, to obtain a set $\mathbf{U}_{n-1}$. A new Voronoi Diagram is computed as many times as the desired order of approximation. The final Voronoi Diagram corresponds to the control points associated to the surface, such that essential variables of the scheme are associated to this set $\mathbf{U}_{n-1}$.

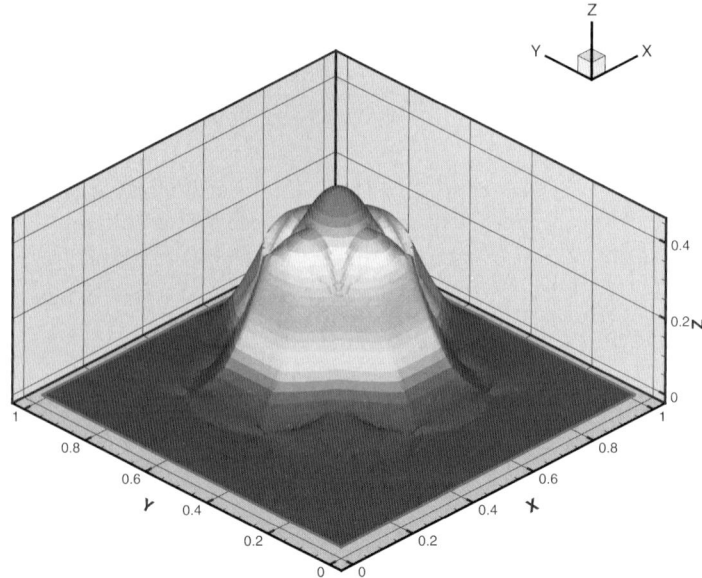

Fig. 10 2D shape function of a quadratic approximation using splines over iterated Voronoi diagrams

In order to evaluate the approximation, when an evaluation point x is introduced into the original cloud of points $\mathbf{U}_0$, it generates a new Voronoi diagram, $\mathbf{V}_{n-1} = \mathscr{V}(\mathbf{U}_{n-1} \cup x)$. The vertices in $\mathbf{V}_{n-1}$ are then approximated using Sibson's interpolation, in terms of the points in $\mathbf{U}_{n-1}$.

We define as control points the vertices of the last Voronoi diagram, associated to the set of points $\mathbf{U}_{n-1}$ ($\mathbf{U}_1$ in the quadratic case). To these $\mathbf{U}_1$ points we associate the essential variables z^0.

Using the same shape functions, the approximation at the evaluation point v will be $z(v) = \sum_i \phi_i z_i^1$, where $z^1 = \sum_j \phi_j z_j^0$ and z^0 represent the essential variables at the control points.

Figure 10 represents the 2D shape function obtained by using splines over iterated Voronoi diagrams.

4.2.2 Properties

As remarkable properties of the quadratic surfaces thus proposed we could cite:

- Linear precision: if $z_i = l(\mathbf{u}_i)$ and l is a bivariate linear function, then $z^2(\mathbf{v}) = l(\mathbf{v})$. This follows since Sibson's interpolant has that property.
- Quadratic precision is assumed, although there exists no formal proof.
- Local support.

- Continuity: $z^2(\mathbf{v})$ is $\mathscr{C}^2$ except at the $\mathbf{u}_i$ where it is $\mathscr{C}^0$. There is no analytical proof of this behaviour, but computational experiments confirm this.
- No knot insertion: if a point $\mathbf{v}$ is inserted into U_0 and refined control mesh is formed, it will not describe the same surface as the original. Reason: Sibson's interpolant does not have the idempotence property.

Although the desired improvement in consistency seems to be achieved, this kind of approximation has an important computational cost associated to the iterative computation of Voronoi diagrams, and renders a continuous approximation only, see Fig. 10. This fact encouraged us to seek for a simpler, smoother, if possible, approximation, that eventually could conserve the desired properties at a reasonable computational cost. This new type of approximation is described next.

4.3 B-Spline Surfaces Constructed over Natural Neighbour Interpolation

We have proposed recently [16] a novel technique based on the application of the well-know de Boor's algorihtm [12] to natural neighbour, instead of linear, interpolation. This algorithm is the same used in the context of B-spline curves to generate higher-order curves from linear combinations of linear interpolations. For a detailed description of the de Boor's algorithm the interested reader can consult [16].

In the following development we employ Sibson coordinates, although the proposed algorithm can also be applied to Laplace and Hiyoshi-Sugihara interpolants, as will be shown later. Consider again, for simplicity, a set of nodes $N = \{\mathbf{x}_1, \mathbf{x}_2, \ldots, \mathbf{x}_M\} \subset \mathbb{R}^2$ and a quadratic surface (the extension to three or higher dimensions and higher-order surfaces is straightforward). From now on, we will work in non-parametric form, since it is extremely hard to find the two-dimensional counterparts of the intervals U_i^r for irregularly scattered sites. Then, we define a new class of surfaces constructed in the way:

$$s(\mathbf{x}) = \sum_{I=1}^{n} \sum_{J=1}^{n^I} N_{IJ}(\mathbf{x}) d_{IJ}, \quad \text{with } d_{IJ} = d_{JI} \tag{15}$$

where n represents the number of neighbours of the point $\mathbf{x}$. In addition,

$$N_{IJ}(\mathbf{x}) = \phi_I(\mathbf{x}) \varphi_J^I(\mathbf{x}) \tag{16}$$

and d_{IJ} represent the control points in B-spline terminology (i.e., the degrees of freedom). $\phi_I(\mathbf{x})$ represents the natural neighbour (Sibson) coordinate of the point $\mathbf{x}$ with respect to site I. Functions $\varphi_J^I(\mathbf{x})$ represent the natural neighbour coordinates of point $\mathbf{x}$ with respect to site J, in the original cloud of points, but without the I-th site (see Fig. 11), in the sense described by the previous section. Finally, n^I is the number of natural neighbours of the point $\mathbf{x}$ when we eliminate the site I, similarly to the de Boor's algorithm. Note that the number of degrees of freedom of the proposed

Fig. 11 Schematic representation of the proposed
algorithm. (**a**) Set of sites
$\{I,\ldots,N\}$. We consider an
evaluation point $\boldsymbol{x}$, whose
neighbours are depicted as
filled circles. The support of
the function ϕ_I is highlighted.
(**b**) After eliminating site I,
the support of function φ_J^I
is highlighted. Note the new
set of neighbouring sites,
$\{J,K,L,M\}$

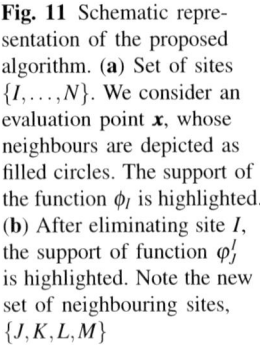
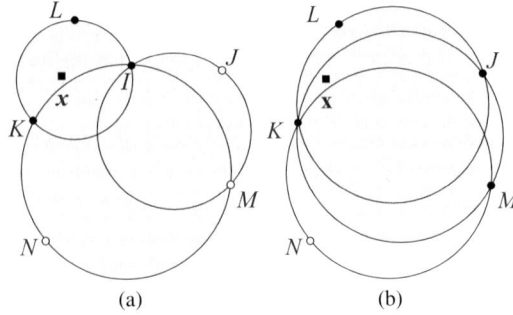

(a)   (b)

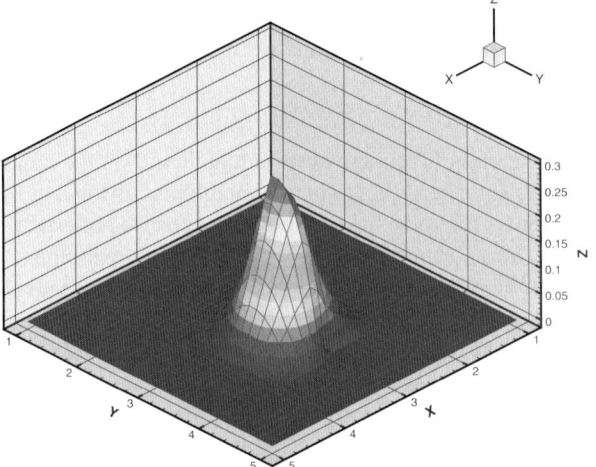

Fig. 12 Shape of a typical function N_{IJ} for a set of irregularly distributed sites

approximation is much less than $M^2/2$, since the sums in Eq. (15) extend only over
natural neighbors of each node.

The typical shape of the functions N_{IJ} described before is shown in Fig. 12 for a
general set of irregularly distributed sites.

4.3.1 Properties of the Proposed Surfaces

Surfaces defined after Eq. (15) posses some similarities with standard B-spline
curves. The following properties are demonstrated in [16].

- The functions N_{IJ} are always positive.
- The functions N_{IJ} form a partition of unity, i.e.,

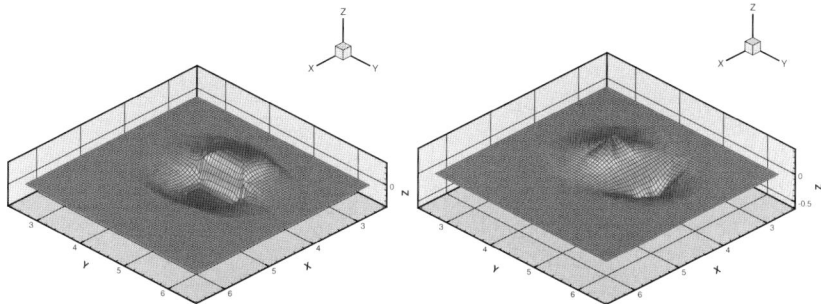

Fig. 13 x- and y derivatives of the function $N_{IJ} + N_{JI}$

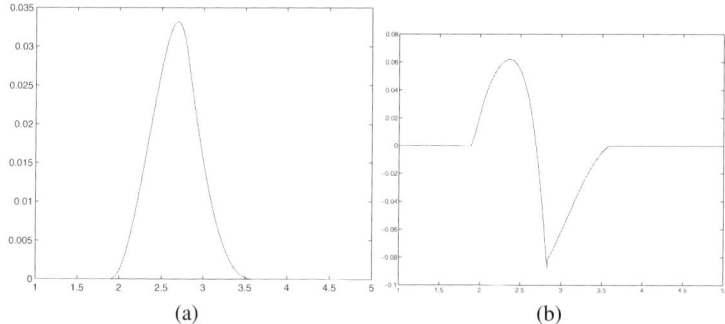

Fig. 14 (**a**) $N_{IJ} + N_{JI}$ value along the line joining two neighbouring sites. (**b**) Derivative of the same expression. Note the small jump in the derivatives near $x \simeq 2.8$

$$\sum_{I=1}^{n} \sum_{J=1}^{n^I} N_{IJ}(\boldsymbol{x}) = 1 \quad \forall \boldsymbol{x} \in \mathbb{R}^2. \tag{17}$$

- The basis functions $N_{IJ}(\boldsymbol{x})$ span the space of linear polynomials.
- The basis functions $N_{IJ}(\boldsymbol{x})$ span the space of quadratic polynomials.
- The proposed surfaces are $\mathscr{C}^{p-1}$, where p stands for the order of the surface, except at the nodes, in the direction joining two neighbouring nodes, where they are $\mathscr{C}^0$.

The x- and y-derivatives of the function pairs $N_{IJ} + N_{JI}$ are shown in Fig. 13. A single function N_{IJ} is depicted in Fig. 12. Note that a single function N_{IJ} is clearly not $\mathscr{C}^1$. Note also that a high gradient exists in this particular case, but not a discontinuity. In fact, the closer sites I and J are located, the higher is the gradient in the derivative.

If, on the contrary, the derivative is computed along the line joining two neighbouring sites, the pair $N_{IJ} + N_{JI}$ value takes the form depicted in Fig. 14(a), with the derivative shown in Fig. 14(b).

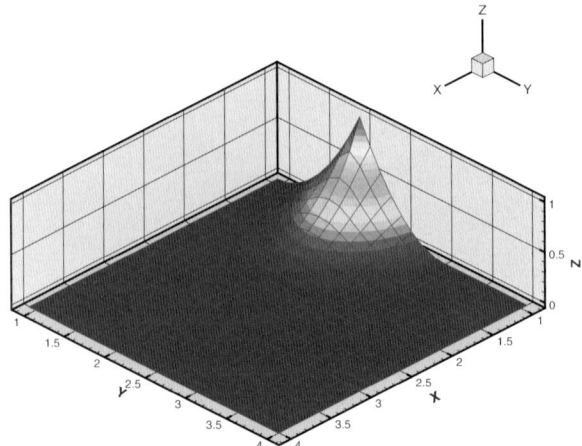

Fig. 15 Interpolating function $N_{II}(x)$ obtained by repeating knots in the boundary of the convex hull of the data sites. A discontinuity on the derivative appears at the site location

4.3.2 The Case of Repeated Knots

It is well-known that in the case of B-splines, the continuity of the sequence of curves can be controlled by the use of repeated knots. In the limit of knots infinitely close (repeated knots) a jump in the derivative appears, thus giving a $\mathscr{C}^0$ surface that, in addition, interpolates the data at the site location. In general, if a knot I is of multiplicity r, the surface of order n thus obtained possesses continuity $\mathscr{C}^{n-r}$, except at the directions of the lines joining neighbour nodes, where it remains $\mathscr{C}^0$. Proofs of this assertion can be found at [16].

By repeating the knots on the boundary of the convex hull of the data sites, for instance, one can make the surface to be (piecewise) quadratically interpolant along the boundary. The resulting function N_{II} for a site on the boundary is depicted in Fig. 15.

This last property has again a tremendous importance when using this kind of approximation in the context of Galerkin procedures, as in the Finite Element method. The use of repeated knots ensures interpolation (without loss of consistency of the approximation) and thus an easy imposition of essential boundary conditions and consistent interpolation along the boundary by simply fixing the value of the approximation at the node (data site).

5 Numerical Examples

We perform some numerical examples concerning the ability of the proposed surfaces to approximate data over irregularly scattered knots. All examples employ quadratic approximation only, but, as mentioned before, extension to higher degrees of consistency is straightforward.

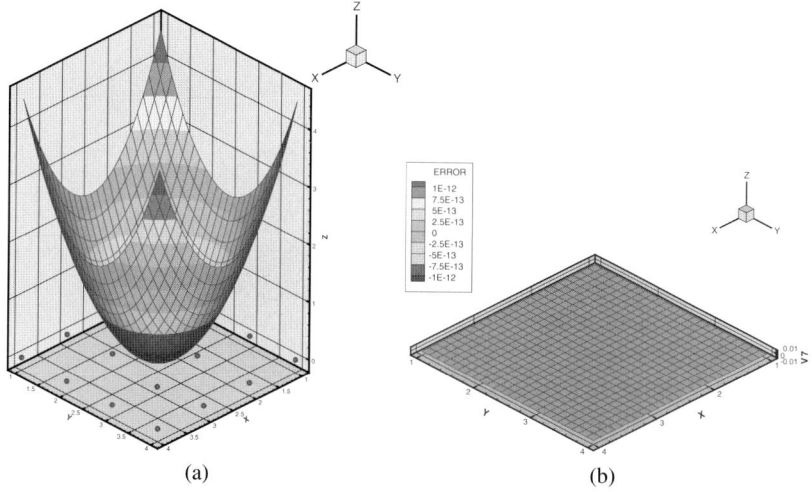

(a)   (b)

Fig. 16 (**a**) Surface obtained by a least squares fitting of the polynomial $(x-2.5)^2 + (y-2.5)^2 = 0$. Knots are depicted as red spheres. Boundary knots are repeated. (**b**) Error obtained in the L_2-fit

5.1 L_2-Fit of a Quadratic Polynomial

The first numerical example is the least squares fit of the quadratic polynomial $(x-2.5)^2 + (y-2.5)^2 = 0$, in order to show that, effectively, the proposed technique spans quadratic polynomials. Boundary knots are repeated in order to obtain an interpolant surface.

The obtained surface is shown in Fig. 16(a). As expected, the error obtained vanish identically everywhere in the domain, see Fig. 16(b).

The same L_2-fit has been performed with the derivative of the mentioned surface. With the coefficients obtained before, the approximation of the derivative of the surface through the derivatives of the N_{IJ} basis functions is achieved. The error in the approximation of the derivative is again within machine precision (less than 10^{-14}), see Fig. 17.

5.2 Infinite Plate with a Hole

The theoretical solution to this well-known problem can be found in [26], among other classical books. Material parameters were Young's modulus $E = 1.0$ and Poisson coefficient $v = 0.25$. μ represents the shear modulus and κ is the Kolosov constant, defined as

$$\kappa = 3 - 4v \tag{18}$$

$$\kappa = \frac{3-v}{1+v} \tag{19}$$

respectively, for plane strain and plane stress.

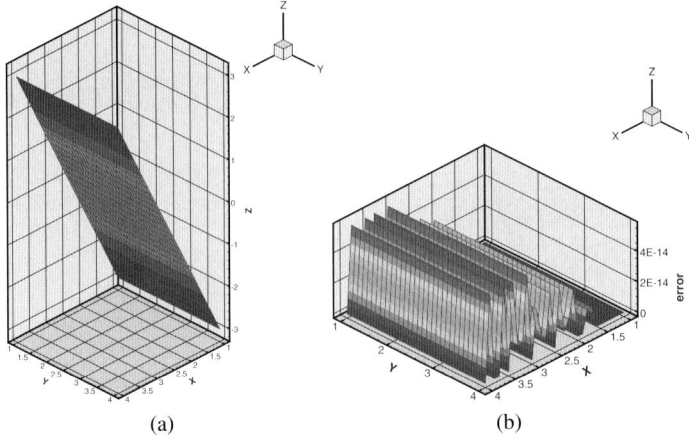

(a)   (b)

Fig. 17 (**a**) Approximation of the x-derivative of the surface $(x - 2.5)^2 + (y - 2.5)^2 = 0$. (**b**) Error obtained in the L_2-fit of the derivatives

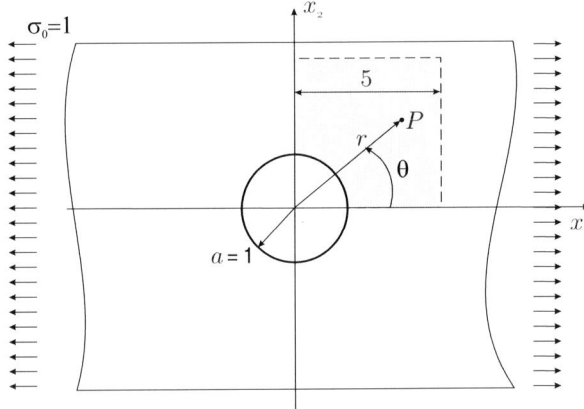

Fig. 18 Geometry of the problem of an infinite plate with a hole under traction

Applying symmetry conditions, only one quarter of the plate was modelled, and exact tractions were applied at the boundary of the model. The geometry of the model is shown in Fig. 18. The plate was discretized with quadratic consistency approximants and, again, repeated nodes were employed along the essential boundary. The position of the knots (following the standard B-spline notation, note that they are not "nodes" in the Finite Element sense) is shown in Fig. 19.

The obtained convergence rate ($R = 2.96$, see Fig. 20) is very close to the theoretical one, for quadratic consistency.

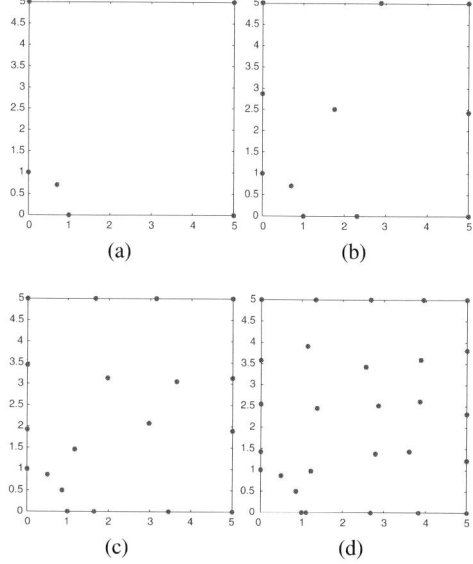

Fig. 19 Discretization of the plate with a hole problem: (**a**) 36 d.o.f. (**b**) 100 d.o.f. (**c**) 200 d.o.f. and (**d**) 322 d.o.f. Note that only the knots (not the nodes or "control points") have been represented

Fig. 20 Convergence of the results in L_2 norm for the plate with a hole problem

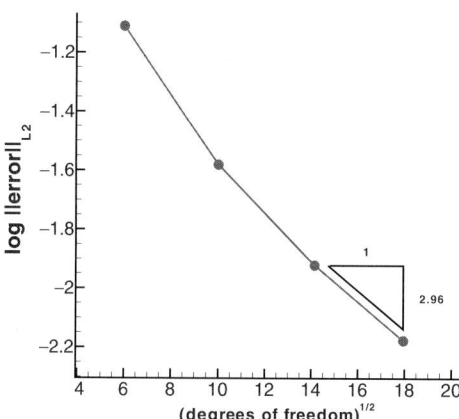

6 Conclusions

In this paper we have reviewed several schemes to generate natural neighbour approximations with high-order consistency and smoothness.

The Partition of Unity method provides an excellent tool to enrich the natural neighbour interpolant in order to verify the requirements of the inf-sup or LBB condition in mixed approximations. It has been demonstrated that the enrichment with the set 1,xy, among other possibilities, provides stable results and seems to verify the LBB condition.

The last method presented is able to generate arbitrary order consistency approximations, conserving many of the interesting properties of the Natural Element Method, notably the interpolant character along the boundary. This is achieved, as in B-spline curves, by repeating the boundary knots defining the surface. Conversely to the B-spline case, continuity is not improved if standard Sibson interpolants are used, but this can be solved by employing Hiyoshi-Sugihara interpolants, which are $\mathscr{C}^1$-continuous at least, but only linearly complete.

That method [16] resembles the *isogeometric* Finite Element Method by T. Hughes, in the sense that a new approximation method has been proposed that shares many of its properties with the B-spline curves. However, the obtention of B-spline surfaces is quite rigid, since they are constructed by tensor-product patches. In the proposed method, there is no need of a tensor-product structure, and hence we would speak about an *isogeometric* meshless method.

The other objective of the method, namely the generation of arbitrary order consistency in the context of the NEM, has been achieved. This opens the possibility of *p*-like adaptive simulations with NE methods. Obtaining smooth approximations with simpler structure than that of Hiyoshi and Sugihara [17] is now part of the authors' current effort of research.

References

1. I. Alfaro, D. Bel, E. Cueto, M. Doblaré, and F. Chinesta. Three-dimensional simulation of aluminium extrusion by the alpha-shape based natural element method. *Computer Methods in Applied Mechanics and Engineering*, 195(33–36):4269–4286, 2006.
2. I. Alfaro, J. Yvonnet, F. Chinesta, and E. Cueto. A study on the performance of natural neighbour-based Galerkin methods. *International Journal for Numerical Methods in Engineering*, accepted for publication, 2006.
3. I. Alfaro, J. Yvonnet, E. Cueto, F. Chinesta, and M. Doblaré. Meshless methods with application to metal forming. *Computer Methods in Applied Mechanics and Engineering*, 195(48–49):6661–6675, 2006.
4. D. N. Arnold, F. Brezzi, and M. Fortin. A stable finite element for the Stokes equations. *Calcolo*, 21:337–344, 1984.
5. V. V. Belikov, V. D. Ivanov, V. K. Kontorovich, S. A. Korytnik, and A. Yu. Semenov. The non-Sibsonian interpolation: A new method of interpolation of the values of a function on an arbitrary set of points. *Computational Mathematics and Mathematical Physics*, 37(1):9–15, 1997.
6. J. Braun and M. Sambridge. A numerical method for solving partial differential equations on highly irregular evolving grids. *Nature*, 376:655–660, 1995.
7. D. Chapelle and K. J. Bathe. The inf-sup test. *Computers and Structures*, 47(4–5):537–545, 1993.
8. J.-S. Chen, C.-T. Wu, S. Yoon, and Y. You. A stabilized conforming nodal integration for galerkin mesh-free methods. *International Journal for Numerical Methods in Engineering*, 50:435–466, 2001.
9. E. Cueto, M. Doblaré, and L. Gracia. Imposing essential boundary conditions in the Natural Element Method by means of density-scaled α-shapes. *International Journal for Numerical Methods in Engineering*, 49:519–546, 2000.

10. E. Cueto, N. Sukumar, B. Calvo, M. A. Martínez, J. Cegoñino, and M. Doblaré. Overview and recent advances in natural neighbour Galerkin methods. *Archives of Computational Methods in Engineering*, 10(4):307–384, 2003.

11. C. de Boor. Bicubic spline interpolation. *Journal of Mathematics and Physics*, 41:212–218, 1962.

12. G. Farin. *Curves and surfaces for CAGD*. Morgan Kaufmann, San Francisco, CA, 2002.

13. D. Gonzalez, E. Cueto, M. A. Martinez, and M. Doblare. Numerical integration in natural neighbour Galerkin methods. *International Journal for Numerical Methods in Engineering*, 60(12):2077–2104, 2004.

14. D. González, E. Cueto, F. Chinesta, and M. Doblaré. A natural element updated Lagrangian strategy for free-surface fluid dynamics. *Journal of Computational Physics*, 223(1):127–150, 2007.

15. D. González, E. Cueto, and M. Doblaré. Volumetric locking in natural neighbour Galerkin methods. *International Journal for Numerical Methods in Engineering*, 61(4):611–632, 2004.

16. D. González, E. Cueto, and M. Doblaré. Higher-order natural element methods: towards an isogeometric meshless method. *International Journal for Numerical Methods in Engineering*, DOI: 10.1002/nme.2237, 2007.

17. H. Hiyoshi. *Study on Interpolation based on Voronoi Diagrams*. Ph.D. thesis, Tokyo University, 2000.

18. L. Illoul, J. Yvonnet, F. Chinesta, and S. Clenet. Application of the natural-element method to model moving electromagnetic devices. *IEEE Transactions on Magnetics*, 42(4):727–730, 2006.

19. R. Sibson. A vector identity for the Dirichlet tesselation. *Mathematical Proceedings of the Cambridge Philosophical Society*, 87:151–155, 1980.

20. R. Sibson. A brief description of natural neighbour interpolation. In *Interpreting Multivariate Data. V. Barnett (Editor)*, pages 21–36. Wiley, New York, 1981.

21. N. Sukumar. *The Natural Element Method in Solid Mechanics*. Ph.D. thesis, Northwestern University, 1998.

22. N. Sukumar and B. Moran. C^1 natural neighbour interpolant for partial differential equations. *Numerical Methods for Partial Differential Equations*, 15(4):417–447, 1999.

23. N. Sukumar, B. Moran, and T. Belytschko. The natural element method in solid mechanics. *International Journal for Numerical Methods in Engineering*, 43(5):839–887, 1998.

24. N. Sukumar, B. Moran, A. Yu Semenov, and V. V. Belikov. Natural neighbor Galerkin methods. *International Journal for Numerical Methods in Engineering*, 50(1):1–27, 2001.

25. A. H. Thiessen. Precipitation averages for large areas. *Monthly Weather Report*, 39:1082–1084, 1911.

26. S. Timoshenko and J. N. Goodier. *Teoría de la Elasticidad*. Editorial Urmo, Spain, 1972.

27. L. Traversoni. Natural neighbour finite elements. In *International Conference on Hydraulic Engineering Software. Hydrosoft Proceedings*, pages 291–297. Computational Mechanics publications, Southampton, 1994.

28. J. Yvonnet, D. Ryckelynck, P. Lorong, and F. Chinesta. A new extension of the natural element method for non-convex and discontnuous problems: the Constrained Natural Element method. *International Journal for Numerical Methods in Enginering*, 60(8):1452–1474, 2004.

A Partition of Unity-Based Multiscale Method

Michael Macri and Suvranu De(✉)

Abstract In this paper we present an octree partition of unity method (OctPUM) developed in the context of a multiscale environment with an enrichment technique for the modeling of heterogeneous media in the presence of singularities such as cracks which overcome long-standing problems associated with the assumption of local periodicity in traditional asymptotic homogenization methods. In order to compute the microscopic fields near the crack edge within the macroscale computations, a structural enrichment-based homogenization method is introduced in which the approximation space of the OctPUM at the macroscopic scale is enriched by functions generated at the microscopic scale using the asymptotic homogenization technique.

Keywords: Multiscale method · Partition of Unity · Octree Partition of Unity Method

1 Introduction

Modeling of heterogeneous materials poses a significant computational challenge. A full finite element solution at the macroscopic level, taking into account the detailed micro-structural features is computationally infeasible. Hence robust and reliable multiscale computational techniques are necessary. The goal of multiscale modeling is to take into account the interconnectivity of the essential phenomena occurring at multiple length and time scales preserving macroscopic conservation principles.

The objective of recent multiscale methods, which distinguishes them from traditional ones such as multigrid, wavelet, fast multipole or adaptive mesh refinement techniques, is to capture the macro-scale behavior with a cost which is sub-linear

S. De and M. Macri

Advanced Computational Research Laboratory, Department of Mechanical, Aerospace and Nuclear Engineering, Rensselaer Polytechnic Institute, 110, 8th Street, Troy, NY 12180, USA

e-mail: des@rpi.com

A.J.M. Ferreira et al. (eds.) *Progress on Meshless Methods, Computational Methods in Applied Sciences.*

compared to the cost of a full micro-scale solver [1]. Scale linking is performed using hierarchical, concurrent or a combination of these schemes [2].

The mathematical theory of asymptotic homogenization [3–5] which uses asymptotic expansions of field variables about macroscopic values is a well-known hierarchical multiscale technique [6–11]. The asymptotic homogenization method provides overall effective properties as well as microscopic stress and strain values. It has been shown in [10] that the use of a homogenization approach, with periodic boundary conditions on the representative volume element (RVE), results in more accurate estimates of stiffness as well as local strain than the unit cell based micro-structural models.

However, the homogenization method suffers from a major limitation stemming from its basic assumptions, viz. (a) uniformity of the macroscopic fields within each RVE and (b) local spatial periodicity of the RVE. Hence, this method breaks down in critical regions of high gradients such as cracks. To extend the homogenization method to non-periodic problems, techniques such as the s-version of the finite element method [12, 13], various multigrid-like bridging scale methods [14, 15], an adaptive global-local method based on the Voronoi cell finite element method (VCFEM) [16], a homogenized Dirichlet projection method (HDPM) [17, 18] and a generalized finite element method with "mesh-based handbook functions" [19] have been developed.

The s-version of the finite element method is based on the principle of superimposed meshes [13] and requires cumbrous quadrature techniques. Unstructured multigrid-like methods require remeshing to capture evolving fine scale features. The VCFEM is an assumed stress hybrid method in which each heterogeneity is embedded in a Voronoi cell which is treated as a finite element.

To overcome the drawbacks of the existing methods we propose an enrichment method based on the principles of partition of unity [20] which allow enrichment of the approximation space in localized subdomains using specialized functions that may be generated based on *a priori* information regarding asymptotic expansions of local stress fields and microstructure. The use of such enrichment functions has been popularized in the context of meshfree [21–25] and generalized/extended finite element methods [26–29]. To take full advantage of the partition of unity properties we have developed an octree partition of unity method [30] which is in-between meshfree methods and finite element methods in character; combining advantages of both techniques. OctPUM uses hierarchical partitioning of space to discretize the geometry. Approximation functions are compactly supported on n-dimensional cubes and are generated using the partition of unity approach. However, the integration domains are much more regular than in meshfree methods resulting in more efficient numerical integration.

To overcome the problem of predicting the correct displacement response at the microscopic scale, we have developed a new structural enrichment based method in which we enrich OctPUM in the vicinity of the cracks with specialized functions derived using asymptotic homogenization theory which allow macro-scale computations to be performed with the micro-structural features explicitly considered.

We begin by reviewing the OctPUM in Section 2. In Section 3 we present a structural enrichment based method and in Section 4 we demonstrate the effectiveness of our approach including a realistic microstructure of particulate composites.

2 The Octree Partition of Unity Method (OctPUM)

We use the well-known octree data-structure [31] to discretize the computational domain. To keep our discussion simple, we will concentrate on two-dimensional examples where the corresponding data structure is known as a "quadtree". We begin by enclosing the computational domain (Ω) within a square known as the "root quadrant" which is then subdivided recursively into four quadrants along the Cartesian directions until the terminal quadrants satisfy user-defined size requirements.

We will initially assume that the domain Ω is square and coincides with the root quadrant (Fig. 1). In [30] we discuss how domains which are geometrically more complex may be discretized.

Computational nodes are placed at the corners of the terminal quadrants as shown in Fig. 1(b). Let $l_{IJ} = \|\mathbf{x}_J - \mathbf{x}_I\|_0$ denote the distance between nodes 'I' and 'J' located at $\mathbf{x}_I$ and $\mathbf{x}_J$, respectively. The "support" of node 'I', $\overline{B}_I$, is chosen to be a square, centered at node 'I,' of side $2l_I = 2\min\{l_{ij}\}_{J\in\zeta}$ where ζ is the set of all nodes 'J' that are directly connected to node 'I.' The shape functions generated at node 'I' using the partition of unity approach (next section) are compactly supported on $\overline{B}_I$. Some of the supports are shown in Fig. 1(b). The advantage of this choice is that by construction $\Omega \subset \cup_{I=1}^{N}\overline{B}_I$ where N nodes are used to discretize the domain. Moreover, the area of overlap of supports as well as the integration subdomains is straightforward to compute.

The shape functions are generated using the partition of unity paradigm [20] based on the Shepard partition of unity functions [32]. We define, at each node 'I', a weighting function W_I which is compactly supported on $\overline{B}_I$. Where

$$W_I(x,y) = W(s_{xI})W(s_{yI}) \tag{1}$$

and $s_{xI} = \|x - x_I\|_0/l_I$, $s_{yI} = \|y - y_I\|_0/l_I$ and x_I and y_I are the coordinates of node 'I'. In our work we have chosen a quartic spline weighting function of the form:

$$W(s_{aI}) = \begin{cases} 1 - s_{aI}^2 + 8s_{aI}^3 - 3s_{aI}^4, & s_{aI} \leq 1 \\ 0, & s_{aI} > 1 \end{cases} \quad \text{for } a \in \{x, y\} \tag{2}$$

We define the Shepard partition of unity function at each node 'I', using a simple normalization procedure

$$\phi_I^0(\mathbf{x}) = \frac{W_I(\mathbf{x})}{\sum_{J=1}^{N} W_J(\mathbf{x})} \quad I = 1, 2, ... N \tag{3}$$

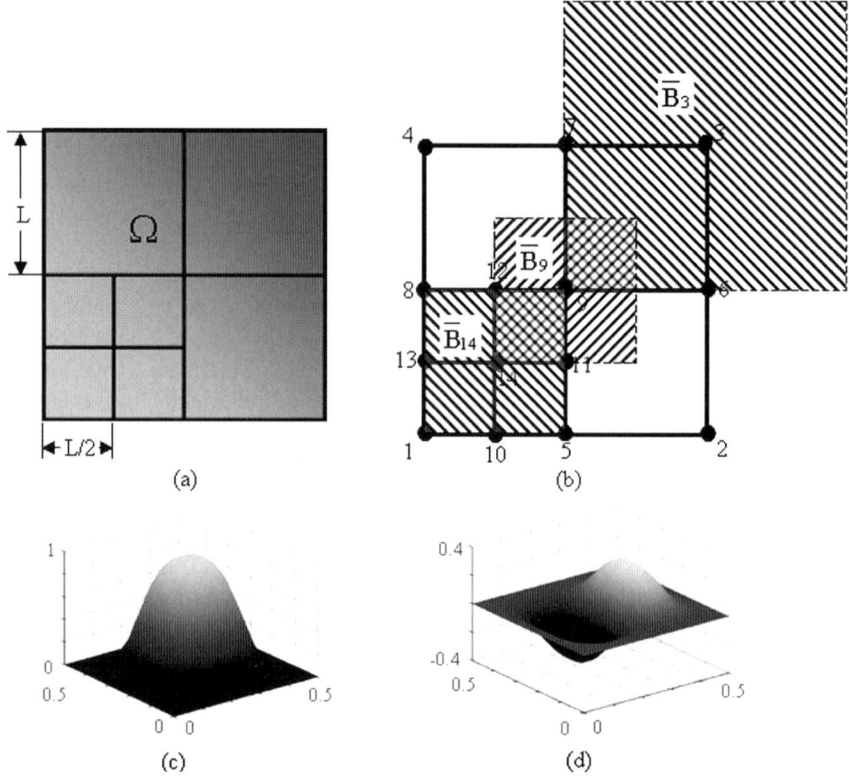

Fig. 1 Construction of the octpum for a square domain (Ω). (**a**) The domain is enclosed within the root quadrant which is then recursively subdivided along the Cartesian directions. Nodes are placed at the corners of the terminal quadrants. (**b**) The supports of nodes 3, 9 and 14 are shown. Some shape functions generated using the partition of unity method are shown in (**c**) and (**d**)

which ensures $\sum_{J=1}^{N} \phi_J^0(\mathbf{x}) = 1 \forall \mathbf{x} \in \Omega$ i.e., rigid body modes are exactly satisfied. We develop global approximation spaces with higher order consistency by defining, at each node 'I', a local approximation space

$$V_I^{h,p} = span_{m \in \zeta} \{p_m(\mathbf{x})\} \subset H^1(\bar{B}_I \cap \Omega) \tag{4}$$

where h is a measure of the size of the element, p is the polynomial completeness, ζ is an index set, H^1 is the first order Hilbert space, and $p_m(\mathbf{x})$ is a polynomial or other function. The global approximation space is then generated as follows

$$V^{h,p} = \sum_{I=}^{N} \phi_I^0 V_I^{h,p} \subset H^1(\Omega) \tag{5}$$

Hence, any function $v^{h,p} \in V^{h,p}$ can be written as

$$v^{h,p}(\mathbf{x}) = \sum_{I=1}^{N} \sum_{m \in \zeta} h_{Im}(\mathbf{x}) \alpha_{Im} \tag{6}$$

where

$$h_{Im}(\mathbf{x}) = \phi_I^0 p_m(\mathbf{x}) \tag{7}$$

is the shape function at node I corresponding to the mth degree of freedom. Figure 1(c) shows the Shepard partition of unity function $h_{140} = \phi_{14}^0$ at node 14. In Fig. 1(d), the shape function $h_{141} = \phi_{14}^0 \frac{x - x_{14}}{l_{14}}$ is shown.

3 Enrichment Techniques for the OctPUM

The advantage of using the OctPUM to model the response of heterogeneous materials such as particulate composites is that, due to the partition of unity property it provides a route to local enrichment using specialized functions in a straightforward manner. To ensure that correct displacement fields are predicted at the microscale, a new structural enrichment-based OctPUM technique is presented in Section 3.2 where functions generated using the asymptotic homogenization method are used for enrichment at the macroscale. We begin by briefly reviewing asymptotic homogenization.

3.1 Asymptotic Homogenization

In the theory of asymptotic homogenization, a macroscopic scale and a microscopic scale are designated and represented by coordinates x and y, respectively (Fig. 2) and a function f^γ implies

$$f^\gamma(\mathbf{x}) = f(\mathbf{x}, \mathbf{y}(\mathbf{x})) \tag{8}$$

where the coordinates x_i and y_j are related by $\mathbf{y} = \frac{\mathbf{x}}{\gamma}$, γ being a constant scale factor that denotes the ratio of length scales at the two scales.

The assumption of "Y-periodicity" of the representative volume element (RVE) ensures

$$f(x_i, y_j) = f(x_i, y_j + kY_j) \tag{9}$$

where Y_j is the dimension of the RVE in the jth direction. Partial derivatives of the function $f(x_i, y_j)$ may be computed using the formula

$$\partial_{x_i} f(x_i, y_j) = \partial_{x_i} f + \partial_{y_j} f \frac{\partial y_j}{\partial x_i} = \partial_{x_i} f + \gamma^{-1} \partial_{y_j} f \tag{10}$$

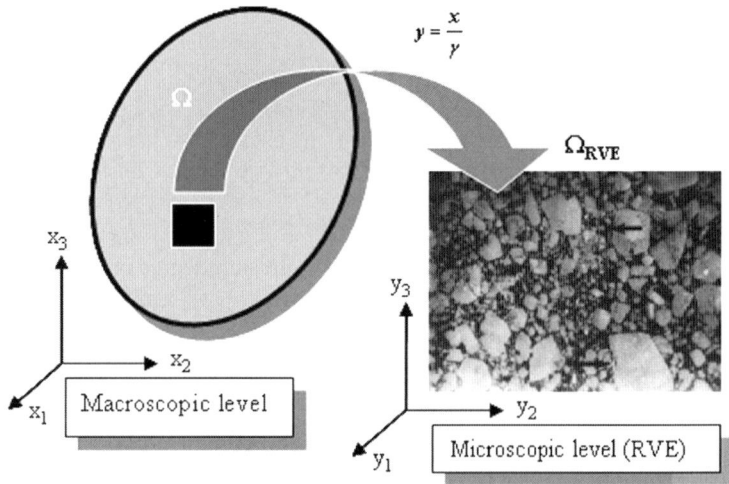

Fig. 2 Relationship between macro (Ω) and micro (Ω_{RVE}) coordinate systems

In the mathematical homogenization method, one assumes an asymptotic expansion of the displacement field in powers of the scale factor γ:

$$u_i^\gamma(\mathbf{x},\mathbf{y}) = u_i^0(\mathbf{x},\mathbf{y}) + \gamma u_i^1(\mathbf{x},\mathbf{y}) + O(\gamma^2) \quad i \in \{1,2,3\} \tag{11}$$

The expansions of the strains (ε_{ij}^γ) and stresses (σ_{ij}^γ) are

$$\varepsilon_{ij}^\gamma(\mathbf{x},\mathbf{y}) = \gamma^{-1}\varepsilon_{ij}^{-1}(\mathbf{x},\mathbf{y}) + \varepsilon_{ij}^0(\mathbf{x},\mathbf{y}) + \gamma\varepsilon_{ij}^1(\mathbf{x},\mathbf{y}) + O(\gamma^2) \quad \{i,j\} \in \{1,2,3\} \tag{12}$$

$$\sigma_{ij}^\gamma(\mathbf{x},\mathbf{y}) = \gamma^{-1}\sigma_{ij}^{-1}(\mathbf{x},\mathbf{y}) + \sigma_{ij}^0(\mathbf{x},\mathbf{y}) + \gamma\sigma_{ij}^1(\mathbf{x},\mathbf{y}) + O(\gamma^2) \quad \{i,j\} \in \{1,2,3\} \tag{13}$$

where

$$\sigma_{ij}^v = C_{ijkl}\varepsilon_{kl}^v \tag{14}$$

is the constitutive relationship for the constituents of the RVE. We will assume that the constitutive behavior of each constituent of the RVE is linear elastic and that

$$\varepsilon_{ij}^v = \varepsilon_{ijx}^v + \varepsilon_{ijy}^{v+1}$$
$$\varepsilon_{ijx}^n = \frac{1}{2}(u_{i,x_j}^n + u_{j,x_i}^n) \tag{15}$$
$$\varepsilon_{ijy}^n = \frac{1}{2}(u_{i,y_j}^n + u_{j,y_i}^n)$$

Substituting the asymptotic expansions into the linear elastic equilibrium equations and rearranging the terms produces the following results. First, that

$$u_i^0(\mathbf{x},\mathbf{y}) = u_i^0(\mathbf{x}) \tag{16}$$

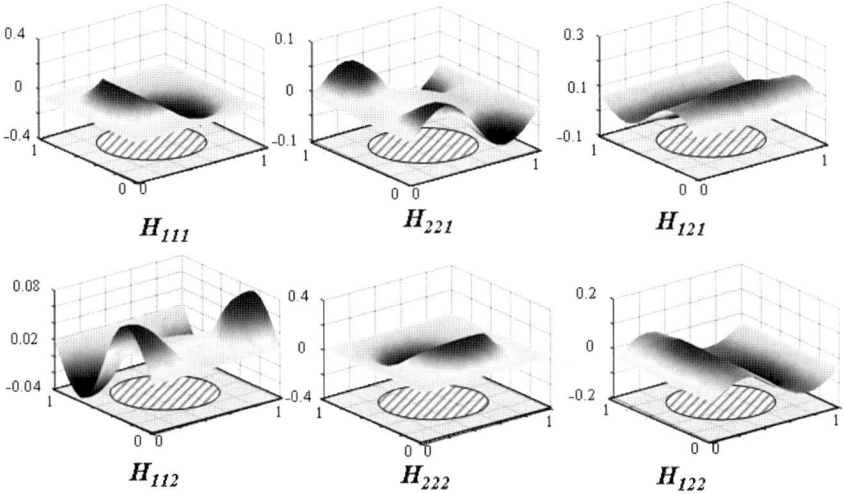

Fig. 3 The six components of, Hijk for the microstructure

Next, the RVE problem is reduced to:

$$(C_{ijkl}(I_{klmn} + \Psi_{mnkl})),_{y_j} = 0 \tag{17}$$

where

$$\Psi_{mnkl} = \frac{1}{2}(H_{mnl,y_k} + H_{mnk,y_l}) \tag{18}$$

and $H_{mnk}(\mathbf{y})$ is a Y-periodic third rank structural tensor that relates the strains at the macroscopic level to the first order displacement perturbation term:

$$u_k^1 = H_{mnk}(\mathbf{y})\varepsilon_{mnx}^0(\mathbf{x}) \tag{19}$$

Some examples of the $H_{mnk}(\mathbf{y})$ functions for varying RVEs are shown in Fig. 3. We have used the finite element method to solve the RVE problem with periodic boundary conditions.

Finally, the "effective" equation at the macroscopic scale is

$$(\tilde{C}_{ijmn}\varepsilon_{mnx}^0),_{x_j} + \tilde{b}_i = 0 \tag{20}$$

where the effective moduli and body force terms are

$$\tilde{C}_{ijmn} = \frac{1}{Y}\int_Y C_{ijmn}(I_{mnkl} + \Psi_{mnkl})dY \tag{21}$$

and

$$\tilde{b}_i = \frac{1}{Y}\int_Y b_i dy \tag{22}$$

3.2 Structural Enrichment-Based Homogenization

We recall from Eqs. (11) and (16) that the displacement field on the RVE is a constant $u_i^\gamma(\mathbf{x}, \mathbf{y}) = u_i^0(\mathbf{x}, \mathbf{y})$. To resolve this problem, especially in the vicinity of the singularities, we present a structural enrichment based homogenization technique where functions, derived using asymptotic homogenization, are used to enrich the macroscale approximation space.

In the OctPUM, if the following approximation is used for the entire domain:

$$u_j^0(\mathbf{x}) = \sum_{I=1}^{N} \sum_{m \in \zeta} h_{Im}(\mathbf{x}) \alpha_{Im}^j \tag{23}$$

and only a portion $\Omega_{enriched} \subset \Omega$ in the vicinity of high solution gradients is to be enriched, then we may define an enriched approximation in $\Omega_{enriched}$ of the form

$$u_j^{enriched}(x) = u_i^0(x) + \sum_{J=1}^{N^{enriched}} \sum_{n \in \zeta} \sum_{k,q,r \in \{1,2,3\}} h_{Jn}(x) H_{kqr} \beta_{Jnkqr}^j \tag{24}$$

Where β_{jnkqr}^j is an unknown degree of freedom. Some enriched shape functions are shown in Figs. 5 and 6. The multiscale enrichment based partition of unity method developed in [33] is similar in approach in the context of the extended finite element method.

4 Numerical Example: Center Cracked Tension (CCT) Specimen Composed of a Heterogeneous Material

We consider a center cracked tension (CCT) specimen in plane stress. The specimen is of unit thickness and is loaded as shown in Fig. 4(a). Since the problem is symmetric, we only examine the upper right quarter of the model. We will consider two types of microstructures shown in Fig. 4(b, c). In all the cases we use a quadratic basis for nodes that are not enriched.

4.1 Microstructure of Type 1

First, we consider a simple RVE consisting of a circular inclusion embedded in a matrix (Fig. 4(b)). We compare all results to a 'fine mesh solution' generated using the commercial finite element program FEMLAB® using approximately 57,000 six-noded triangular elements. We assume a finite size of the RVE such that the modeled domain consists of 16×16 unit cells. The strain energy corresponding to the fine mesh solution is calculated to be 50.971786 *in-lb*. Asymptotic homogenization is used to compute the effective properties as $C_{11} = 1.630429\, psi$, $C_{12} = 0.393643\, psi$

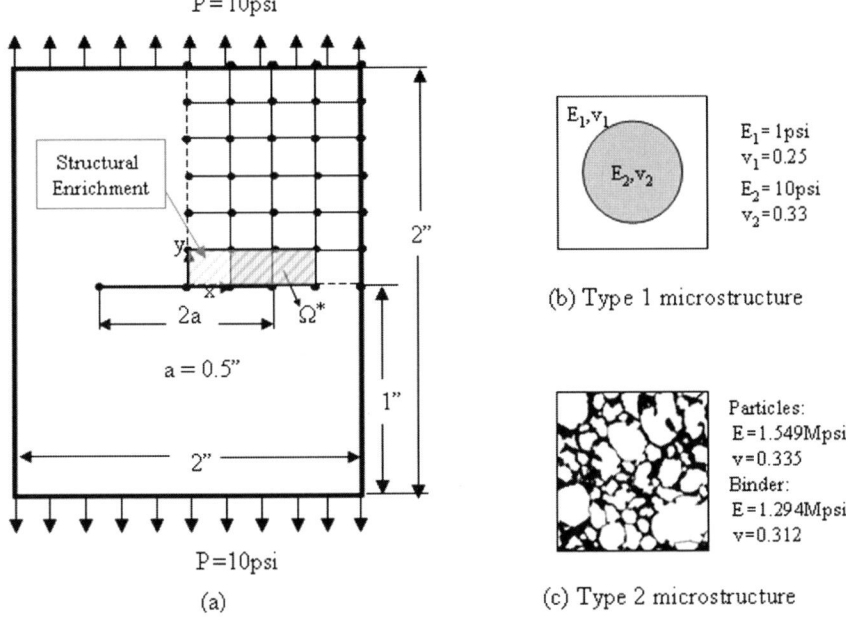

Fig. 4 (**a**) A center cracked tension (CCT) specimen with two types of microstructures as shown in (**b**) and (**c**)

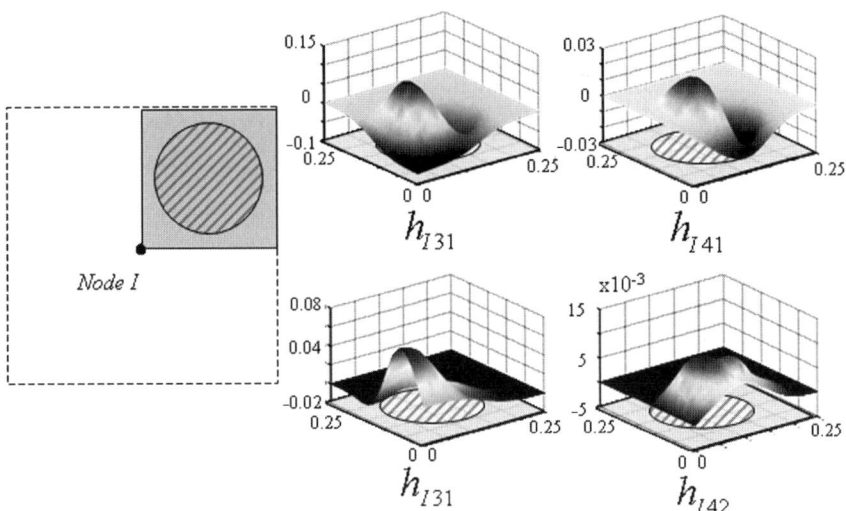

Fig. 5 Enriched shape functions for a 2D analysis are shown when a quarter of the support for node I coincides with a single RVE

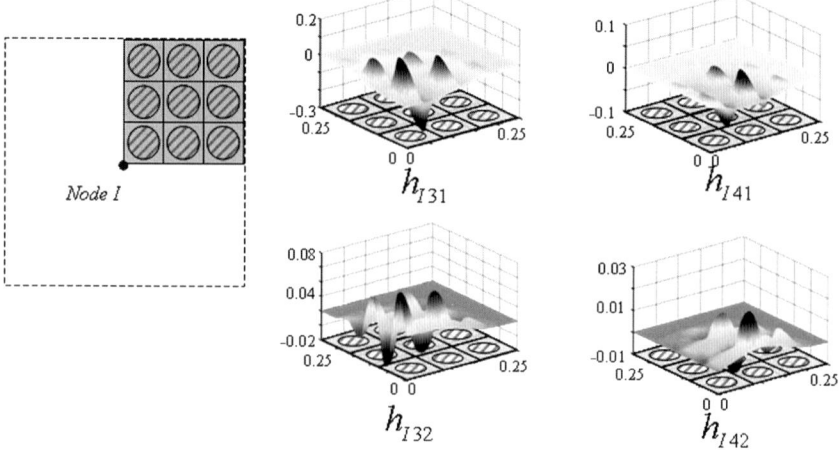

Fig. 6 An enriched shape functions for a 2D analysis are shown when a quarter of the support for node I coincides with nine RVEs in 2D

and $C_{33} = 0.560592\,psi$, where C_{11}, C_{12}, and C_{33} are the components of the elasticity matrix.

The structural enrichment-based homogenization OctPUM method has been developed to compute the actual displacement field within the microstructure as part of the global solution. In this technique, only a single layer of octants adjacent to the crack surface is enriched. Figure 3 shows the third rank tensor H_{ijk} for the microstructure in Fig. 4(b) computed using a finite element mesh of 1,000 six-noded triangular elements.

The following enriched shape functions are used

$$u_j(\mathbf{x}) = \sum_{I=1}^{N} \sum_{m=0}^{11} h_{Im}^j(\mathbf{x}) \alpha_{Im}^j \quad j = 1, 2$$

$$
\begin{aligned}
&h_{I0}^1 = h_{I0}^2 = \phi_I^0 \qquad h_{I1}^1 = h_{I1}^2 = \phi_I^0(x - x_I)/l_I \qquad h_{I2}^1 = h_{I2}^2 = \phi_I^0(y - y_I)/l_I \\
&h_{I3}^1 = \phi_I^0 H_{111} \qquad h_{I4}^1 = \phi_I^0 H_{111}(x - x_I)/l_I \qquad h_{I5}^1 = \phi_I^0 H_{111}(y - y_I)/l_I \\
&h_{I3}^2 = \phi_I^0 H_{112} \qquad h_{I4}^2 = \phi_I^0 H_{112}(x - x_I)/l_I \qquad h_{I5}^2 = \phi_I^0 H_{112}(y - y_I)/l_I \qquad (25) \\
&h_{I6}^1 = \phi_I^0 H_{221} \qquad h_{I7}^1 = \phi_I^0 H_{221}(x - x_I)/l_I \qquad h_{I8}^1 = \phi_I^0 H_{221}(y - y_I)/l_I \\
&h_{I6}^2 = \phi_I^0 H_{222} \qquad h_{I7}^2 = \phi_I^0 H_{222}(x - x_I)/l_I \qquad h_{I8}^2 = \phi_I^0 H_{222}(y - y_I)/l_I \\
&h_{I9}^1 = \phi_I^0 H_{121} \qquad h_{I10}^1 = \phi_I^0 H_{121}(x - x_I)/l_I \qquad h_{I11}^1 = \phi_I^0 H_{121}(y - y_I)/l_I \\
&h_{I9}^2 = \phi_I^0 H_{122} \qquad h_{I10}^2 = \phi_I^0 H_{122}(x - x_I)/l_I \qquad h_{I11}^2 = \phi_I^0 H_{122}(y - y_I)/l_I
\end{aligned}
$$

where the structurally enriched shape functions are direction dependent requiring the addition of the superscript j to represent the direction the shape function is applied to. Several examples of the shape functions are shown in Figs. 5 and 6 where the support of node I contains 4 and 36 RVEs, respectively.

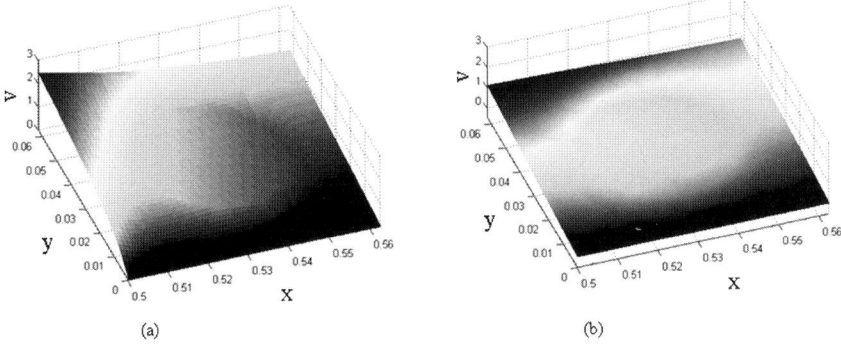

Fig. 7 For the RVE in Fig. 4(b) located at the crack tip, the displacement in the y-direction (v) is plotted for (**a**) the fine mesh and (**b**) the octpum with structural enrichment-based homogenization

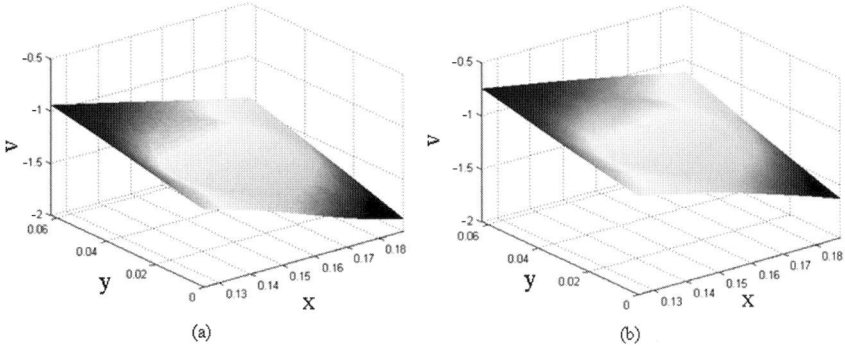

Fig. 8 For the RVE in Fig. 4(b) located along the crack edge, the displacement in the x-direction (u) is plotted for (**a**) the fine mesh and (**b**) the octpum with structural enrichment-based homogenization

The displacement in the y-direction (v) is plotted within the RVE located at the crack tip in Fig. 7 and the displacement in the x-direction (u) is plotted within the RVE located at the crack edge in Fig. 8. The OctPUM with structural enrichment-based homogenization is seen to capture the correct variation of the displacement field within the RVE. In Fig. 9, we show that the normal strain (ε_{yy}) in the RVE located at the crack tip is also similar to what the fine mesh solution provides.

4.2 Microstructure of Type 2

Next we consider the microstructure of a particulate composite shown in Fig. 3(c) composed of crystalline particles embedded in a polymer matrix. To develop a finite element model from the microstructure shown in Fig. 10(a) image processing techniques have been employed. The contours of the particles are identified using the

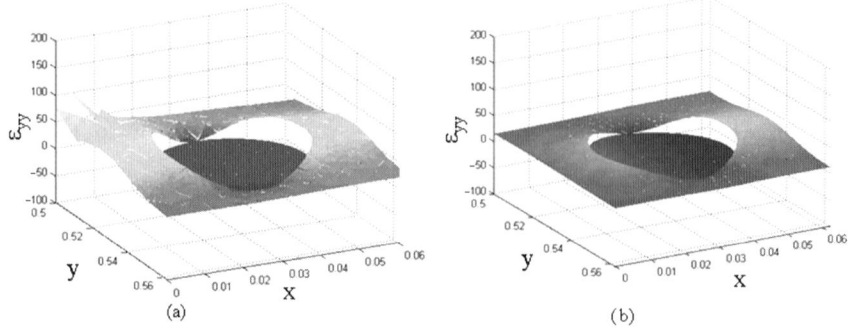

Fig. 9 For the RVE in Fig. 4(b) located at the crack tip, strain (ε_{yy}) is plotted for (**a**) the fine mesh and (**b**) the octpum with structural enrichment-based homogenization

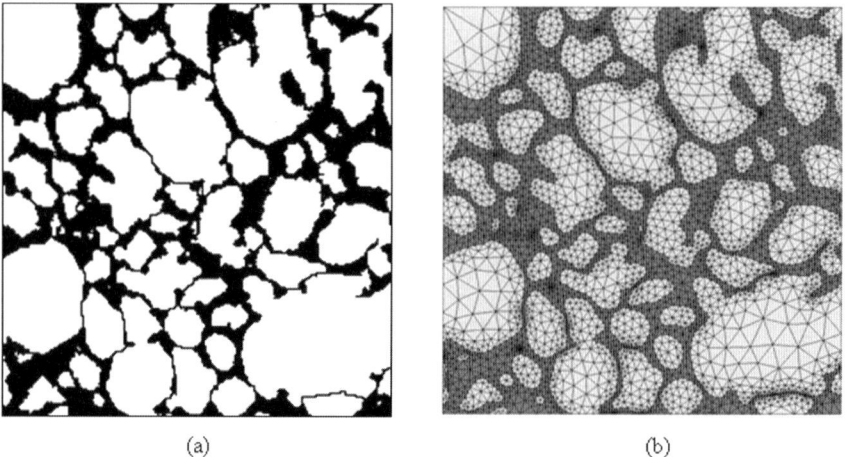

Fig. 10 (**a**) Binary image of PBX 9501 microstructure. (**b**) Finite element discretization

image processing toolbox of Matlab®. Finally, the commercially available code FEMLAB® is used to generate the finite element mesh (Fig. 10(b)). Notice that in the final model very small particles are lumped together with the matrix, generating what is known as the "dirty binder". The smallest grains captured are approximately 10 μm while the max grain size is approximately 300 μm. We use the properties $E = 1549\,ksi$ and $v = 0.335$ for the particles and $E = 1294\,ksi$ and $v = 0.312$ for the binder. The effective properties are $C_{11} = 1632.6\,ksi$, $C_{12} = 531.1\,ksi$ and $C_{33} = 590.2\,ksi$. We compute the displacement field in the y-direction (v) and the normal stress (σ_{yy}) within the RVE located at the crack tip in Fig. 11(a, b) respectively. This is an excellent example where the fine mesh solution is extremely expensive to compute. However, we may still compute the displacement field within the RVE in desired locations within the model using structural enrichment-based OctPUM.

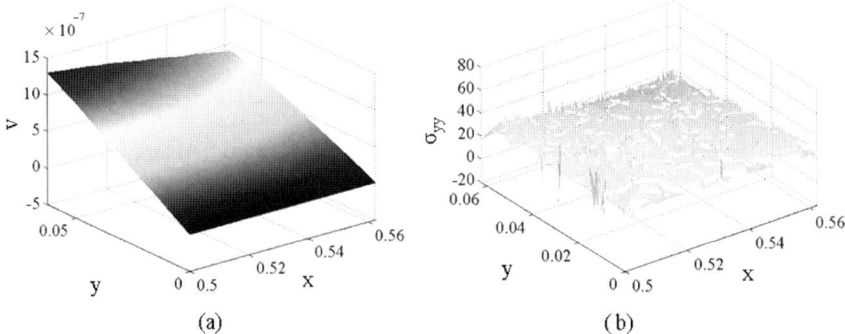

Fig. 11 The (**a**) displacement in the y direction (v) and the normal stress (σ) at the micro structural level for the RVE in Fig. 4(c) located at the crack tip

5 Concluding Remarks

A structural enrichment-based homogenization method has been developed, in the context of a new octree partition of unity (OctPUM) method which is in-between finite element and mesh free methods, to compute the response at the microscopic scale in localized regions of the computational domains such as along the crack edge while performing macroscale simulations. In this technique the enrichment functions are derived from the asymptotic homogenization method.

References

1. Weinan E, Engquist B, Li X, Ren W, Vanden-Eijnden E. The heterogeneous multiscale method: A review (Preprint), 2006.
2. Yip S. Handbook of materials modeling (Ed. XXIX). Springer, 2005.
3. Bakhalov N, Panasenko G. Homogenization: Averaging process in periodic media. Dordrecht: Kluwer, 1989.
4. Bensoussan A, Lions JL, Papanicolaou G. Asymptotic analysis for periodic structures. Amsterdam: Noth-Holland, 1978.
5. Sanchez-Palencia E, Zaoui A. Homogenization techniques for composite media. Springer, 1985.
6. Tolenado A, Murakami H. A high-order mixture model for periodic particulate composites. International Journal of Solids and Structures 1987; 23:989–1002.
7. Guedes JM, Kikuchi N. Preprocessing and post-processing for materials based on homogenization method with adaptive finite element methods. Computer Methods in Applied Mechanics and Engineering 1990; 83:143–198.
8. Hollister SJ, Fyhrie DP, Jepsen KJ, Goldstein SA. Analysis of trabecular bone micromechanics using homogenization theory with comparison to experimental results. Journal of Biomechanics 1989; 22(10):1025.
9. Hollister SJ, Fyhrie DP, Jepsen KJ, Goldstein SA. Application of homogenization theory to the study of trabecular bone mechanics. Journal of Biomechanics 1991; 24(9):825–839.
10. Hollister SJ, Kikuchi N. A comparison of homogenization and standard mechanics analyses for periodic porous composites. Computational Mechanics 1992; 10:73–95.

11. Fish J, Yu Q, Shek K. Computational damage mechanics for composite materials based on mathematical homogenization. International Journal for Numerical Methods in Engineering 1999; 45:1657–1679.

12. Fish J. The s-version of the finite element method. Computers and Structures 1992; 43(3):539–547.

13. Fish J, Wagiman A. Multiscale finite element method for heterogeneous medium. Computational Mechanics 1993; 12:1–17.

14. Fish J, Belsky V. Multigrid method for periodic heterogeneous medium. Part I: Convergence studies for one-dimensional case. Computer Methods in Applied Mechanics and Engineering 1995; 126:1–16.

15. Fish J, Belsky V. Multigrid method for periodic heterogeneous medium. Part II: Multiscale modeling and quality control in multidimensional case. Computer Methods in Applied Mechanics and Engineering 1995; 126:17–38.

16. Ghosh S, Lee K, Raghavan P. A multilevel computational model for multiscale damage analysis in composite and porous materials. International Journal of Solids and Structures 2001; 38:2335–2385.

17. Oden JT, Zohdi TI. Analysis and adaptive modeling of highly heterogeneous elastic structures. Computer Methods in Applied Mechanics and Engineering 1997; 148:367–391.

18. Oden JT, Vemaganti K, Moes N. Hierarchical modeling of heterogeneous solids. Computer Methods in Applied Mechanics and Engineering 1999; 172:3–25.

19. Strouboulis T, Zhang L, Babuska I. p-version of the generalized FEM using mesh-based handbooks with applications to multiscale problems. International Journal for Numerical Methods in Engineering 2004; 60:1639–1672.

20. Yosida K. Functional analysis (Ed. 5). Berlin/Heidelberg: Springer, 1978.

21. Fleming M, Chu Y, Moran B, Belytschko T. Enriched element-free Galerkin methods for crack tip fields. International Journal for Numerical Methods in Engineering 1997; 40:1483–1504.

22. Belytschko T, Fleming M. Smoothing, enrichment and contact in the element-free Galerkin method. Computers & Structures 1999; 71:173–195.

23. Rao BN, Rahman S. An enriched meshless method for non-linear fracture mechanics. International Journal for Numerical Methods in Engineering 2004; 59:197–223.

24. Rao BN, Rahman S. An efficient meshless method for fracture analysis of cracks. Computational Mechanics 2000; 26:398–408.

25. Macri M, De S. Enrichment of the method of finite spheres using geometry independent localized scalable bubbles. International Journal for Numerical Methods in Engineering 2006; 69:1–32.

26. Dolbow J. An extended finite element method with discontinuous enrichment. Evanston, IL: Northwestern University, 1999.

27. Bellec J, Dolbow J. A note on enrichment functions for modeling crack nucleation. Communications in Numerical Methods in Engineering 2003; 19:921–932.

28. Strouboulis T, Babuska I, Copps K. The design and analysis of the generalized finite element method. Computer Methods in Applied Mechanics and Engineering 2000; 181:43–69.

29. Laguardia JJ, Cueto E, Doblaré M. A natural neighbor Galerkin method with quadtree structure. International Journal for Numerical Methods in Engineering 2005; 63:789–812.

30. Macri M, De S. An octree partition of unity method (OctPUM) with enrichments for multiscale modeling of heterogeneous media. Computers & Structures 2008 (in press).

31. Samet H. The design and analysis of spatial data structures. Reading, MA: Addison-Wesley, 1990.

32. Shepard D. A two-dimensional interpolation function for irregularly spaced data. Proceedings of 23rd National Conference ACM, 517–524, 1968.

33. Fish J, Yuan Z. Multiscale enrichment based partition of unity. International Journal for Numerical Methods in Engineering 2005; 62:1341–1359.

Application of Smoothed Particle Hydrodynamics Method in Engineering Problems

Matej Vesenjak(✉) and Zoran Ren

Abstract Computational simulations are becoming increasingly important engineering tool in recent years. One of the new computational techniques are the meshless methods, covering several application fields in engineering. In this paper the meshless Smoothed Particle Hydrodynamics method (SPH) is introduced and its implementation in the explicit code LS-DYNA is discussed. Then the application of the SPH method is presented with two practical examples. The first example deals with the modelling of sloshing problem in a reservoir, where the fluid has been analysed using different numerical techniques. The computational results were compared to and validated with the experimental measurements. The second example describes the fluid flow through the open network of cellular structure with the purpose to study its influence on capability of the impact energy absorption. In both examples the SPH method proved to be an effective and reliable tool.

Keywords: SPH · Smooth Particle Hydrodynamics Method · Explicit Code

1 Introduction

Computational simulations have become an indispensable tool for solving complex problems in engineering and science. They offer insight and help to interpret complex physics phenomena and additionally support analytical solutions and experimental testing. Various computational methods are widely applied in computational solid and fluid mechanics, solving also multi-physical problems [1].

Despite the fact that mesh-based numerical methods are the primary computational methodology in engineering computational mechanics, they still have limited application efficiency in many complex problems (e.g. free surface problems, large

M. Vesenjak and Z. Ren
University of Maribor, Faculty of Mechanical Engineering, Smetanova 31, SI-2000 Maribor, Slovenia, e-mail: m.vesenjak@uni-mb.si

A.J.M. Ferreira et al. (eds.) *Progress on Meshless Methods, Computational Methods in Applied Sciences.*

deformations). The major drawbacks can be attributed to the use of mesh, since the entire formulation and its results are based and depend on the mesh and its quality [1, 2]. Therefore recent research activities have been focused on development of computational methods, able to avoid the mesh dependence. This research resulted in development of meshfree methods, promising to be superior in regard to conventional mesh-based numerical methods in several engineering applications. One of the attractive meshless formulations is the Smoothed Particle Hydrodynamics (SPH), which is represented by a set of particles containing individual material properties and moving according to the general governing conservation equations.

This paper covers the theoretical background of SPH, its implementation in the explicit finite element code LS-DYNA and two practical examples of engineering applications. The first example describes a fluid-structure interaction problem of fluid sloshing in a reservoir. Different modelling approaches and solving formulations (Lagrangian, Eulerian, ALE and SPH) were compared and evaluated for the fluid part of the problem using the explicit code LS-DYNA. The computational results have also been compared to available experimental measurements of the sloshing problem. The second example analyses the behaviour of light-weight cellular materials with viscous fluid pore fillers to increase the energy absorption capabilities. These materials have been increasingly used as energy absorbing components and their development is valuable in modern engineering applications. The cellular structure has been modelled with the finite element method, while the fluid filler flow was modelled with the meshless SPH. Fully coupled fluid-structure interaction between the cellular structure base material and the fluid filler was also considered.

2 Meshless Methods – SPH

The basic idea of meshless methods is to provide accurate, reliable and stable computational solutions for integral equations or partial differential equations with various boundary conditions and a set of arbitrary distributed particles without any mesh connectivity between them. The meshless methods can be divided into three main groups: (i) methods based on strong formulations: are computationally efficient and completely meshless, but often unstable and less accurate, (ii) methods based on weak formulations (Element Free Galerkin – EFG, Meshless Local Petrov-Galerkin – MLPG, Point Interpolation Method – PIM): are very stable and accurate, but there is a need of a background mesh and (iii) particle methods (Molecular Dynamics – MD, Monte Carlo – MC, Smoothed Particle Hydrodynamics – SPH): are similar to the methods based on weak formulations and are stable for arbitrary distributed nodes and excellently cope with large deformation, where the accuracy mostly depends on the smoothing function [1].

The advantage of the particle meshless methods comparing to the conventional mesh-based methods are: (i) the analysed domain is discretised with particles that are not connected with a mesh, allowing for simple and accurate solution at large deformations, (ii) the discretisation of complex geometries is less complicated and (iii) the physical values and paths of the particles are easy to follow and evaluate, consequently it is also simple to determine the free surface of movable interfaces or deformable boundaries.

In the Smoothed Particle Hydrodynamics method, the state of the system is represented by a set of particles, which possess individual material properties and move according to the governing conservation equations. SPH as a meshfree, Lagrangian, particle method was developed by Lucy, Gingold and Monaghan, initially to simulate astrophysical problems [1–8]. Later the SPH was extensively studied and extended to dynamic response with material strength as well as dynamic fluid flows with large deformations. It has some special advantages over the traditional mesh-based numerical methods. The most significant is the adaptive nature of the SPH method, which is achieved at the very early stage of the field variable (i.e. density, velocity, energy) approximation that is performed at each time step based on a current local set of arbitrarily distributed particles. Because of the adaptive nature of the SPH approximation, the formulation of the SPH is not affected by the arbitrariness of the particle distribution. Therefore, it can handle problems with extremely large deformations very well. Another advantage of the SPH method is the combination of the Lagrangian formulation and particle approximation.

Unlike the meshfree nodes in other meshfree methods, which are only used as interpolation points, the SPH particles also carry material properties, functioning as both approximation points and material components. These particles are capable of moving in space, carry all the computational information, and thus form the computational frame for solving the partial differential equations describing the conservation laws. The numerical solution procedure of the SPH formulation consists of the following steps [1]:

- Generation of the meshless numerical model: problem domain is discretised with a finite number of arbitrary located particles without mesh connectivity between them.
- Integral representation (kernel approximation): is used for the field function approximation.
- Particle approximation: is replacing the integration in integral representation with summations for the values of neighbouring particles in the support domain.
- Adaptation: the particle approximation is performed at every time step and depends on the current local particle distribution, however it does not depend on the particle locations in the previous time step.
- Dynamic analysis: the direct time integration of the governing equation is performed with explicit integration scheme allowing for fast time stepping and obtaining time history for all field variables of particles assembling the problem domain.

Basically the SPH method consists of two key tasks. The first represents the integral representation and the second the particle approximation. The concept of the integral representation of the function $f(\mathbf{x})$, used in SPH method, is based on the following presumption

$$f(\mathbf{x}) = \int f(\mathbf{x}') \cdot \delta(\mathbf{x} - \mathbf{x}') \cdot d\mathbf{x}', \tag{1}$$

where f is the function of a three-dimensional position vector $\mathbf{x}$ and $\delta(\mathbf{x} - \mathbf{x}')$ is the Dirac delta function [1,2,9,10]. From the Eq. (1) it is evident that any function $f(\mathbf{x})$ can be written in an integral form. The Dirac delta function can be substituted with a smoothing function

$$f(\mathbf{x}) \approx \int f(\mathbf{x}') \cdot W(\mathbf{x} - \mathbf{x}', h) \cdot d\mathbf{x}', \qquad (2)$$

where W is the smoothing function and h is the smoothing length determining the influence area of the smoothing function. It should be noted that the integral form in Eq. (2) is only an approximation with second order accuracy when the smoothing function is not equal to Dirac's delta function. The smoothing function has to satisfy the following conditions: (i) normalization (unity) condition, (ii) Delta function property condition, (iii) compact condition and (iv) positivity condition [1]. Additionally, the smoothing function has to be symmetric, continuous and uniform, yet its value has to monotonically decrease with increasing the distance to the observed particle.

The computational SPH model consits of a finite number of mass particles spread over certain space, which is achived by the introduction of particle approximation. The countinous approximative integral form (Eq. (2)) has to be transformed into a discretised form of particle sumations in the influence area (Fig. 1).

An infinitive small volume $d\mathbf{x}'$ in the integration equations at the position of the arbitrary particle j can be written as a finite volume ΔV_j with particle mass m_j

$$m_j = \Delta V_j \cdot \rho_j, \qquad (3)$$

where ρ_j is the density of particle j. The continuous SPH integral representation of $f(\mathbf{x})$ can be as discretised particle approximation written as

$$f(\mathbf{x}) \approx \int f(\mathbf{x}') \cdot W(\mathbf{x} - \mathbf{x}', h) \cdot d\mathbf{x}' \approx \sum_{j=1}^{N} f(\mathbf{x}_j) \cdot W(\mathbf{x} - \mathbf{x}_j, h) \cdot \Delta V_j, \qquad (4)$$

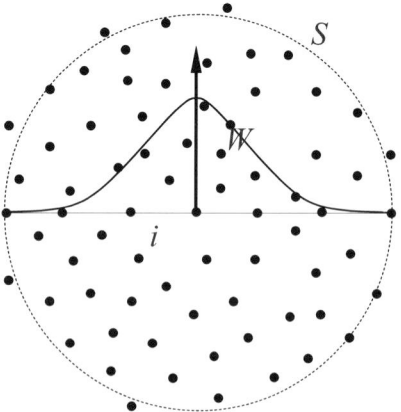

Fig. 1 Particle approximation (central particle i) within the influence area (s) of the smoothing function W [1]

where N is the number of particle in the influence area of the smoothing function W. Combining the Eqs. (3) and (4) se the approximated function for particle i [8]

$$f(\mathbf{x}_i) \approx \sum_{j=1}^{N} \frac{m_j}{\rho_j} f(\mathbf{x}_j) \cdot W_{ij}, \tag{5}$$

where

$$W_{ij} = W(\mathbf{x}_i - \mathbf{x}_j, h). \tag{6}$$

Using the particles sum for integral approximation is of crucial importance and ensures that the SPH method does not dependent on any background mesh during the numerical integration, gaining the advantage comparing to some other meshless methods (e.g. EFG). From the Eq. (5) it can be observed that particle's mass and density are introduced in the basic equation, making the SPH method very attractive for hydro dynamical problems, where the density is one of the most important variables.

According to the Eqs. (2) and (5) the governing equations of mass, momentum and energy conservation can be written in the following form

$$\frac{\partial \rho_i}{\partial t} = \sum_{j=1}^{N} m_j \cdot v_{ij} \cdot \frac{\partial W_{ij}}{\partial x_i}, \tag{7}$$

$$\frac{\partial v_i}{\partial t} = \sum_{j=1}^{N} m_j \cdot \left(\frac{\sigma_i}{\rho_i^2} + \frac{\sigma_j}{\rho_j^2} \right) \cdot \frac{\partial W_{ij}}{\partial x_i}, \tag{8}$$

$$\frac{\partial u_i}{\partial t} = \sum_{j=1}^{N} m_j \cdot \frac{\sigma_i \cdot \sigma_j}{\rho_i \cdot \rho_j} \cdot v_{ij} \cdot \frac{\partial W_{ij}}{\partial x_i}, \tag{9}$$

where σ_i and σ_j are components of stress tensor in particle i and j, respectively, determined with the constitutive equation and v_{ij} is the component of the relative velocity vector between particle i and particle j [1, 2, 8, 11].

Despite all the described advantages, the SPH method has still to cope with some numerical difficulties, like particle inconsistency, inaccuracy at domain boundaries and instabilities at tensile stress state [3]. However, SPH method accuracy and stability depend also on the particle number (particle density) in the influence area and the time step of the time integration scheme.

In application, the principal potential advantage of the SPH method is that there is no need for connectivity between particles with a conventional mesh avoiding elements distortion problems at large deformations. In comparison with Eulerian description, the SPH method offers higher efficiency in terms of domain modelling, since only the material domains have to be discretised and not also the areas through which the material might move during the simulation. However, the SPH method is relatively new in comparison with standard Lagrangian and Eulerian methods, still having some difficulties in the area of stability, consistency and fulfilling the conservation equations [1, 4, 7, 8].

With recent improvement and development the Smoothed Particle Hydrodynamics method definitely became a reliable tool providing adequate accurate and stabile results and excellent adaptivity achieving a high level for implementation in many commercial computational packages and application in several engineering areas.

3 SPH in LS-DYNA

Due to its high efficiency and applicability the SPH method has been recently, implemented and used in several computational codes. One of the engineering finite element codes which also include the SPH method is LS-DYNA [12, 13]. The LS-DYNA was primarily developed for solving structural dynamic problems with explicit time integration scheme. Through the years it became one of the leading computational codes for crash tests evaluation and it spread from explicit to implicit time-integration using different formulations of mesh-based as well as meshfree methods. The SPH method in LS-DYNA is very efficient at high strain rate and large deformations problems.

The SPH particles describe the Lagrangian motion of mass points which simultaneously represent also the interpolation points and are approximated with the cubic B-spline function [3, 4]. A schematic loop of solving the SPH integration cycle is shown in Fig. 2.

During the entire computational simulation it is important to know, which particles interact with each other and which particles are in the influence area, therefore the neighbour search is of crucial importance for the analysis. The influence area (spherical or ellipsoidal shape) is defined by the radius of $2h$ [3]. The search of neighbouring particles in a model with N particles requests to perform for only one particle N-1 distance calculations. This leads to $N(N-1)$ calculations for the distances between all the particles, which consequently increases computational time. This can be overcome by application of methods for search of neighbour particles, similar to the methods being used in solving contact problems. One of the most effective methods is the bucket sort method, which is based on splitting the analysed

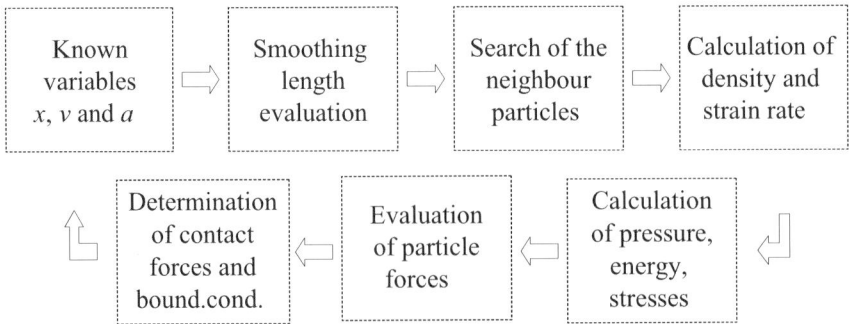

Fig. 2 Schematic loop of solving the SPH integration cycle

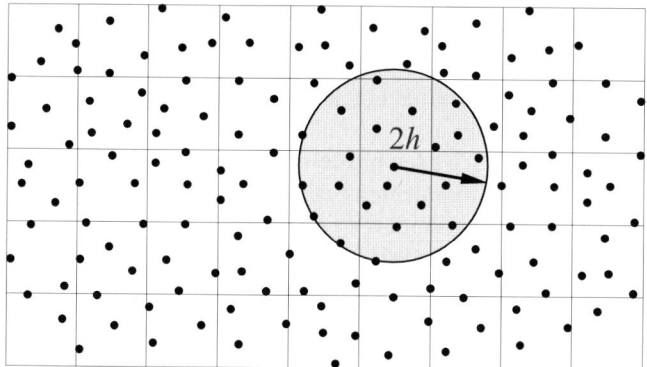

Fig. 3 Bucket sort and neighbour search [3]

domain into several boxes of a given size. The neighbour search accounts only for particles in the same box and the neighbouring boxes regarding the influence area (Fig. 3). After a list of possible neighbours is known, the distances between them are computed. With the bucket sort technique it is possible to reduce the number of distance calculations ($\approx N \log N$) which consequently enormously reduces the computational time and increases its efficiency [3, 8].

Another difficulty that may appear using SPH at large deformations is the change of particle's number in the influence area. If the smoothing length remains constant during the entire simulations, the number of particles in this area significantly depends on the type of loading (e.g. at compressive loading the number of particles increases, resulting in much longer computational times, at tensile loading the number of particles decreases, resulting in low accuracy and stability problems). Therefore, it is reasonable to use a variable smoothing length h, changing in time and space. Its advantage is to maintain approximate the same number of particles in the influence area. The default equation for evaluation of the variable smoothing length is defined as

$$\frac{dh}{dt} = \frac{1}{3} \cdot h \cdot div(v), \tag{10}$$

where is $div(v)$ divergence of speed. The objective of the smoothing length evaluation is to maintain a constant number of particles in the influence area. The smoothing length increases, if the distances between parts become larger (tensile loading) and decreases, if the distances between are getting smaller (compressive loading). However, the value of the smoothing length has to remain in certain limits to assure the numerical stability

$$h_m \cdot h_0 < h < h_v \cdot h_0, \tag{11}$$

where h_0 represents the initial value of the smoothing length and the default values for h_m and h_v are 0.2 and 2, respectively. In case where $h_m = h_v = 1$ the smoothing length becomes constant in time and space [3, 8].

4 Practical Examples

In the following section the use and efficiency of the SPH method, as implemented in the LS-DYNA code, is discussed and illustrated with two practical engineering application examples. The first example describes the modelling of fuel sloshing in a reservoir, where different formulations, using mesh-based and meshless methods, are compared and evaluated according to experimental measurements. The second example describes the impact analysis of a cellular structure, where the influence of viscous fluid pore filler flow has been studied.

4.1 First Example: Modelling of Sloshing in a Reservoir

This example presents a computational model for simulation of a fuel reservoir deformation under impact loading conditions, considering also the fuel motion. For this purpose different methods describing the fluid motion were evaluated using a simplified reservoir problem, analysed with the explicit dynamic code LS-DYNA [13]. Computational results were then compared with previously published experimental observations [14].

4.1.1 Problem Definition

The analysed reservoir consists of a closed PMMA container box (Fig. 4) with 30 mm wall thickness, which is 60% filled with water and 40% with air, and is subjected to gravitation (negative z direction) and longitudinal time-dependent acceleration (negative x direction) with a peak acceleration of approximately 30 g [14].

The box was modelled with the four-noded Belytschko-Tsay shell elements with three integration points through the thickness. The elastic material model is used for the box container with material data corresponding to the PMMA material ($\rho = 1{,}180\,\mathrm{kg/m^3}$, $E = 3{,}000\,\mathrm{MPa}$ and $\nu = 0.35$).

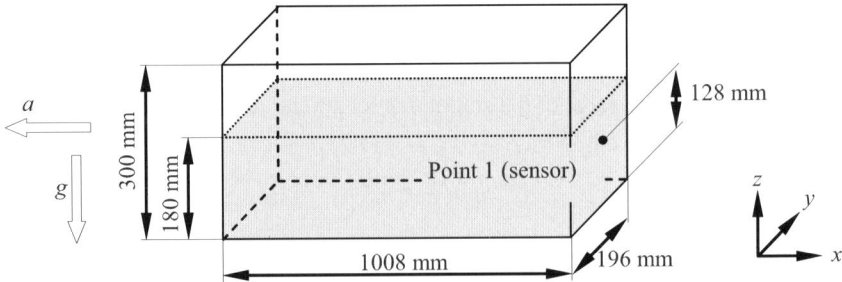

Fig. 4 Dimensions of reservoir and initial conditions

4.1.2 Domain Descriptions and Solution Techniques for the Fluid

The modelling of the fluid domain and its interaction with structure can be in LS-DYNA analysed using different types of domain description and solution techniques. Four of them have been evaluated in this example in order to simulate the fluid motion in the reservoir:

- Mesh-based Lagrangian formulation: usually used for describing solid mechanics problems, where the motion of every single mass point is being observed in space and time. The Lagrangian formulation is very simple and easy to use, although the method becomes very complicated and complex for description of large number of mass particles [15]. One finite element represents the same part of the material throughout the course of the analysis [13, 16].
- Mesh-based Eulerian formulation: most commonly used for solving the fluid dynamics problems, where fluid flow through the fixed mesh in a space is observed. The problem is being observed at one point in space which does not follow the motion of the single particle. In one time step Δt several mass particles may pass through the observed point [15, 17, 18].

Although the Eulerian mesh in LS-DYNA appears not to move or deform during the analysis, it does actually change its position and form, but only within the single time steps, since Lagrangian formulation is initially used within that time step. After that the following two approximations are performed, before entering the next time step: (i) mesh smoothing (the nodes are moved to their original position) and (ii) advection (the internal variables of the translated nodes are recomputed) [17,18];

- Mesh-based Arbitrary Lagrange-Eulerian formulation (ALE): the mesh partly moves and deforms because it follows the material (Lagrangian formulation), while at the same time the material can also flow through the mesh (Eulerian formulation) [17, 18].

The ALE solving procedure in LS-DYNA is similar to Eulerian procedure. The only difference is the mesh smoothing. In the Eulerian formulation the nodes are moved back to their original positions, while in the ALE formulation the positions of the moved nodes are defined according to the average distance to the neighbour nodes [17, 18];

- Meshless Smoothed Particle Hydrodynamics method (SPH): as described in the previous two sections.

Solid finite elements and particle elements were used for the water and air discretisation in the observed problem, depending on the applied method. A special material model was used for water ($\rho = 1,000\,\text{kg/m}^3$ at 293 K) and air ($\rho = 1\,\text{kg/m}^3$ at 293 K) modelling. The air was considered only in Eulerian and ALE model. A penalty based interaction between fluid and structure was applied in all computational models.

Explicit dynamic analyses were carried out by using all four described techniques. The models have been solved with LS-DYNA Linux Version 970. The computational time frame was set to 80 ms and the smallest time step of the simulation was automatically adjusted by the code to ensure the stability and convergence of results.

4.1.3 Computational Results

The motion of the fluid during the acceleration was recorded with high-speed camera during the experimental testing [14]. The recorded fluid free surface shape at the time instance $t = 38$ ms (dotted line) was compared with results of the fluid free surface shape obtained with different computational models, Fig. 5. It is obvious that the Lagrangian and SPH models are only good for approximating the fluid motion at the right side wall, since in reality the fluid would not retain the form of the container which is the case observed in simulations at the left side wall. However, this observation must be considered in view of required computational results. In case where only the impulse of the fluid motion towards the reservoir wall is needed, the deformations and deflections on the opposite side could be neglected, until they do not influence the required results. Eulerian and ALE formulations performed much

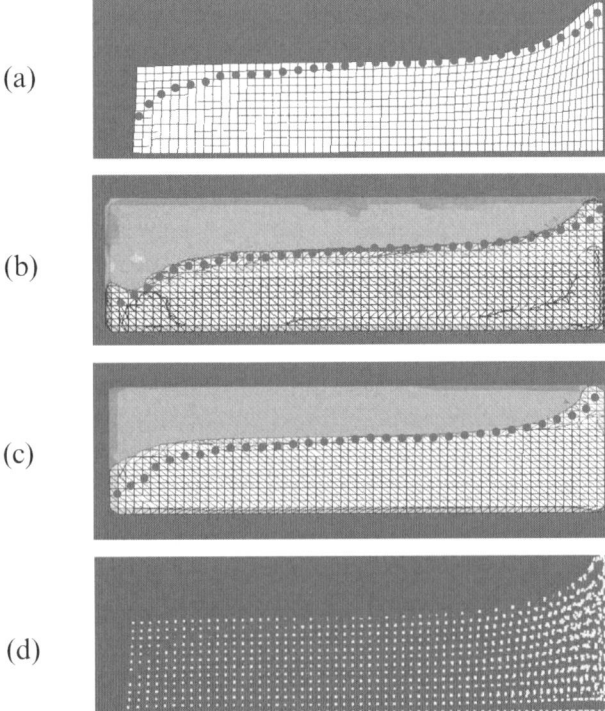

(a)

(b)

(c)

(d)

Fig. 5 The fluid free surface shape at the time of $t = 38$ ms: (**a**) Lagrangian model; (**b**) ALE model; (**c**) Eulerian model; (**d**) SPH model

better in describing the form of the fluid free surface. However, this is only achieved by substantial increase in calculation times, which is not always acceptable. It is important to observe that by using the Lagrangian formulation results in significant element distortion and consequently large computational errors. This once again confirms the fact that the Lagrangian formulation is unsuitable for modelling large deformations.

The fluid pressure acting on the reservoir surface at Point 1 (Fig. 4) due to the acceleration was also measured during the experimental testing and evaluated with different modelling approaches in LS-DYNA. The computational results have been determined by two different approaches: (i) in the Lagrangian and SPH model the pressure at Point 1 (Fig. 4) was measured with contact forces which appeared at the observed point [19] and (ii) for the Eulerian and ALE model the pressure was determined by the leakage control, i.e. by determining the force that is needed for establishing equilibrium in every observed element on the boundary between the fluid and the box wall [20]. The comparison of the pressure at Point 1 for the computational and experimental results is shown in Fig. 6.

The best agreement with the experimental results was achieved by using the Lagrangian formulation and SPH method, whereby the simulation with the Lagrangian model at some point failed due to large element deformation resulting in too distorted elements. The SPH formulation provided excellent results, especially when taking into account that using this formulation results in fast and uncompli-cated analyses, since the mesh consists only of SPH particles. The pressure drop, observed by the Eulerian and ALE formulation, is attributed to the air which obvi-ously dampened the fluid motion in the reservoir and consequently reduced the pressure measured during the simulation. The model size and the required CPU times for each analysis are listed in Table 1 to illustrate the required computational effort for solving the chosen problem with different approaches.

From Table 1 it can be observed that the Lagrangian and SPH solution techniques are the most efficient considering the computational time. Computationally the most

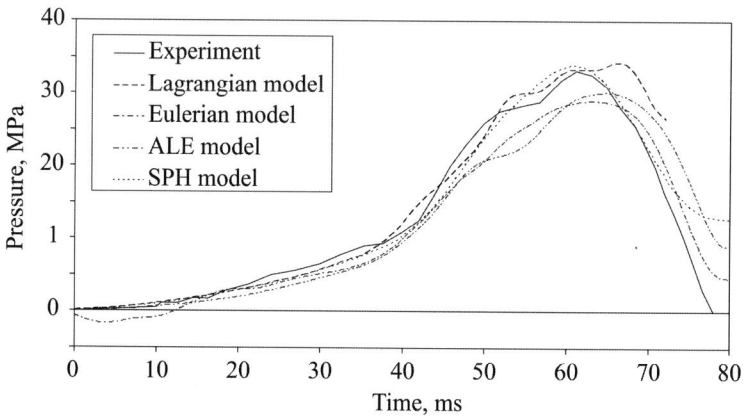

Fig. 6 Comparison of the pressure time-variation at Point 1 (Fig. 4)

Table 1 CPU time comparison

Model	Total number		CPU time [min]
	Nodes/particles	Elements	
Lagrangian	2,898	2,420	16
Eulerian	10,162	8,706	225
ALE	7,462	6,396	260
SPH	2,898	–	13

extensive simulations have been observed using the Eulerian and ALE formulations, which results in use of the performed approximations during the solution procedure in order to reduce or to overcome the motion and deformation of the mesh.

4.1.4 Conclusions

Four different approaches and solution techniques have been used to analyse the fluid motion and its flow in a reservoir box, with the purpose to evaluate and compare their efficiency and applicability. Computational simulations have shown that the fluid motion can be properly described by applying different alternative formulations in LS-DYNA. The fluid motion in regard to the fluid free surface prediction can be best described with the ALE and Eulerian methods, while the Lagrangian and SPH models provide better predictions of fluid forces acting on the reservoir structure. However, the main advantage of using the SPH model is in the short pre-processing and computational time. Additionally, these models are also very economical and suitable for use in simulations of large scale and advanced engineering problems.

4.2 Second Example: Fluid Flow Through Cellular Structure

Cellular structures have an attractive combination of physical and mechanical properties and are being increasingly used in modern engineering applications [21, 22]. Research of their behaviour under quasi-static and high strain rates is valuable for engineering applications such as those related to impact and energy absorption problems. A logical solution to increase the stiffness and energy absorption of open-cell cellular materials is by filling the cellular structure with viscous fluid. The fluid offers certain level of flow resistance during collapse of cellular structure due to its viscosity, which in turn increases the structure stiffness. Preliminary investigations have shown that in combination with high strain-rate loading this results in substantial increase of energy absorption [23, 24]. This example shows the results of parametric computational simulations of cellular structures with open-cell morphology under impact loading conditions accounting for fluid filler flow, modelled with the meshless SPH method using the finite element code LS-DYNA [13, 20].

Fig. 7 Analysed open-cell
cellular structure

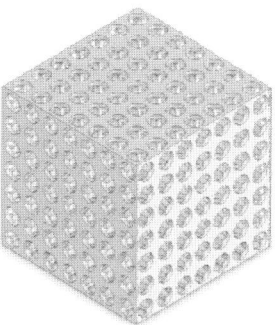

4.2.1 Computational Model

The open-cell cellular material with a regular structure (Fig. 7) was modelled
with three relative densities $\rho/\rho_0 = 0.37, 0.27$ and 0.16. This corresponds to the
basic geometry dimensions: hole diameter of $d = 3$ mm and the intercellular wall
thickness of $c = 1.5, 1.0$ and 0.5 mm.

The polymer FullCure M730 was used as the base material with the following
material properties: $E = 2{,}323$ MPa, $v = 0.3$, σ_y(tensile) $= 49$ MPa and σ_y
(compressive) $= 91$ MPa. The strain rate effects were also considered by implementing the Cowper-Symonds constitutive relation [13, 24–27]. The cellular structure
base material was discretised with eight-noded fully integrated quadratic solid elements. With additional parametric analyses the proper mesh density ($l \approx 0.1$ mm)
and time step size ($\Delta t \approx 0.04\,\mu$s) have been determined to assure adequate precision
of computational results [24, 28].

Water ($\rho = 1{,}000$ kg/m^3 at 293 K) was chosen as a viscous fluid filler, which was
modelled with the SPH particles. The relationship between the change of volume
and pressure in this study has been represented with the Mie-Grüneisen equation
of state [13, 24]. An optimal distance between the SPH particles ($l \approx 0.112$ mm)
and mass of single particles ($m_i \approx 1.42\,\mu$g) has been determined with separate
parametric simulations [24].

Fully coupled fluid-structure interaction between the cellular structure's base
material and the fluid filler was considered. The upper surface of the cellular structure has been subjected to a uniaxial compressive impact loading, with displacement
controlled compressive load at a strain rate of $1{,}000\,\text{s}^{-1}$. Symmetry boundary conditions have been applied due to regular geometry of the structure [24, 29]. A single
LS-DYNA analysis run of the model with 16 cells lasted approximately 12 h on a
PC-cluster of 4 units with Intel Pentium IV 3,200 MHz processors and 1 GB RAM
each.

Initial parametric simulations of the liquid filler outflow have been also performed with the ANSYS CFX [30] code in order to evaluate and validate the SPH
fluid models. Figure 8 illustrates the comparison between the computational results
obtained with the LS-DYNA code using the SPH model and the ANSY CFX code
using the finite volume method. Very good agreement of results from both codes

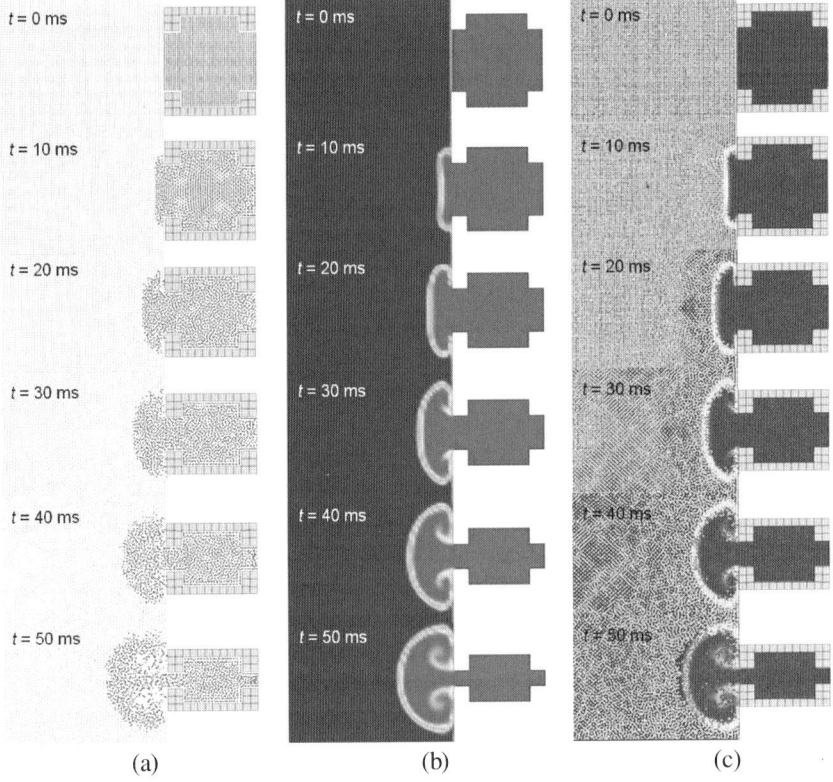

Fig. 8 Fluid filler outflow simulations: (**a**) LS-DYNA (SPH model); (**b**) ANSYS CFX (finite volume method); (**c**) comparison between (**a**) and (**b**)

can be observed, which in turn validates suitability of the SPH model to accurately simulate the filler flow through the cellular material.

4.2.2 Computational Results

Figure 9 shows the deformation of the cellular structure with liquid filler outflow under impact loading at different time sequences.

Figure 10 illustrates the influence of the relative density and the filler. As already observed in the previous investigations, the stiffness increases with increasing the relative density [24]. Computational simulations have shown that the filler influences more the behaviour of cellular structure with a higher relative density than the cellular structure with a lower relative density. The reason for this effect can be explained by smaller pore sizes in a cellular structure with high relative density, which offers higher resistance during the filler outflow, which consequently

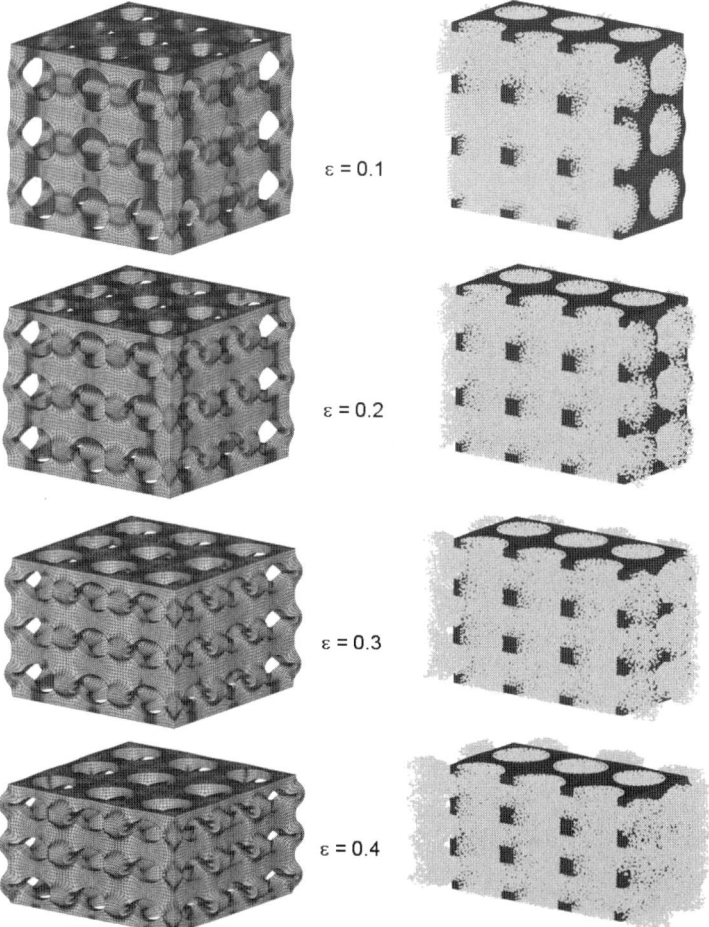

Fig. 9 Behaviour of the cellular structure with fluid filler under impact loading

contributes to the increase of cellular structure macroscopic stiffness and its energy absorption capacity.

The influence of the relative density, the pore filler and the cellular structure size on the macroscopic yield stress is shown in Fig. 11. From the figure it can be observed that the macroscopic yield stress increases with increasing the cellular structure size. Again the influence of the pore filler can be observed, which is more pronounced at cellular structures with higher cell number, due to additional energy dissipation during the pore filler outflow.

Further computational simulation considering different filler types (e.g. oil, silicon) with various viscosities have shown that the increase of the filler's viscosity results in increase of cellular structure macroscopic stiffness and higher energy absorption capacity.

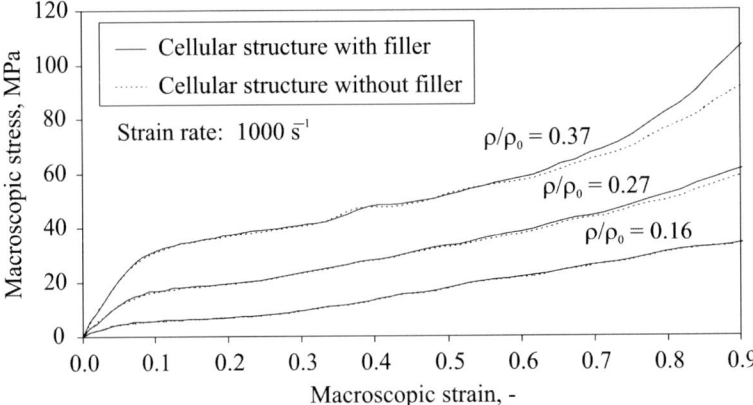

Fig. 10 Influence of the pore filler

Fig. 11 Influence of the relative density and the pore filler on the macroscopic yield stress

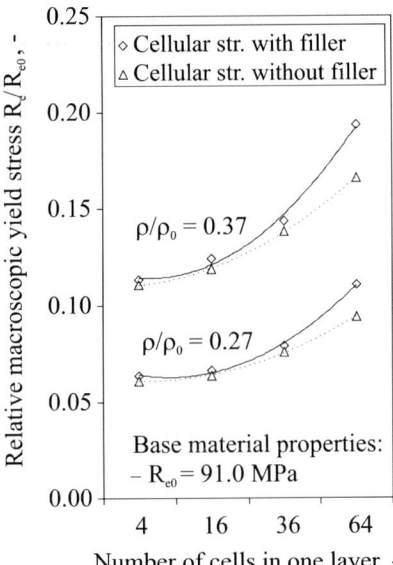

4.2.3 Conclusions

The computational simulations of cellular structures with fluid fillers have shown the practical applicability of the SPH method for solving engineering problems. The conducted simulations have shown that the fluid filler influence is more pronounced in cellular structure with higher relative density than in cellular structures with lower relative density. With further computational simulations it was also determined that the increase of the filler viscosity results in increase of cellular structure stiffness which contributes to higher capability of deformational energy absorption.

5 Conclusions

The paper describes the meshless Smoothed Particle Hydrodynamics method and its implementation in the explicit code LS-DYNA.

The practical application of the SPH method and the comparison to other available solution techniques and domain descriptions was presented with analysing two engineering examples. The first example described the modelling of fluid motion problem in a reservoir, where different modelling approaches and solution formulations were evaluated and compared. The second example described the fluid flow through open-cell cellular materials and its influence on the cellular structure behaviour, where fully coupled fluid-structure interaction between the cellular structure base material and the fluid filler was considered.

Although the SPH method is relatively new computational method and still suffers from some numerical difficulties, its application into the LS-DYNA code proved to be very successful. The SPH method thus became a reliable and efficient tool, providing adequate accurate and stabile results and excellent adaptivity, especially for solving large scale and advanced engineering problems.

References

1. Liu GR, Liu MB (2003) Smoothed particle hydrodynamics: a meshfree particle method. World Scientific, Singapore
2. Li S, Liu WK (2004) Meshfree particle methods. Springer, Berlin
3. Lacome JL (2001) Smoothed particle hydrodynamics. Livermore Software Technology Corporation, Livermore, CA
4. Schwer L (2004) Preliminary Assessment of Non-Lagrangian Methods for Penetration Simulation. Proceedings 8th International LS-DYNA Users Conference
5. Gray JP, Monaghan JJ, Swift RP (2001) SPH elastic dynamics. Comput Method Appl M 190:6641–6662
6. Idelsohn SR, Onate E, Del Pin F (2003) A Lagrangian meshless finite element method applied to fluid-structure interaction problems. Comput Struct 81:655–671
7. Buyuk M, Kan CDS, Bedewi N, et al. (2004) Moving Beyond the Finite Elements, A Comparison Between the Finite Element Methods and Meshless Methods for a Ballistic Impact Simulation. Proceedings 8th International LS-DYNA Users Conference
8. Vignjevic R (2002) Review of development of the smooth particle hydrodynamics (SPH) method. Cranfield University, Cranfield
9. Hitoshi M, Quantum mechanics I: Dirac delta function. http://hitoshi.berkley.edu/221A. Accessed 17. 1. 2006
10. Kurt B (2001) The Dirac delta function. Rose-Hulman Institute of Technology, Terre Haute
11. Trina RM (1995) Physically-based fluid modelling using smoothed particle hydrodynamics. University of Illinois, Chicago, IL
12. LS-DYNA. http://www.lstc.com/. Accessed 4. 4. 2007
13. Hallquist JO (1998) LS-DYNA theoretical manual. Livermore Software Technology Corporation, Livermore, CA
14. Meywerk M, Decker F, Cordes J (1999) Fuel Sloshing in Crash Simulations. Proceedings EuroPAM 99
15. Škerget L (1994) Fluid mechanics (in Slovene). Technical Faculty Maribor and Faculty of Mechanical Engineering Ljubljana, Ljubljana

16. Christon MA, Cook GO (2001) LS-DYNA's incompressible flow solver, user's manual. Livermore Software Technology Corporation, Livermore, CA
17. Olovsson L (2004) LS-DYNA training class in ALE and fluid-structure interaction. Livermore Software Technology Corporation, Livermore, CA
18. Olovsson L, Soulli M, Do I (2003) LS-DYNA - ALE capabilities, fluid-structure interaction modeling. Livermore Software Technology Corporation, Livermore, CA
19. Müllerschön H (2004) Contact modeling in LS-DYNA. DYNAmore, Stuttgart
20. Hallquist JO (2003) LS-DYNA keyword user's manual. Livermore Software Technology Corporation, Livermore, CA
21. Gibson LJ, Ashby MF (1997) Cellular solids: structure and properties. Cambridge University Press, Cambridge
22. Ashby MF, Evans A, Fleck NA, et al. (2000) Metal foams: a design guide. Elsevier Science, Burlington, MA
23. Lankford J, Dannemann KA (1998) Strain Rate Effects in Porous Materials. Proceedings Materials Research Society Symposium, pp. 103–108.
24. Vesenjak M (2006) Computational modelling of cellular structure under impact conditions, Ph.D. thesis, Faculty of Mechanical Engineering, Maribor
25. Ren Z, Vesenjak M, Öchsner A (2008) Behaviour of cellular structures under impact loading: a computational study. Mater Sci Forum 566:53–60
26. Altenhof W, Ames W (2002) Strain rate effects for aluminum and magnesium alloys in finite element simulations of steering wheel armature impact tests. Fatigue Fract Eng M 25:1149–1156
27. Bodner SR, Symonds PS (1962) Experimental and theoretical investigation of the plastic deformation of cantilever beams subjected to impulsive loading. J Appl Mech 29:719–728
28. Vesenjak M, Öchsner A, Hriberšek M, et al. (2007) Behaviour of cellular structures with fluid fillers under impact loading. Int J Multiphys 1:101–122
29. Ochnser A (2003) Experimentelle und numerische Untersuchung des elasto-plastischen Verhaltens zellularer Modellwerkstoffe, Ph.D. thesis, University Erlangen, Nuremberg
30. CFX Users Manual 10.0. www.ansys.com. Accessed 4. 4. 2007

Visualization of Meshless Simulations Using Fourier Volume Rendering

Andrew Corrigan(✉), John Wallin, and Matej Vesenjak

Abstract Fourier volume rendering is a volume visualization technique first applied to regular grid data. We adapt this technique to deal directly with meshless data, with the intended application of visualizing simulations which use meshless methods to solve the underlying equations of the simulation. Because we consider a general class of meshless data, the technique is applicable to the data produced by many meshless methods such as Kansa's method, symmetric collocation, and smoothed particle hydrodynamics. We discuss the technique's implementation on graphics hardware, and demonstrate its usefulness in visualizing data produced by both astrophysical and fluid dynamics simulations.

Keywords: Fourier Volume method · Visualization of Meshless Simulations · Fourier Volume Rendering

1 Introduction

Volume visualization is an important tool for understanding three-dimensional simulation data. Fourier volume rendering (FVR) [6, 18, 19] is a particular volume visualization technique. Previous work on FVR has so far only adapted this technique to deal directly with regular grid data or wavelet data [12], including an indirect adaptation to irregularly sampled data [28]. By indirect it is meant that the data is first sampled onto a three-dimensional grid before volume visualization is performed. As pointed out by others [4, 15, 23, 25], in the context of meshless data this could result in the loss of important detail at feasible grid resolutions, and

A. Corrigan and J. Wallin
George Mason University, Department of Computational and Data Sciences, MS 642, 4400 University Drive, Fairfax, VA 22030-4444, USA, e-mail: acorriga@gmu.edu

M. Vesenjak
University of Maribor, Faculty of Mechanical Engineering, Smetanova 31, SI-2000 Maribor, Slovenia, e-mail: m.vesenjak@uni-mb.si

A.J.M. Ferreira et al. (eds.) *Progress on Meshless Methods, Computational Methods in Applied Sciences.*
291

furthermore would leave meshless data less efficient to visualize than grid data. The purpose of this paper is to avoid such issues and adapt FVR to deal directly with a general class of meshless data in the form of a summation of N integrable functions $\Phi_k : \mathbb{R}^3 \to \mathbb{R}$ with coefficients $\alpha_k \in \mathbb{R}$

$$s(x) = \sum_{k=1}^{N} \alpha_k \Phi_k(x). \tag{1}$$

2 Related Work

Most related to this work are techniques which also deal directly with meshless data. One such technique is direct slice-based volume visualization which was considered in the context of meshless data in [16] with further improvements made in [32]. The technique presented in those works for visualizing meshless data, in particular radial basis function interpolants, involves sampling the meshless data over an array of two-dimensional sampling planes in a view-dependent manner. They use a spatial data structure to only evaluate terms whose support is relevant to regions of a given slice. While this is not required, a loss of performance is to be expected otherwise. When visualizing static data, building this data structure can be treated as a pre-processing step. However, this is not always possible, such as when the meshless data is time-varying with moving data sites. In contrast, FVR requires no spatial data structure since it deals with the data's Fourier transform which tends to be centered at the origin. A related technique for slice-based isosurface visualization of meshless data is presented in [5].

Another such technique is splatting. This technique was considered in the context of meshless data in [25]. When using an orthogonal projection, splatting attempts to compute the same image as does FVR. The difference is that splatting computes the image by directly approximating the integral in the spatial domain. This technique involves computing a so-called footprint using numerical integration, which is accumulated at different points in the image for each term in the meshless data. FVR and splatting can be considered complementary since FVR performs most efficiently for data with low frequency content, typically when each term has large spatial support, while splatting performs most efficiently for data with small spatial support, typically resulting in high frequency content. Also, given a fixed cutoff frequency, FVR becomes cheaper when zooming in, due to the associated larger frequency step size decreasing the required amount of sampling. On the other hand, splatting becomes cheaper when zooming out, due to less resolution being required of the footprint. Relatively recent work on splatting meshless data includes multiresolution [20] and hierarchical [15] techniques for handling enormous meshless data sets, and a technique for splatting meshless data with elliptical basis functions [22].

The traditional computer graphics technique of ray-tracing for visualizing isosurfaces [13] is general enough to deal directly with an arbitrary function, and therefore can deal directly with meshless data. Another approach to isosurface visualization [4], uses local tetrahedrizations to compute an isosurface while avoiding globally sampling the meshless data onto a grid. The SPH visualization code SPLASH [23] is also related to this work. This code implements a number of visualization techniques which deal directly with meshless data, such as volume visualization using splatting, cross section visualization, and surface visualization based on optical depth.

3 Direct Adaptation of FVR to Meshless Data

Fourier volume rendering computes from a function $s : \mathbb{R}^3 \to \mathbb{R}$ a two-dimensional image $I : \mathbb{R}^2 \to \mathbb{R}$ defined as

$$I(u,v) = \int_{\mathbb{R}} s\left(tw_t + uw_u + vw_v\right) dt \tag{2}$$

where $\{w_t, w_u, w_v\}$ is an orthonormal basis of $\mathbb{R}^3$. In words, for each point on the image plane spanned by $\{w_u, w_v\}$, FVR integrates the function s through the point along the image plane's normal w_t. What distinguishes Fourier volume rendering from other volume visualization techniques is that it computes the image via its Fourier transform [6, 18, 19]

$$\hat{I}(u,v) = \hat{s}\left(uw_u + vw_v\right). \tag{3}$$

Therefore, to perform Fourier volume rendering with meshless data its Fourier transform is required. For grid-based Fourier volume rendering this is obtained by computing a three-dimensional FFT. For meshless data such an expensive computation is not required, since its Fourier transform can be obtained analytically. The inverse Fourier transform can then be approximated using a two-dimensional inverse FFT.

3.1 The Fourier Transform of Meshless Data

To adapt FVR, the Fourier transform of the image (3) is obtained by applying properties of the Fourier transform. By linearity, the Fourier transform of the meshless data is

$$\hat{s}(f) = \sum_{k=1}^{N} \alpha_k \widehat{\Phi_k}(f) \tag{4}$$

which reduces the problem to computing the Fourier transform of each function Φ_k. We use the Fourier transform convention

$$\hat{\Phi}(f) = \int_{\mathbb{R}^3} \Phi(x) e^{-2\pi i x^T f} dx \tag{5}$$

which is defined when Φ is integrable: $\Phi \in L^1\left(\mathbb{R}^d\right)$. Therefore if each Φ_k is integrable then the Fourier transform of the meshless data is defined. An important special case is when each Φ_k is defined in terms of a radial function Φ, in which case (5) specializes to

$$\hat{\Phi}(f) = \frac{2}{\|f\|} \int_0^\infty t\phi(t)\sin(2\pi\|f\|t)\,dt \tag{6}$$

where $\Phi(x) = \phi(\|x\|)$ [33, Theorem 5.26].

In Section 4 the Fourier transform of Φ_k is obtained for different types of mesh-less data. This allows for the direct evaluation of the Fourier transform of the meshless data, and therefore by (3) the Fourier transform of the image.

3.2 Approximation of the Inverse Fourier Transform

Having obtained the Fourier transform of the image, the next step in FVR is to compute the inverse Fourier transform of $\hat{I}$ to obtain the image I.

$$I(u,v) = \int_{-\infty}^{\infty}\int_{-\infty}^{\infty} \hat{I}(f_u, f_v)\,e^{i2\pi(uf_u+vf_v)}\,df_u df_v \tag{7}$$

For this, a discrete Fourier transform is used

$$I(u,v) \approx \sum_{k_{f_v}=-N_{f_v}}^{N_{f_v}}\sum_{k_{f_u}=-N_{f_u}}^{N_{f_u}} \hat{I}(f_u, f_v)\,e^{i2\pi(uf_u+vf_v)}\Delta_{f_u}\Delta_{f_v} \tag{8}$$

where $u = k_u\Delta_u$, with analogous substitutions made for v, f_u and f_v. Also, k_u and k_v vary over $-N_u\ldots N_u - 1$ and $-N_v\ldots N_v - 1$ respectively.

In image processing terms, by only considering a finite number of samples, an ideal lowpass filter is applied to the image [11, Page 167]. This filter provides a control of performance at the expense of quality. By choosing a sufficiently large cutoff frequency, the effects of the lowpass filter are not apparent. However, choosing a larger cutoff frequency requires more frequency domain sampling, and therefore decreases performance. An insufficiently large cutoff frequency will cause blurring and ringing. While blurring is an acceptable trade-off for increased performance, ringing can be distracting. To decrease the effects of ringing a standard approach is to use a different lowpass filter, such as a Butterworth filter [11, Page 173]. In cases where the image contains low frequency content, a low cutoff frequency can be used without sacrificing image quality. In Fig. 1 the effects of ringing and blur-ring are shown for the bar galaxy data set, while for the sloshing data set it is shown that for almost no loss of quality significant gains in performance can be made. This demonstrates that FVR is well suited for data with low frequency content.

Another effect of this approximation is that the image becomes periodic, due to the use of numerical integration with a finite step size in each dimension. This can lead to aliasing if the period is not sufficiently large. However, one can choose the step size, so it can be left up to the user how much aliasing if any at all can be tolerated. If the functions Φ_k have global support, then there will always be some amount of aliasing. However, these functions are required to be integrable and

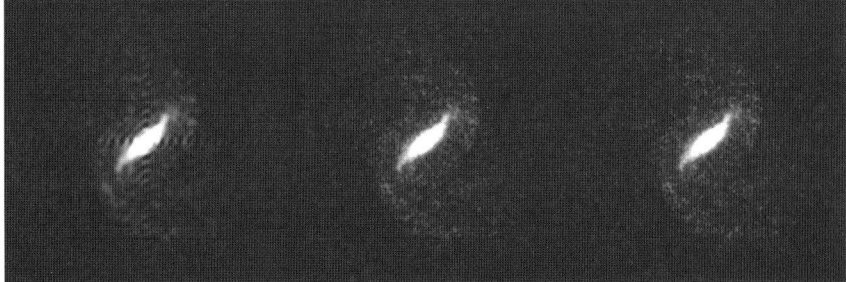

(a) The bar galaxy data set (6,400 particles) contains significant high frequency detail. Therefore there is a loss of quality when the lowpass filter is made too small as in the left image. However, little ringing is present in the center image, resulting in a nice tradeoff between quality and performance. Here we use an integration step size of 0.065 and 512×512 samples in the spatial domain. From left to right the number of samples and performance in frames per second on an NVIDIA Geforce 8600 GTS is respectively 64×32, 128×64, 192×96, and 63, 21, and 10.4

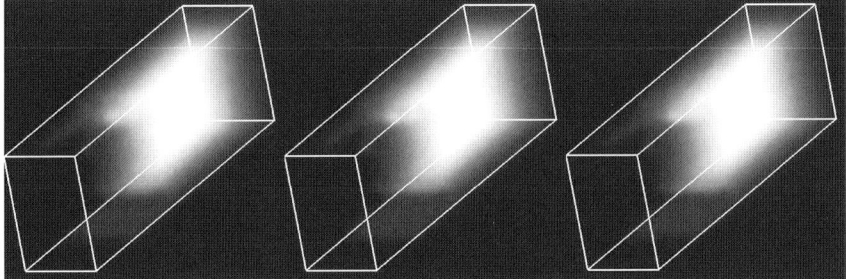

(b) The water sloshing in a reservoir data set (11,895 particles) contains little high frequency detail. Therefore a lowpass filter can be applied without a noticeable loss of quality. Here we use an integration step size of 0.7 and 512×512 samples in the spatial domain. From left to right the number of samples and performance in frames per second on an NVIDIA Geforce 8600 GTS is respectively 32×16, 64×32, and 128×64, and 92, 40, and 12.4

Fig. 1 The above images illustrate the trade-off in performance and quality using the approximation (8)

therefore rapidly decreasing towards zero, and so even in this case where aliasing is theoretically unavoidable, its effect can be made negligible.

4 Particular Meshless Methods

In the preceding section the specific form of Φ_k was not considered, which is necessary for determining its Fourier transform. In this section the form Φ_k takes for various types of meshless data is considered, from which its Fourier transform is obtained. In each case an explicit form of the Fourier transform $\widehat{\Phi_k}$ is given. With this information the image approximation (8) can be sampled directly.

4.1 Kansa's Method

Kansa's method [17, 27] generalizes radial basis function interpolation [3, 33] to solve partial differential equations using collocation. The resulting meshless data is of the form of (1) with

$$\Phi_k = \Phi(\cdot - x_k)$$

where Φ is some function, usually referred to as a radial basis function. Assuming Φ is integrable, each Φ_k is as well. It follows by the shift property [33, Theorem 5.16] that

$$\widehat{\Phi_k}(f) = e^{-i2\pi x_k^T f}\hat{\Phi}(f).$$

In fact, the Fourier transform of the meshless data can be simplified since $\hat{\Phi}$ can be factored outside of the summation

$$\hat{s}(f) = \hat{\Phi}(f) \sum_{k=1}^{N} \alpha_k e^{-i2\pi x_k^T f}. \tag{9}$$

Examples of integrable radial basis functions are given in Table 1. Non-integrable radial basis functions and polynomial terms are sometimes used, examples of which

Table 1 A list of three-dimensional Fourier transforms of various integrable functions used in meshless methods. With the exception of that of the Gaussian, the inverse multiquadric, and the Sobolev spline [33, Theorems 6.10, 6.13, and Page 133], the Fourier transform of each function was computed using (6)

Name	$\Phi(x) = \phi(\|x\|), r = \|x\|$	$\hat{\Phi}(f), r = \pi\|f\|, m = \pi r$
Gaussian	$e^{-\alpha r^2}, \alpha > 0$	$\left(\frac{\pi}{\alpha}\right)^{\frac{3}{2}} e^{-\frac{m^2}{\alpha}}$
Inverse multiquadric	$(c^2 + r^2)^{-\beta}, c > 0, \beta > \frac{3}{2}$	$\frac{2\pi^{3/2}}{\Gamma(\beta)}\left(\frac{m}{c}\right)^{\beta-3/2} K_{3/2-\beta}(2cm)$
Monaghan's M_4	$(2-r)^3 - 4(1-r)^3, 0 \leq r < 1$ $(2-r)^3, 1 \leq r < 2 0, r \geq 2$	$\frac{3\pi}{m^6}(\cos 2m - 1)(\cos 2m + m\sin 2m - 1)$
Sobolev	$\frac{2^{1-\beta}}{\Gamma(\beta)}r^{\beta-3/2}K_{3/2-\beta}(r), \beta > \frac{3}{2}$	$(2\pi)^{3/2}(1+4m^2)^{-\beta}$
Wendland's $\phi_{3,0}$	$(1-r)^2_+$	$\frac{\pi}{2m^5}(m\cos 2m - \frac{3}{2}\sin 2m + 2m)$
Wendland's $\phi_{3,1}$	$(1-r)^4_+(4r+1)$	$\frac{15\pi}{2m^8}\left(\frac{9}{2}m\sin 2m + (6-m^2)\cos 2m + 4m^2 - 6\right)$
Wendland's $\phi_{3,2}$	$(1-r)^6_+(35r^2 + 18r + 3)$	$\frac{315\pi}{2m^{11}}\begin{pmatrix}\left(\frac{315}{2} - 36m^2\right)\sin 2m \\ +(4m^2 - 123)m\cos 2m \\ +32m(m^2-6)\end{pmatrix}$
Wendland's $\phi_{3,3}$	$(1-r)^8_+(32r^3 + 25r^2 + 8r + 1)$	$\frac{10395\pi}{4m^{14}}\begin{pmatrix}\left(60m^3 - \frac{2295}{2}m\right)\sin 2m \\ -(4m^4 + 1440 - 375m^2)\cos 2m \\ +1440 - 960m^2 + 64m^4\end{pmatrix}$

include the multiquadric and the thin-plate spline. Our adaptation of FVR does not apply to data using such radial basis functions. Splatting is also unable to deal with such radial basis functions for the same reason of non-integrability. While slice-based volume visualization can deal with such radial basis functions, it does so much slower compared to integrable radial basis functions, due to it being required that every term is evaluated at every point, instead of only a few local terms.

4.2 Meshless Symmetric Collocation

Meshless symmetric collocation is another method for solving partial differential equations, which uses generalized interpolation [8–10, 33]. For concreteness the Poisson equation with Dirichlet boundary conditions is considered. In this case the functions Φ_k are

$$
\Phi_k(x) = \begin{cases} \Delta\Phi(x - x_k), & 1 \le k \le N_L \\ \Phi(x - x_k), & 1 \le k - N_L \le N_B \end{cases} \tag{10}
$$

where $x_1 \ldots x_{N_L}$ are in the interior and $x_{N_L+1} \ldots x_{N_L+N_B}$ are on the boundary of the domain. Applying the shift and differentiation properties [33, Theorem 5.16] of the Fourier transform shows that

$$
\widehat{\Phi}_k(f) = \begin{cases} -(2\pi\|f\|)^2 e^{-i2\pi x_k^T f}\hat{\Phi}(f), & 1 \le k \le N_L \\ e^{-i2\pi x_k^T f}\hat{\Phi}(f), & 1 \le k - N_L \le N_B \end{cases} \tag{11}
$$

The factor $\hat{\Phi}(f)$ can be factored outside of the summation.

$$
\hat{s}(f) = \hat{\Phi}(f)\left(\sum_{k=1}^{N_L}\alpha_k e^{-i2\pi x_k^T f} - (2\pi\|f\|)^2 \sum_{k=N_L+1}^{N_L+N_B}\alpha_k e^{-i2\pi x_k^T f}\right) \tag{12}
$$

For this problem, the function Φ should be in C^4. For example, the Gaussian or the Wendland function $\phi_{3,2}$ [33, Page 129] could be used. These functions and others appear along with their Fourier transforms in Table 1.

4.3 Smoothed Particle Hydrodynamics

Smoothed particle hydrodynamics (SPH) is a particle-based meshless method for simulating fluid dynamics [21]. Underlying this method is a kernel approximation

$$
A_s(x) = \sum_{k=1}^{N} m_k \frac{A_k}{\rho_k} W(x - x_k, h_k) \tag{13}
$$

where A is a function being approximated such as density, and $A_k, m_k, \rho_k, r_k, h_k$ are respectively the function value, mass, density, position, and smoothing-length of each particle. Letting $\alpha_k = m_k \frac{A_k}{\rho_k}$ and $\Phi_k(x) = W(x - x_k, h_k)$ shows that Eq. (13) is in a form consistent with (1). A common choice for the kernel is

$$\Phi_k(x) = \frac{1}{4\pi h_k^3} M_4 \left(\frac{\|x - x_k\|}{h_k} \right). \tag{14}$$

which, by the shift and scaling properties [33, Theorem 5.16], has a Fourier transform of

$$\widehat{\Phi_k}(f) = \frac{1}{4\pi} \widehat{M_4}(h_k \|f\|) e^{-i2\pi x_k^T f} \tag{15}$$

The function M_4 is listed in Table 1.

5 Implementation

This technique consists of two major computations: the first step is to sample the Fourier transform, and the second step is to apply an inverse FFT to those samples. The Fourier transform sampling is simple to implement since it merely involves sampling each term in a small window around zero, without the use of any complicated data structures. The second step merely involves applying an existing FFT code to the Fourier transform samples. Each of these steps are overall straightforward to implement, but there are a few details of the implementation worth mentioning.

The Fourier transforms of the compactly-supported polynomials are not defined at zero as written and are difficult to sample accurately near zero due to round-off error. One solution to this problem is to compute the limit of these functions at zero using l'Hospital's rule and to then perform linear or quadratic interpolation between zero and points sufficiently far from zero where the function can be sampled without significant round-off error. This approximation is reasonable since the derivatives of these functions are typically quite small near zero.

For many types of meshless data the Fourier transform of the radial basis function used can be factored out of the summation as in (9) and (12), resulting in less computation per term. For the graphics hardware implementation described in the next section, factoring out the Fourier transform modestly increases the performance results by 15–45 percent.

Because the meshless data's Fourier transform is conjugate symmetric it is only required to sample it over a half plane since the rest of the samples can be obtained by complex conjugation. This halves the amount of sampling required.

Table 2 Performance measured in frames per second when visualizing various data sets interactively. In each case the M_4 function is used with an image size of 512×512. The number of partial sums was chosen such that the number of thread blocks was at least the number of multiprocessors

Data set		Number of Fourier transform samples				
8,600 GTS	# of terms	32×16	64×32	128×64	256×128	512×256
Ring galaxy	13,107	89	36	11	2.8	0.8
Bar galaxy	6,400	124	63	21	5.8	1.4
Sloshing	11,895	92	40	12.4	3.6	1.2
Cellular structure	161,973	12.8	4	1.2	0.4	n/a
8,800 GTS						
Ring galaxy	13,107	170	86	29	8	2.4
Bar galaxy	6,400	210	133	54	16	4.4
Sloshing	11,895	178	92	32	9.2	2.4
Cellular structure	161,973	30	10	2.8	0.8	0.4

5.1 Implementation on Graphics Hardware

We have implemented this technique on NVIDIA graphics hardware[1] using their CUDA interface.[2] A brief description of this implementation is as follows. The first step, sampling of the Fourier transform, is implemented as a CUDA kernel. This kernel is executed over a grid of threads where each thread is responsible for computing a partial sum of one sample. Within this grid, threads are organized into thread blocks, where each thread within a block has access to fast shared memory. Since each thread accesses the same parameters from memory, if a thread block is of length N, then each thread loads into shared memory the parameters of one of N terms. Once all of the parameters are loaded into shared memory, each thread accumulates the N terms before moving on to the next batch of terms. After sampling is performed the image is computed from its Fourier transform samples using a complex-to-real FFT provided by the CUDA FFT library.

Table 2 gives frame rates achieved for visualizing various data sets. These performance measurements indicate that this technique is quite capable of interactively displaying meshless data, and is scalable with respect to the data size, the number of Fourier transform samples, and available hardware. Furthermore, these results indicate that the 512×512 complex-to-real FFT is not a bottleneck with respect to achieving interactive frame rates.

The code has been released as the open source library *libMeshlessVis* and is available for download.[3]

---

[1] http://www.nvidia.com/page/geforce8.html

[2] http://developer.nvidia.com/cuda/

[3] http://code.google.com/p/libmeshlessvis

6 Applications

We have applied this technique to meshless-based simulations in two application domains. We observe in each case that Fourier volume rendering produces
animations or images with information useful for understanding the simulation.

6.1 Astrophysical Data

The implementation of FVR was used to visualize the meshless data generated
by MASS99 [1], a code which couples N-body gravity with SPH-based hydrodynamics.

The first sequence of images in Fig. 2 shows the time evolution of the distribution
of gas in a collisional ring galaxy model based on the AM0644-741 galaxy. This
use of FVR reveals features such as the high density regions in the ring, which
correspond to regions of star formation seen in this galaxy. Other features revealed
include the appearance of a double ring and the density changes along the main ring,
which are also observed in optical and radio images of this galaxy [14].

The second sequence of images in Fig. 2 shows the time evolution of the distribution of gas in a bar galaxy model. The rotation of the bar is clearly visible in the
sequence, along with the changing structure of the spiral arms. Using this model,
the authors tested the Tremaine-Weinberg method of determining pattern speeds in
a system that has a well characterized angular speed [24]. Doing this type of analysis with real observational data is impossible, since the rotational period of these
systems is millions of years long.

6.2 CFD Data

The Implementation of FVR was used to visualize the fluid sloshing in a reservoir [31] and the fluid flow through cellular structure subjected to compressive
loading [26, 30]. The first example (Fig. 3(a)) shows the water sloshing in a rectan-
gular reservoir ($1{,}008 \times 196 \times 300$ mm) introduced by longitudinal time-dependent
reservoir acceleration with a peak of approximately 30 g [31]. The water has been
modeled with SPH particles and the moving reservoir has been defined as rigid
using a penalty based contact algorithm to prevent any penetration. From Fig. 3(a)
the free surface during the sloshing can be observed as well as the distribution of the
SPH particles in the reservoir. The second example focuses on the energy absorption capabilities of cellular structures [26, 30]. In order to increase the capability of
impact energy absorption the regular cellular structure made of polymer FullCure
M730 has been filled with fluid filler. Figure 3(b) depicts computational results of
the fluid flow through cellular structure subjected to dynamic compressive loading. During the cellular material deformation the fluid flow additionally dissipates

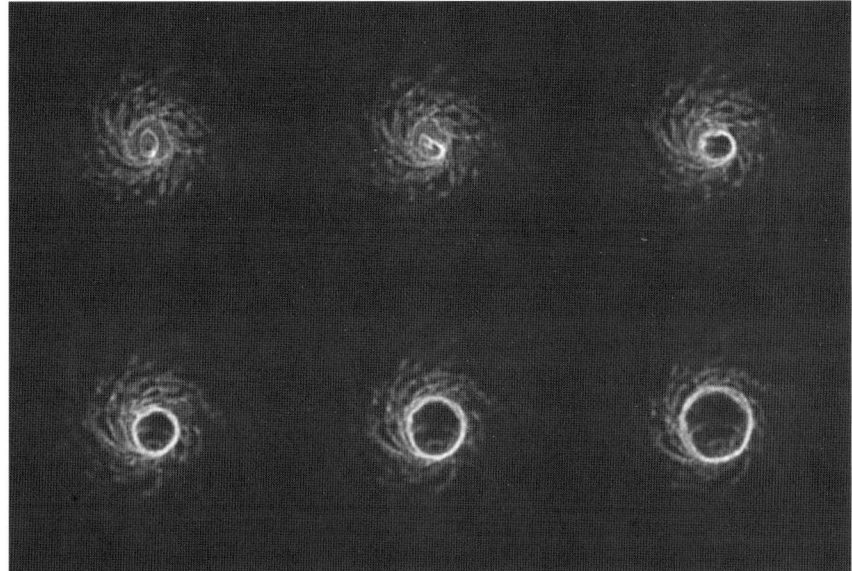

(a) A collisional ring galaxy

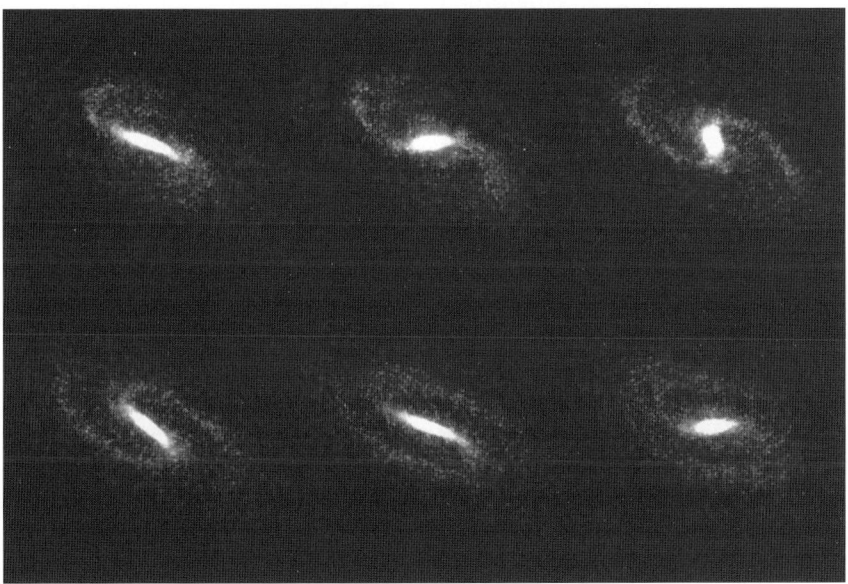

(b) A bar galaxy

Fig. 2 Visualization of meshless astrophysical data sets

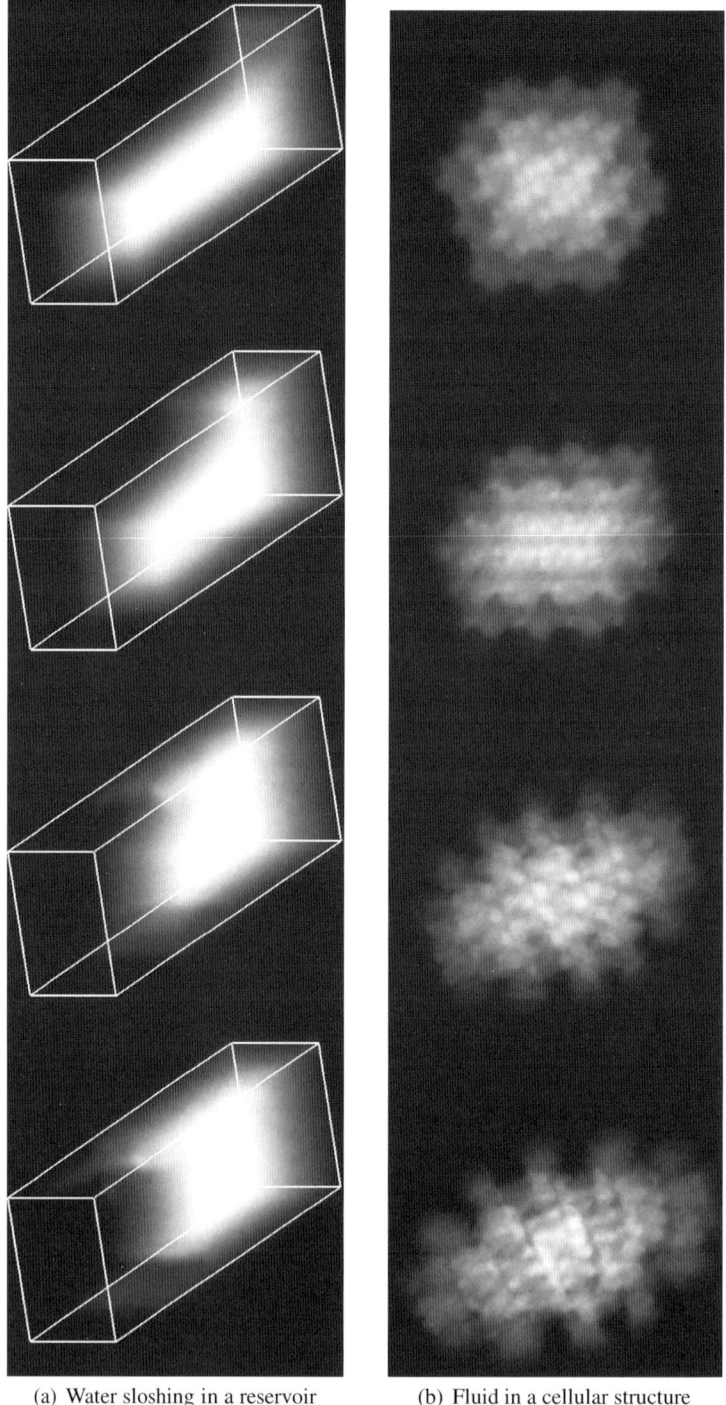

(a) Water sloshing in a reservoir   (b) Fluid in a cellular structure

Fig. 3 Visualization of meshless fluid dynamics data sets

energy and contributes to better material characteristics. Figure 3(b) shows the visualization of filler particles flow at different time instances using the FVR. From the figure the outflow of the SPH particles can be clearly observed, as well as their distribution and concentration.

7 Future Work

We are currently investigating combining Fourier volume rendering and splatting together as a single hybrid technique. This could be done by decomposing the meshless data based on the frequency content of each basis function. In particular, splatting will deal with basis functions with high frequency content (i.e. low spatial support), or that need to be spatially filtered to avoid aliasing, while Fourier volume rendering will deal with low frequency content. Although we have focused on meshless data sets, we are also investigating the application of this approach to unstructured grid FEM data sets, which are also of the form (1).

Fourier volume rendering has been enhanced to include spatial depth cues and lighting effects [7, 29], which could possibly be used with this adaptation of FVR. We have also left as future work this technique's application to data resulting from meshless methods such as moving least squares [2] where the meshless data is defined at each point by a local least squares fit with respect to a polynomial basis, since it is not clear what the Fourier transform of this type of data is.

8 Conclusions

We have presented an adaptation of Fourier volume rendering which for the first time enables it to deal directly with meshless data. Because the adaptation is direct, it avoids the prohibitive computational and memory costs and quality loss of discretizing the meshless data into a three-dimensional grid. Other advantages of this adaptation of Fourier volume rendering of meshless data include a trade-off between image quality and performance via lowpass filtering. While lowpass filtering will cause ringing in images with high frequency content, it is otherwise a highly effective approximation. Due to the technique dealing with the Fourier transform of the meshless data, no spatial data structures are needed. Therefore, the technique is indifferent to whether or not the meshless data's geometry is static or dynamic. Because a general form of meshless data was considered this technique is applicable to a number of types of meshless data. Finally, we demonstrated the technique's usefulness in visualizing different meshless-based simulations.

In general we believe that it is important that visualization techniques which deal directly with meshless data are available, so that meshless methods are not at a disadvantage when being considered for use in scientific simulation. Therefore, we encourage further work on adapting visualization techniques to deal directly with meshless data.

Acknowledgments The authors gratefully acknowledge feedback offered by fellow participants at the conference. A portion of this work was performed while the first author was under the advisement of H. Quynh Dinh at Stevens Institute of Technology, whose advice is gratefully acknowledged, along with advice given by Greg Slabaugh, Klaus Mueller, Neophytos Neophytou, and Yi Li. The financial support of the Slovenian Research Agency is greatly appreciated.

References

1. A. Antunes and J. Wallin. Convergence on N-body plus SPH. In *Bulletin of the American Astronomical Society*, pages 1433, December 2001.
2. T. Belytschko, Y. Krongauz, D. Organ, M. Fleming, and P. Krysl. Meshless methods: An overview and recent developments. *Computer Methods in Applied Mechanics and Engineering*, 139:3–47, 1996.
3. M. Buhmann. *Radial Basis Functions: Theory and Implementations*. Cambridge University Press, New York 2003.
4. Christopher S. Co and Kenneth I. Joy. Isosurface generation for large-scale scattered data visualization. In G. Greiner, J. Hornegger, H. Niemann, and M. Stamminger, editors, *Proceedings of VMV 2005*, pages 233–240, 2005.
5. A. Corrigan and H. Q. Dinh. Computing and rendering implicit surfaces composed of radial basis functions on the GPU. In *Poster proceedings of the International Workshop on Volume Graphics*, June 2005.
6. S. Dunne, S. Napel, and B. Rutt. Fast reprojection of volume data. In *Proceedings of the First Conference on Visualization in Biochemical Computing*, pages 11–18, 1990.
7. A. Entezari, R. Scoggins, T. Möller, and R. Machiraju. Shading for Fourier volume rendering. In *VVS '02: Proceedings of the 2002 IEEE symposium On Volume Visualization and Graphics*, pages 131–138, 2002.
8. G. Fasshauer. Solving partial differential equations by collocation with radial basis functions. In *Surface Fitting and Multiresolution Methods*, pages 131–138. Vanderbilt University Press, Nashville, TN, 1997.
9. C. Franke and R. Schaback. Solving partial differential equations by collocation using radial basis functions. *Applied Mathematics and Computation*, 93(1):73–82, 1998.
10. P. Giesl and H. Wendland. Meshless collocation: Error estimates with application to dynamical systems. Preprint Göttingen/München 2006, to appear in *SIAM Journal on Numerical Analysis*, 2006.
11. R. Gonzalez and R. Woods. *Digital Image Processing*. Prentice-Hall, second edition, 2002.
12. M. H. Gross, L. Lippert, R. Dittrich, and S. Häring. Two methods for wavelet-based volume rendering. *Computers and Graphics*, 21(2):237–252, 1997.
13. J. Hart. Ray tracing implicit surfaces. In *Siggraph 93 Course Notes: Design, Visualization and Animation of Implicit Surfaces*, pages 1–16, 1993.
14. J. L. Higdon and J. F. Wallin. Wheels of fire. III. Massive star formation in the "double-ringed" ring galaxy AM0644-741. *Astrophysical Journal*, 474:686–700, 1997.
15. M. Hopf and T. Ertl. Hierarchical splatting of scattered data. In *Proceedings of IEEE Visualization*, pages 433–440, 2003.
16. Y. Jang, M. Weiler, M. Hopf, J. Huang, D. Ebert, K. Gaither, and T. Ertl. Interactively visualizing procedurally encoded scalar fields. In O. Deussen, C. Hansen, D. Keim, and D. Saupe, editors, *Proceedings of EG/IEEE TCVG Symposium on Visualization VisSym'04*, 2004.
17. E. J. Kansa. Multiquadrics - A scattered data approximation scheme with applications to computational fluid dynamics. *Computers & Mathematics with Applications*, 19:147–161, 1990.
18. M. Levoy. Volume rendering using the Fourier projection-slice theorem. In *Proceedings of the Conference on Graphics Interface'92*, pages 61–69, Morgan Kaufmann, San Francisco, CA, USA, 1992

19. T. Malzbender. Fourier volume rendering. *ACM Transactions Graphics*, 12(3):233–250, 1993.
20. J. Meredith and K.-L. Ma. Multiresolution view-dependent splat based volume rendering of large irregular data. In *Proceedings of IEEE Symposium on Parallel and Large Data Visualization and Graphics*, 2001.
21. J. J. Monaghan. Smoothed particle hydrodynamics. *Reports on Progress in Physics*, 68(8):1703–1759, 2005.
22. N. Neophytou, K. Mueller, K. T. McDonnell, W. Hong, X. Guan, H. Qin, and A. Kaufman. GPU-accelerated volume splatting with elliptical RBFs. In *Proceedings of the Joint Eurographics - IEEE TCVG Symposium on Visualization 2006*, May 2006.
23. D. J. Price. SPLASH: An interactive visualisation tool for SPH data. Accepted to the Publications of the Astronomical Society of Australia, 2007.
24. R. J. Rand and J. F. Wallin. Pattern speeds of BIMA SONG galaxies with molecule-dominated interstellar mediums using the Tremaine-Weinberg method. *Astrophysical Journal*, 614:142–157, 2004.
25. R. Rau and W. Strasser. Direct volume rendering of irregular samples. In *Visualization in Scientific Computing'95*, pages 72–80, 1995.
26. Z. Ren, M. Vesenjak, and A. Öchsner. Behaviour of cellular structures under impact loading: A computational study. *Materials Science Forum*, 566:53–60, 2008.
27. R. Schaback. Convergence of unsymmetric kernel-based meshless collocation methods. *SIAM Journal on Numerical Analysis*, 45(1):333–351, 2007.
28. P. Stark. Fourier volume rendering of irregular data sets. Master's thesis, Simon Fraser University, 2002.
29. T. Totsuka and M. Levoy. Frequency domain volume rendering. In *SIGGRAPH'93: Proceedings of the 20th Annual Conference on Computer Graphics and Interactive Techniques*, pages 271–278, 1993.
30. M. Vesenjak, A. Öchsner, M. Hribersek, and Z. Ren. Behaviour of cellular structures with fluid fillers under impact loading. *International Journal of Multiphysics*, 1:101–122, 2007.
31. M. Vesenjak, Z. Ren, H. Mulerschon, and S. Matthaei. Computational modelling of fuel motion and its interaction with the reservoir structure. *Journal of Mechanical Engineering*, 52, 2006.
32. M. Weiler, R. Botchen, S. Stegmaier, T. Ertl, J. Huang, Y. Jang, D. S. Ebert, and K. P. Gaither. Hardware-assisted feature analysis and visualization of procedurally encoded multifield volumetric data. *IEEE Computer Graphics and Applications*, 25(5):72–81, 2005.
33. H. Wendland. *Scattered Data Approximation*. Cambridge University Press, Cambridge 2005.